Ulrike Lindequist, Timo Niedermeyer, Eberhard Teuscher

Natural Poisons and Venoms

Also of interest

Natural Poisons and Venoms.
Volume 1: Plant Toxins: Terpenes and Steroids
Eberhard Teuscher, Ulrike Lindequist, 2023
ISBN 978-3-11-072472-1, e-ISBN (PDF) 978-3-11-072473-8,
e-ISBN (EPUB) 978-3-11-072486-8

Natural Poisons and Venoms.
Volume 2: Plant Toxins: Polyketides, Phenylpropanoids and Further Compounds
Eberhard Teuscher, Ulrike Lindequist, 2025
ISBN 978-3-11-072851-4, e-ISBN (PDF) 978-3-11-072853-8,
e-ISBN (EPUB) 978-3-11-072864-4

Natural Poisons and Venoms.
Volume 3: Plant Toxins: Alkaloids and Lectins
Eberhard Teuscher, Ulrike Lindequist, 2025
ISBN 978-3-11-112740-8, e-ISBN (PDF) 978-3-11-113621-9,
e-ISBN (EPUB) 978-3-11-113692-0

Natural Poisons and Venoms.
Volume 4: Animal Toxins
Jan-Peter Hildebrandt, Eberhard Teuscher, Ulrike Lindequist, 2023
ISBN 978-3-11-072854-5, e-ISBN (PDF) 978-3-11-072855-2,
e-ISBN (EPUB) 978-3-11-072862-0

Set: Natural Poisons and Venoms.
Volume 1–5
Eberhard Teuscher, Ulrike Lindequist, Jan-Peter Hildebrandt,
Timo Niedermeyer, 2025
ISBN 978-3-11-160199-1

Ulrike Lindequist, Timo Niedermeyer,
Eberhard Teuscher

Natural Poisons and Venoms

Volume 5
Fungal and Microbial Toxins

Founded by
Eberhard Teuscher and Ulrike Lindequist

DE GRUYTER

Authors
Prof. Dr. Ulrike Lindequist
Lise-Meitner-Str. 6a
17491 Greifswald
Germany
lindequi@uni-greifswald.de

Prof. Dr. Timo Niedermeyer
Freie Universität Berlin
Institut für Pharmazie
Königin-Luise-Str. 2+4
14195 Berlin
Germany
timo.niedermeyer@fu-berlin.de

Prof. Dr. Eberhard Teuscher
Goethestr. 9
07950 Zeulenroda-Triebes
Germany
teuscher-triebes@t-online.de

ISBN 978-3-11-072856-9
e-ISBN (PDF) 978-3-11-072857-6
e-ISBN (EPUB) 978-3-11-072865-1

Library of Congress Control Number: 2025944262

Bibliographic information published by the Deutsche Nationalbibliothek
The Deutsche Nationalbibliothek lists this publication in the Deutsche Nationalbibliografie; detailed bibliographic data are available on the internet at http://dnb.dnb.de.

Cover image: grafxart8888/iStock/Getty Images Plus
Typesetting: Integra Software Services Pvt. Ltd.

www.degruyterbrill.com
Questions about General Product Safety Regulation:
productsafety@degruyterbrill.com

Preface

Living beings produce pharmacologically active substances to ensure their survival. In many cases, such substances can harm or kill other living creatures or can trigger allergies. Some of these compounds pose a danger not only to human beings but also to animals.

Our concern is to try to acquaint the readers – especially physicians, veterinarians, pharmacists, biologists, chemists, food chemists, biochemists, students of these disciplines, but also interested laymen – with poisonous microorganisms, fungi, plants, and animals, in words and pictures.

The focus of this volume is on the toxic ingredients of macrofungi and microfungi, cyanobacteria, and selected bacteria pathogenic to humans and animals. The reader will be informed about the biology of the poisonous fungi and bacteria and the chemical structures, pharmacological effects, and modes of action of the contained poisonous compounds.

We report on the numerous possibilities to poison, which are, e.g., confusion of poisonous mushrooms with edible ones, accidental tasting of poisonous mushrooms, testing the edibility of unknown fungi, misuse as intoxicants, for attempted suicides or homicides, misapplication of intoxicants as therapeutics in medicine, folk medicine and folk customs, ingestion of toxin-contaminated food or water, or infections of human beings by pathogenic microorganisms.

We inform about the symptoms of the poisonings, in many cases shown by case reports, and we give hints on poisoning prevention and about measures to treat the poisonings. However, the books are not intended as a guide to medical practice in cases of poisoning.

For poisonings of animals, especially of farm and domestic animals, by fungi or bacteria, we characterize the conditions of poisonings and describe the major symptoms and the possible treatment of these poisonings as well as their prevention.

We discuss not only poisonous organisms and their active ingredients which in earlier and current times were/are used as therapeutics, but we present also compounds that may be starting points for the development of new medicines.

Over the decades, our knowledge in the field of natural poisons has expanded rapidly. Reasons for this include, e.g., modern molecular biological and chemical methods available today. The easier accessibility to exotic sources through globalization and the easier accessibility of living beings from the marine habitat have fostered the interest in investigations of these sources.

We have always tried to see the toxins not only as a danger to humans and animals but also as an admirable variety of substances that have been evolved in the course of evolution to protect the living products of evolution.

The five volumes in the series *Natural Poisons and Venoms* are based on our book *Biogene Gifte*, third edition, published in 2010 by Wissenschaftliche Verlagsgesellschaft Stuttgart in German language. Because of the enormous increase in knowledge in the

https://doi.org/10.1515/9783110728576-202

field of natural poisons and venoms, to reach a larger readership, and to fit into the STEM series of the De Gruyter publishing house, the topic is now covered in five volumes in English language.

After a fundamental chapter on toxins (definition of terms used in toxicology, history, chemistry, biology, first aid, and general and clinical toxicology), Volumes 1 to 3 are devoted to poisonous plants and their active substances. Volume 4 summarizes the current knowledge about poisonous animals, their poisons, their attacks, and the measures after an attack. Volume 5 deals with poisonous fungi and bacteria, including cyanobacteria, and their ingredients. The volumes are staggered in time, so that the highest possible topicality is guaranteed. The volumes can be purchased independently, but the purchase of the set containing all volumes at a reduced rate is recommended.

For the scientific nomenclature of the kingdom of fungi, we follow the 'Index Fungorum'; for the English language vernacular names, we follow predominantly the names of the British Mycological Society or from Wikipedia.

Despite the extensive bibliographies at the end of each chapter, the cited literature represents only a selection. For reasons of space, the information is as short as possible. Preference was given to review papers. Further literature is summarized at the end of the volume (cross-chapter literature and further reading).

We would like to thank everyone who helped us with the creation of the books. We would especially like to thank Mr. Dipl.-Phys. Karl-Heinz Lichtnow, Greifswald (Germany), for drawing the drafts of structural formulas and schemes and his readily available help with critical computer problems of any kind. A big thanks goes to Dr. Markus Scholler, Karlsruhe (Germany), for his advice on all mycological matters. We would like to thank all image authors, Getty images, and Wikimedia for the provision of the photos. Our thanks go to De Gruyter publishing house, especially Mrs. Karin Sora, for making these books possible, Dr. Bettina Noto for the always pleasant and helpful cooperation in the completion of the volumes and her careful editing work, and Mrs. Anne Stroka and her staff for the preparation of the manuscript to the printable form.

We are always grateful for critical information on errors and suggestions for improvement.

Greifswald, Triebes, and Berlin, July 2025
Ulrike Lindequist, Eberhard Teuscher, and Timo Niedermeyer

A Short Glance on Volume 5

Whereas Volumes 1, 2, and 3 of this book series *Natural Poisons and Venoms* focus on poisonous constituents of plants and Volume 4 on those of animals, the focus of this Volume 5 is on poisons of macroscopic and microscopic fungi and of selected bacteria including those of cyanobacteria. It is arranged according to the producing organisms, and inside these large chapters according to the biogenetic origin/chemical groups of the toxic ingredients. Volume 5 also contains an index of the Volumes 1–4.

Poisoning by macrofungi is mainly caused by incorrect identification and confusion with edible species. *Amanita phalloides*, the death cap, is responsible for about 90% of all fatal mushroom poisonings worldwide. In recent years, several fungal species have been identified as poisonous for the first time. In some cases, the structure of the responsible ingredients could be elucidated. For example, we know today that certain amino acids are responsible for the strong cardiac toxicity of *Trogia venenata*, little white mushroom, that caused several hundred death cases of apparently healthy people in Southern China. *Pleurocybella porrigens*, angel wing mushroom, caused fatalities of patients suffering from kidney problems in Japan. Because of the low stability of its active constituent pleurocybellaziridin, it needed a long time to clarify its structure and mode of action. Toxicity is assumed for some species, e.g., *Tricholoma equestre*, but the substances responsible are still unknown. Otherwise, this species was considered edible in the past.

In contrast to poisoning by macrofungi, intoxications by microfungi (molds) are mostly caused by toxins that enter the human or animal organism accidentally via contaminated food. Toxins of the molds are commonly called mycotoxins. While acute poisoning by toxins of microfungi is observed almost exclusively in animals, the carcinogenic and mutagenic effects of many mycotoxins are also significant for humans. Aflatoxin B1, produced by several Aspergillus species, is the strongest carcinogen of natural origin known to date. Epidemiological investigations from several parts of the world underline a relation between the consumption of mycotoxins with the nutrition and the incidence of cancer diseases. New strategies have been developed to mitigate the content of mycotoxins in food and feed and to protect humans and animals.

Some highly human pathogenic bacteria, e.g., *Clostridium botulinum, Clostridium tetani*, and *Vibrio cholerae*, exhibit their toxicity by exotoxins that are mostly proteins. They enter the human body, e.g., by contaminated food, drinking water, via wounds, or during medical application, and can lead to fatal outcome. The mechanisms of action of some bacterial exotoxins are now well-known. Although protection against most exotoxins can be achieved through vaccination, serious poisoning occurs time and again. Bacterial endotoxins are lipopolysaccharides that are released after death of the bacteria. They are strong pyrogens and can cause severe sepsis.

https://doi.org/10.1515/9783110728576-203

Cyanobacteria, also called blue-green algae, are phototrophic bacteria. They live in waters of all kinds, on soil or stones, or as endophytes in plants. Under certain circumstances, they form water blooms that can lead to mass poisoning of animals and humans. The toxins produced by cyanobacteria belong to different substance classes and have a broad spectrum of activity. Some compounds are of great interest for medical applications or already used as cancer therapeutics.

Contents

Abbreviations and Icons

ADI	acceptable daily intake
approx.	approximately
BfArM	Federal Institute for Drugs and Medical Devices (Germany)
BfR	Federal Institute for Risk Assessment (Germany)
B.C.	before Christ
B.P.	boiling point
BW	body weight
cAMP	cyclic adenosine monophosphate
CNS	central nerve system
CRP	C-reactive protein
CYP	cytochrome P
DGfM	German Society for Mycology
DW	dry weight
EFSA	European Food Safety Authority
EMA	European Medicines Agency
FAO	Food and Agriculture Organization of the UN
FDA	Food and Drug Administration (USA)
FW	fresh weight
GABA	γ-aminobutyric acid
HAB	harmful algal blooms
HMPC	Herbal Medicinal Products Committee
IARC	International Agency for Research on Cancer
i.m.	intramuscular
i.p.	intraperitoneal
i.v.	intravenous
JECFA	Joint FAO/WHO Expert Committee on Food Additives
LD	lethal dose
LOAEL	lowest observed adverse effect level
MAO	monoamine oxidase
MAPK	mitogen-activated protein kinases
max.	maximal
MDD	mean daily dose
MOE	margin of exposure
M.P.	melting point
Mr	molecular weight
MRT	magnetic resonance tomography
NF-kB	nuclear factor kB
NMDA	*N*-methyl-D-aspartate
NOAEL	no observed adverse effect level
OATP	organic anion transporting polypeptide
p.e.	parenteral
P-gp	P-glycoprotein
p.i.	per inhalationem
p.o.	peroral
resp.	respectively
RKI	Robert Koch Institute, Germany
s.c.	subcutaneous
TCM	Traditional Chinese Medicine
WHO	World Health Organization

https://doi.org/10.1515/9783110728576-205

☠ symptoms of acute poisoning
(☠) low risk of acute poisoning
🚹 case description
🐕 poisoning of animals
✱ allergy
☹ chronic poisoning
(☹) low risk of chronic poisoning
🚑 first aid and potential treatment

Part I: **Fungal Toxins**

1 General

Fungi (Mycota and Eumycota) form a separate kingdom of heterotrophic eukaryotic organisms. The classification depends on the author. We prefer to distinguish six phyla, namely Ascomycota, Basidiomycota, Chytridiomycota, Glomeromycota, Microsporidia, and Zygomycota. Anamorphic fungi with unknown sexual state were formerly placed in the Deuteromycota, but this phylum is not accepted anymore because it is polyphyletic with most species belonging to the sac fungi (Ascomycota). Fungi are often divided into microfungi with small fruit bodies and macrofungi with large fruit bodies. The latter group can be recognized with the naked eye. Macrofungi include most Basidiomycetes and some Ascomycetes and are commonly called mushrooms. The fruit bodies (sporocarps) of mushrooms often possess a fleshy consistence. Poisonous or inedible mushrooms are sometimes referred to as toadstools. Molds and yeasts belong to the microfungi.

The estimated number of fungal species occurring worldwide is between 2.2 and 3.8 million. Only about 120,000 species are scientifically described [11]. The number of poisonous fungi, among them mushrooms, is much lower and cannot be specified exactly. New poisonous fungal species are continually being discovered. To date, more than 100 mushroom toxins and 500 toxins from microfungi have been reported. However, the active substances of many poisonous fungi are not yet or only partially known (overview chemistry: **[182]**, non-peptide mushroom toxins: [16], peptide mushroom toxins: [32, 33]).

All fungal toxins are mycotoxins. In medicine and toxicology, however, the term 'mycotoxin' is preferably used for mold toxins **[18, 76]**. In this book, we use the term 'mycotoxin' in this sense for metabolites of molds that are toxic to humans and animals. While Chapter 2 describes toxins of macrofungi, Chapter 3 is devoted to the toxins of microfungi, molds in particular.

Molecular genetic studies have led to many changes in the taxonomy and classification of fungi [18]. For an unambiguous identification and classification of a fungal species, the combination of molecular, phenotypic, ecological, chorological, and other data is necessary [19].

https://doi.org/10.1515/9783110728576-001

2 Toxins of Macrofungi

2.1 General

2.1.1 Epidemiology of Poisoning by Macrofungi

Global data about mushroom poisoning is not available. Local studies indicate that the absolute number and the incidence of mushroom poisoning may be increasing, with regional differences. At the beginning of the 2000s, between 50 and 100 fatal cases/year were determined in Western Europe. In France, an average of 1,300 poisoning cases/year was registered between 2010 and 2017 [**41**]. The Federal Statistical Office of Germany reports over the period 2000–2018 a total of 4,412 hospitalizations and 22 deaths due to the toxic effects of mushroom consumption in Germany [**177**]. The US National Poison Data system reports a mean of 7,428 fatal mushroom poisonings in the 1999–2016 period, with a low lethality rate (less than 3 lethal cases/year) [2]. In 2024, the China Chinese Center for Disease Control and Prevention (CDC) investigated 599 cases of mushroom poisoning, including 1,486 patients and 13 deaths (case fatality rate: 0.87%). Ninety-seven mushroom species were identified as the cause of the intoxication. The number is in a similar range as in the years before [17]. In Japan, 569 cases of mushroom poisoning, which involved 1,920 patients and 10 deaths, were reported between 2001 and 2010 [34]. Many cases are likely to remain unregistered [2, 7–9].

The incidence of mushroom poisoning varies from region to region, season to season (mostly between August and November in the northern temperate region) and depending on the weather [13]. New poisoning syndromes emerge. Exotic species enter new areas so that known syndromes are being reported outside their previously known regions of incidence [**13, 178**]. Animals, especially pets such as dogs and cats, can also be poisoned by mushrooms [26].

The **main causes of poisoning by mushrooms** are:

- **The misidentification of toxic species, carelessness during mushroom picking, and/or ignorance of the toxicity of the mushrooms**

The most common cause of poisoning is a lack of knowledge in the recognition of mushrooms that leads to confusion of poisonous with edible species or to ignorance of the toxicity [28]. Mushroom intoxication due to misidentification is especially a concern for newly arrived persons in a region, e.g., migrants, accustomed to foraging in their home countries, who might not be familiar with local mushroom occurrence.

For common mushroom pickers, the species identification is based on phenotypical characters. Only mushrooms that are known to be safe should be collected. For the normal collector, these should mostly only be tube mushrooms. For safety reasons, mycologists could be consulted. Well-illustrated books can also be helpful (se-

https://doi.org/10.1515/9783110728576-002

lected examples: [**19, 42, 66, 68, 75, 78, 79, 82, 84, 103, 104, 105, 107, 110, 127, 142, 183**]). Mushroom Apps often do not give exact information and can only be suitable for an initial consultation.

- **Tasting by children**
  Children are at a particularly high risk for accidental poisonings. In the course of exploration of their environment, they try mushrooms grown on or nearby playground, on the way to school, in garden or in a flowerpot. For example, a small child suffered amanitin poisoning from amanitin-containing specimen of *Lepiota elaiophylla* Vellinga & Huijser (see Section 2.10.1) that had grown next to a yucca palm in a flowerpot [27]. To avoid this, mushrooms at such places should be removed. When children are in the forest, they must be well supervised.
- **Misuse as intoxicants**
  Acute illnesses or chronic damage in connection with use of hallucinogenic mushrooms such as *Amanita muscaria* (L.) Lam. or Psilocybe species or with their active components such as psilocybin are regarded as intentional poisonings. The number of cases is increasing. Besides, the confusion of hallucinogenic mushrooms with stronger poisonous ones, e.g., Cortinarius species, during intended collection has been described (see Section 2.8.4).
- **Misuse for criminal purposes or suicides**
  The use of poisonous mushrooms for the purpose of murder occurred already during history [**87**] and has also been reported in the present [10]. Suicides and suicide attempts are also reported.
- **Consumption of conditionally poisonous mushrooms**
  Under special circumstances, edible mushrooms can also cause intoxications [12, 13, 14, 25, **19, 42, 49, 180**]. These conditions include
  - the consumption of raw, incorrectly or too long stored or incorrectly prepared mushrooms;
  - the consumption of usually edible mushrooms that are too old or moldy;
  - the consumption of very large amounts of edible mushrooms in a short time (high fiber content and therefore risk of indigestibility);
  - the concomitant use of some edible fungi with alcohol; and
  - the consumption of edible mushrooms contaminated with microorganisms, heavy metals, poisonous plants, insects, moth and nematodes, pesticides, or radioactivity.
    - For example, the consumption of *Sarcosphaera coronaria* (Jacq) J. Schröt., violet crown cup, enriched with arsenic from the soil, has led to chronic arsen poisoning in Switzerland in 1920 [4, 30]. *Cyanoboletus pulverulentus* (Opat.) Gelardi, Vizzini & Simonini, ink stain bolete, hyperaccumulates arsenic, mainly in the form of dimethylarsinic acid, in the hymenium. The consumption of this edible species should be restricted [6]. Dried porcini (*Boletus edulis* Bull.) from China can be contaminated with nicotine that is/was there used as insecticide [3].

2.1.2 Diagnosis and Characteristics of Poisoning by Macrofungi

The diagnosis of poisoning is based on

- the anamnesis (identification of the ingested mushroom species, based on morphological, chemical or molecular characteristics, best by a mycological specialist),
- the interval between ingestion and onset of symptoms (latency period)
- the symptoms encountered, and
- qualitative and quantitative laboratory tests [**41, 42, 159** = **Vol. 1, 178**].

The latency period is decisive for the course of the poisoning. While the symptoms of mild or easily treatable poisoning tend to set in after a short time, the symptoms of potentially fatal poisoning often only start after several hours (>6 h in the case of amanitin-containing mushrooms) or even days (Cortinarius species). Gastrointestinal symptoms occur in almost all cases of mushroom poisoning. In mild cases, the poisoning is thus overcome. Depending on the toxins responsible, the intoxication symptoms are various. They are described in the following chapters. White et al. [**178**] classify poisonous mushrooms according to the main syndromes that they cause into the following groups:

- mushrooms with cytotoxic effects, e.g., *Amanita phalloides* or Cortinarius species;
- mushrooms with neurotoxic effects, e.g., *Amanita muscaria*, Inocybe species, Psilocybe species;
- mushrooms with myotoxic effects, e.g., *Russula subnigricans*;
- mushrooms with metabolic effects, e.g., *Coprinopsis atramentaria* or *Trogia venenata*;
- mushrooms with gastrointestinal effects, e.g., *Chlorophyllum molybdites* or Lactarius species; and
- mushrooms with miscellaneous activities (allergies, phototoxicity, and other), e.g., *Paxillus involutus* (overviews symptoms of mushroom poisoning: [12, **42, 60, 180**]).

It must be taken into consideration that the clinical signs depend not only on the consumed species but also on the geographical location and the mode of preparation of the fungi and on individual factors of the patients such as age, health condition, different sensitivities, co-medication etc.

2.1.3 Measures in Case of Poisoning or Suspected Poisoning by Macrofungi

If poisoning is suspected, the following measures are necessary:

- immediate contact to an information center for poisoning cases of the respective country (see p. 239) and, depending on the advices, contact to a mycological specialist and/or a medical doctor or a pharmacist;

- storage of mushroom leftovers, from cleaning, food, or vomiting for diagnostics;
- in case of a long latency period or severe symptoms immediate call of the rescue service and referral to a hospital;
- information and observation of all meal participants and, if necessary, presentation to a doctor.

For first aid and clinical measures, see [13, **19, 31, 42, 49, 88, 159 = Vol. 1, 178**] and the following chapters!

To avoid mushroom poisoning, it is necessary to use only the mushrooms that are known as edible, to avoid the consumption of old or damaged or raw mushrooms and to eat the mushrooms only in acceptable amounts.

2.1.4 Allergies Caused by Macrofungi

Allergies can occur after ingesting mushrooms, after skin contact with the mushrooms, or after inhaling the fungal spores. Of practical relevance are especially allergies that are caused by edible mushrooms (overview: [20, 21, 22, 25]).

Allergic reactions from type 1 have been reported, for example, for *Boletus edulis*, porcini, *Flammulina velutipes* (Curtis) Singer, enoki, *Tricholoma matsutake* (S. Ito & S. Imai) Singer, matsutake, *Lentinula edodes* (Berk.) Pegler, shiitake, *Auricularia auricula-judae* (Bull.) Quél., wood ear, and *Agaricus bisporus*, (J.E. Lange) Imbach, cultivated mushroom, portobello, button mushroom. Their symptoms can be confused with gastrointestinal poisoning symptoms. The responsible allergens are presumably heat-stable proteins.

The Paxillus syndrome, caused by consumption of cooked *Paxillus involutus* (Batsch.) Fr., brown roll-rim, poison pax (Ph. 2.1), Paxillaceae, is an autoimmune hemolytic anemia (AIHA) that belongs to type 2 allergic reactions (cytotoxic reactions). *P. involutus* has been widely consumed earlier. Due to a number of fatal intoxications, the species is now classified as dangerously poisonous.

P. involutus is widely distributed throughout the Northern Hemisphere and has been introduced to South Africa, South America, Australia, and other parts of the world. The mushrooms occur in woody and grassy areas in late summer and in autumn. Their reddish–, yellowish–, or olive brown–colored funnel-shaped caps, up to 12 cm in diameter, possess a characteristic depressed center with a distinctive inrolled rim. The gills are decurrent, light in color, and stain brown when touched or bruised [31].

An until now unknown toxin causes the formation of antibodies in pre-sensitized consumers. Subsequent contact with the mushrooms results in the formation of immune complexes consisting of antigen-antibodies that target the surface of human erythrocytes. This leads to activation of the complement system and following to hemolysis in sensitized people.

Ph. 2.1: *Paxillus involutus*, brown roll-rim (source: oliver schulz/iStock/Getty Images Plus).

† A 46-year-old male and his wife oftentimes collected and consumed *P. involutus* mushrooms preceding 10 years. Approx. 3 h after consumption, the man experienced severe pain around the lumbar spine. Increasing upper abdominal pain occurred, but no vomiting or diarrhea. Signs of inflammation, liver failure, and thrombocytopenia increased. Petechial hemorrhages appeared on the patient's skin. Sudden hypertension occurred. The patient died 4 days after consuming the mushrooms following multiorgan failure. Disseminated intravascular coagulation was identified as cause of death. The wife remained asymptomatic [31].

☠ Symptoms begin within 15 min to 2 h after ingestion. They may include jaundice, abdominal pain, collapse, dark-colored urine, and, 1–2 days later, icterus. In severe cases, liver and kidney failure may occur. Typically, not all participants in the meal are affected, but only those sensitized by previous contact with the mushrooms [21, 31].

Independently of AIHA, raw or insufficiently cooked roll-rim can cause gastrointestinal symptoms. Several phenolic compounds, e.g., involutine and involutone, have been isolated from the fruit bodies. They derived probably from atromentine, a terphenylquinone, and are in vitro not cytotoxic [16].

Contact eczema and other skin symptoms (type IV allergies) have been observed, for example, after contact with fresh or dried Suillus, Boletus, and Agaricus species, and with *Ramaria flava* (Schaeff.) Quél., yellow coral mushroom. People who process mushrooms professionally or mushroom pickers are affected. The responsible allergens are unknown [22].

External contact with or consumption of *Lentinula edodes*, the widely cultivated shiitake mushroom (Ph. 2.2, Omphalotaceae), can cause the shiitake dermatitis (flagellate erythema). Whether its genesis is immunological or purely toxic is still unclear. After a latency period of 24–48 h, a characteristically arranged skin rash appears. The whiplash-like appearance is typical, especially on the trunk and the extensor sides of the extremities. In addition to the β-glucan lentinan held responsible for the mentioned symptoms, other substances may be involved in the development of shiitake dermatitis [1, 5, 24, 29].

Ph. 2.2: *Lentinula edodes*, shiitake (source: Jan Lelley).

Flagellate erythema has been observed also after consumption of *Boletus edulis* and *Auricularia auricula-judae* [15, 22, 23].

Cases of photosensitive dermatitis caused by ingestion of *Bulgaria inquinans* (Pers.) Fr. or some other Ascomycetes and sun exposition have been reported from China [**60**].

References

For numbers in bold, see cross-chapter literature p. 233.

[1] Adji A et al. (2024) Biodiversitas 25(12): 4790
[2] Balice G et al. (2024) Toxins 16: 265
[3] Berndt S (2018) Z Mykol 84/2: 359, Tintling 24(3): 84
[4] Berndt S (2022) Tintling 139(5): 96
[5] Boels D et al. (2022) Clin Toxicol 60(8): 954
[6] Braeuer S et al. (2018) Food Chem 242: 225
[7] Brandenburg WE, Ward K (2018) Mycologia 110: 637
[8] Diaz JH (2005) Crit Care Med 33: 419
[9] Gummin DD (2018) Clin Toxicol 56(12): 1213
[10] Hawksworth DL, Wiltshire PEJ (2011) Forensic Sci Int 206(1–3): 1
[11] Hawksworth D, Lücking R (2017) Microbial Spectrum 5(4): FUNK-0052-2016
[12] Jo WS et al. (2014) Mycobiology 42(3): 215
[13] Keller SA et al. (2018) Int J Environ Res Publ Health 15: 2855

[14] Lagrange E, Vernoux JP (2020) Toxins 12(8): 482
[15] Lang N et al. (2016) J Dtsch Dermatol Ges 14: 303
[16] Lee S et al. (2022) Nat Prod Rep 39: 512
[17] Li H et al. (2025) China CDC Weekly 7(19): 645
[18] Liimatainen K et al. (2022) Fungal Diversity 112: 89
[19] Loizides M et al. (2022) Mycol Progr 21: 7
[20] Mayser P (2021) Tintling 26 (1): 58
[21] Mayser P (2021) Tintling 26(2): 11
[22] Mayser P (2021) Tintling 26(3): 17
[23] Molin S et al. (2015) Ann Allergy Asthma Immunol 115(3): 254
[24] Nguyen AH et al. (2017) Int J Dermatol 56(6): 610
[25] Nieminen P, Mustonen AM (2020) Toxins 12: 639
[26] Puschner B, Wegenast C (2018) Vet Clin North Am Small Anim Pract 48(6): 1053
[27] Schabel G (2017) Tintling 22(1): 81
[28] Schenk-Jaeger KM et al. (2012) Eur J Intern Med 23(4): e85
[29] Stephany MP et al. (2016) Am J Dermatol 17(5): 485
[30] Stijve T (2008) Sterbeeckia 28: 17
[31] Stöver A et al. (2019) Diagnostics 9(4): 130
[32] Vetter J (2023) Molecules 28: 5932
[33] Walton J (2018) The Cyclic Peptide Toxins of Amanita and Other Poisonous Mushrooms. Springer. Berlin. Heidelberg. New York
[34] Yamaura Y (2013) Chudoku Kenkyu 26(1): 39

2.2 N-Free Organic Acids as Toxins of Macrofungi

The organic acids ustalic acid (see Section 2.2.1) and cycloprop-2-ene-carboxylic acid (see Section 2.2.2) are thought to be the cause of some cases of mushroom poisoning observed in East Asia. Possibly the presence of oxalic acid in *Inonotus obliquus* (Fr.) Pilát, chaga, Hymenochaetaceae, that is often used as medicinal mushroom, is toxicologically relevant with long-term use of the fungi. An oxalic acid content of up to nearly 100 mg/g dry extract (4) or even more (9) has been detected. Cases of nephropathy after long-term intake of larger amounts (10–15 g powder per day for at least 3 months), sometimes in combination with vitamin C (500 mg/day), have been described [7, 9].

2.2.1 Ustalic Acid as Toxin of Tricholoma Species

Ustalic acid (Fig. 2.1) was first isolated by Kawagishi et al. in 2002 from *Tricholoma ustale* (Fr.) P. Kumm., burnt knight (Ph. 2.3, Tricholomataceae [16]). Probably it originates from oxidative cleavage of phlebiarubrone, a red terphenylequinone pigment [17]. Recently, a taxonomic revision of the Japanese *T. ustale* has led to the result that the Japanese specimen are not *T. ustale* but really *Tricholoma kakashimeji*. This new identified species contains ustalic acid and is probably responsible for several cases of mushroom poisoning in Japan [1]. Whether *T. ustale* from Europe or North America contains ustalic acid is unclear.

Ustalic acid

Fig. 2.1: A toxin of Tricholoma species.

Ph. 2.3: *Tricholoma ustale*, burnt knight (source: Geert Schmidt-Stohn).

The content of ustalic acid in dried Japanese fungi was reported to be 41.9–155.7 ppm (=0.0042–0.0156%, [5]). In the leftover of a food poisoning case an amount of 3.7 µg/g ustalic acid was determined [20].

Ustalic acid is an inhibitor of Na^+/K^+-ATPase. About 10 mg/animal is fatal for mice [16].

☠ Ingestion of mushrooms containing ustalic acid leads to vomiting, diarrhea, and gastrointestinal pain.

2.2.2 Cycloprop-2-ene-carboxylic Acid as Toxin of Russula Species

Although poisonings by the Japanese fungus *Russula subnigricans* Hongo, rank russula (Ph. 2.4), Russulaceae, have been observed already in the year 1954, the responsible toxin cycloprop-2-ene-carboxylic acid (Fig. 2.2) was first isolated in 2009 [12]. This substance is unstable, volatile, and polymerizes easily. The polymerized product lacks toxicity. Additionally, *R. subnigricans* contains cytotoxic chlorinated phenylether, the russuphelins (Fig 2.2 [18]), and the unusual amino acids (*S*)-(–)-baikain, (2*S*,3*R*)-(–) hydroxybaikain, and L-pipecolic acid (6, Fig. 2.2).

Cycloprop-2-ene carboxylic acid | Russuphelin A | Baikain | Pipecolic acid

Fig. 2.2: Possible toxins of *Russula subnigricans.*

Ph. 2.4: *Russula subnigricans*, rank russula (source: Zhuo-Hong Chen, provided by Guo-Jie Li).

Cycloprop-2-ene-carboxylic acid causes severe rhabdomyolysis, but not through direct interaction with myocytes but by triggering some other biochemical reactions. The LD_{100} value of cycloprop-2-ene carboxylic acid in mice is 2.5 mg/kg BW, p.o. Serum creatine phosphokinase activity is increased [12, 13].

Intoxications by *R. subnigricans* have been reported from Japan [6, 8], Taiwan, South Korea [2, 3, 14], and China [10, 15]. They were mainly due to confusion with *R. adusta* (Pers.) Fr., blackening brittlegill, a limited edible species [10], that contains the ellagic acid derivative nigricanin [19].

☠ Poisoning by *Russula subnigricans* begins with gastrointestinal effects like nausea, vomiting, and diarrhea usually within 2 h after ingestion, in rare cases later. Most cases resolve without documented rhabdomyolysis over 24 h. In a minority of cases, rhabdomyolysis (peak creatine kinase > 200,000 IU/L), myalgias, hypertension, renal failure, hyperkalemia, and cardiovascular collapse occur. Speech impairment, convulsions, and loss of consciousness have been observed. Fatal cases have been reported [2, 3, 11]. In one case, the symptoms were misdiagnosed as myocardial infarction [14]. A patient with glucose-6-phosphate dehydrogenase deficiency developed acute hemolysis on the second day after ingestion [15].

References

For numbers in bold, see cross-chapter literature p. 233.

[1] Aoki W et al. (2021) Mycoscience 62: 307
[2] Cho JT, Han JH (2016) J Korean Med Sci 31: 1164
[3] Chun MS et al. (2023) Wilderness Environ Med 34(3): 372
[4] Glamočlija J et al. (2015) J Ethnopharmacol 162: 323
[5] Ito T et al. (2021) J Nat Med 75(3): 688
[6] Kusano G et al. (1987) Chem Pharm Bull (Tokyo) 35: 3482
[7] Kwon O et al. (2022) Medicine (Baltimore) 101(10): e28997
[8] Lee PT et al. (2001) Am J Kidney Dis 38: E17
[9] Lee S et al. (2020) J Korean Med Sci 35(19): e122
[10] Li H et al. (2021) China CDC Weekly 3(3): 41
[11] Lin S et al. (2015) Wilderness Environ Med 26: 380
[12] Matsuura M et al. (2009) Nat Chem Biol 5: 465
[13] Matsuura M et al. (2016) Chem Pharm Bull 64: 602
[14] Min MK et al. (2022) Wilderness Environ Med 33(3): 324
[15] Pu Y et al. (2023) Clin Toxicol (Phila) 61(6): 473
[16] Sano Y et al. (2002) Chem Commun (13): 1384
[17] Sawayama Y et al. (2006) Biosci Biotechnol Biochem 70(12): 2998
[18] Takahashi A et al. (1993) Chem Pharm Bull (Tokyo) 41: 1726
[19] Tan JW et al. (2004) Helv Chim Acta 87: 1025
[20] Yoshioka N et al. (2020) Forensic Sci Int 317: 110554

2.3 Polyynes as Possible Toxins of Macrofungi

Polyynes (polyacetylenes) are aliphatic hydrocarbons with single bonds alternating with triple bonds $(-C \equiv C-)_n$ **[160 = Vol. 2]**.

About 213 polyynes, isolated from Basidiomycota and Ascomycota, have been known in 2023. Most of them possess antibacterial and antifungal activities [5]. They are probably part of the defense strategy of the fungi. It is conceivable that such compounds may be among the substances that make some species of mushrooms toxic when they are ingested in the raw state. Examples are:

- Cinnatriacetins A–C, isocinnatriacetin A and B, ethylcinnatriacetin A (Fig. 2.3, from fruit bodies, [6, 7]), and feldin (Fig. 2.3, from mycelial cultures, [4]) from *Fistulina hepatica* (Schaeff.) With., beefsteak fungus (Ph. 2.5), Fistulinaceae, young edible, possess antifungal activity.
- Diatretynes 1, 2, and 3 from cultures of *Clitocybe diatreta* (Fr.) P. Kumm., pierced funnel, Clitocybaceae [1].
- Diatretyne 2 (nudic acid B, Fig. 2.3) from cultures of *Collybia nuda* (Bull.) Z.M. He & Zhu L. Yang (*Lepista nuda* (Bull.) Cooke, Ph. 2.6), wood blewit, Tricholomataceae, toxic when raw ingested [5].
- 10-Hydroxy-undeca-2,4,6,8-tetraynamide from *Mycena viridimarginata* P. Karst., olive-edge bonnet (Ph. 2.7, Mycenaceae [2]).
- Xerulin, dihydroxerulin, and xerulinic acid (Fig. 2.3) from cultures of *Oudemansiella melanotricha* (Dörfelt) M.M. Moser (*Xerula melanotricha* Dörfelt), Physalacriaceae, inhibitors of 3-hydroxy-3-methylglutaryl-coenzyme A reductase and thus of cholesterol biosynthesis [3].

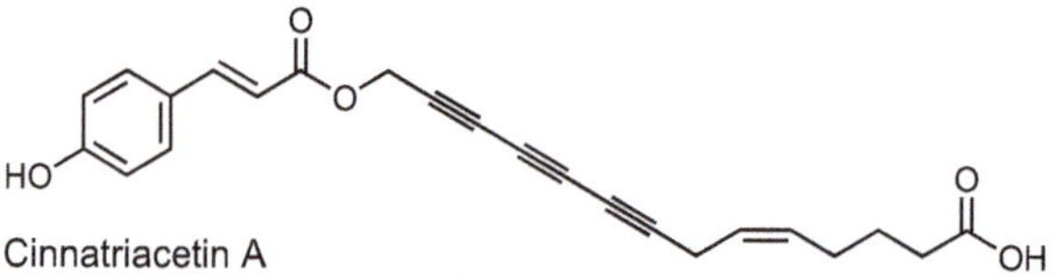

Cinnatriacetin A

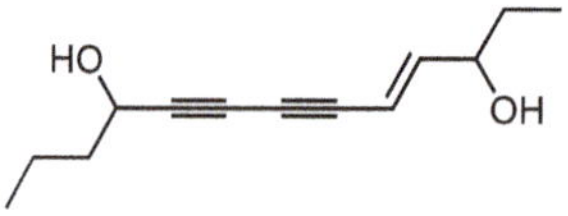

Feldin

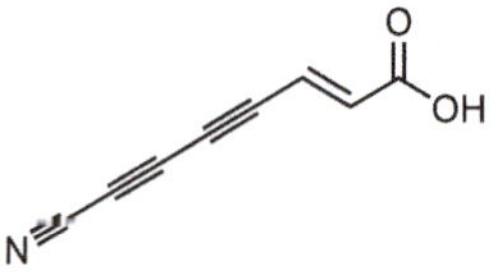

Diatretyne 2

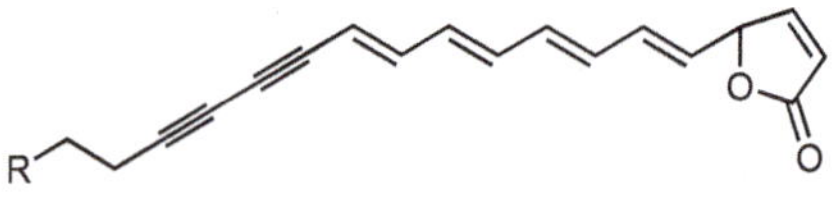

Xerulin R=CH_3

Xerulinic acid R=COOH

Fig. 2.3: Polyynes as possible toxins of Basidiomycetes.

Ph. 2.5: *Fistulina hepatica*, beefsteak fungus (source: Geert Schmidt-Stohn).

Ph. 2.6: *Collybia nuda*, wood blewit (source: Geert Schmidt-Stohn).

Ph. 2.7: Mycena sp. (source: Geert Schmidt-Stohn).

References

For numbers in bold, see cross-chapter literature p. 233.

[1] Anchel M (1955) Science 121 (3147): 607
[2] Bäuerle J et al. (1982) Arch Microbiol 132(2): 194
[3] Kuhnt D et al. (1990) J Antibiot (Tokyo) 43(11): 1413
[4] Lee J et al. (2020) Biomolecules 10: 1502
[5] Qi JS et al. (2023) Nat Prod Bioprospect 13: Art. 50
[6] Tsuge N et al. (1999) J Antibiot Res 52: 578
[7] Whaley AO et al. (2023) Int J Med Mushrooms 25(12): 43

2.4 Sesquiterpenes as Toxins of Macrofungi

2.4.1 General

Terpenes occur in fungi in a great variety. Sesqui- and triterpenes are particularly common. Mono- and diterpenes were found less frequently (overview: [1, 5, 17]).

The isolated sesquiterpenes belong to the drimane, guaiane, hirsutane, illudane, protoilludane, secoilludane, illudalane, lactarane, secolactarane, marasmane, and sterpurane types, among others. So far, about 1,000 sesquiterpenoids have been obtained from Basidiomycetes. They serve as protectors against predators, and have probably further ecological functions (for biosynthesis, see [51]). Most representatives of these groups of substances have been tested only for their antimicrobial and cytostatic activity [9, 43, 51]. Therefore, we know little about their toxicology. However, some sesquiterpenes are responsible for the toxicity or at least contribute to the toxicity of poisonous representatives of the genera Lactarius, Lactifluus, and Russula (see Section 2.4.2), Armillaria (see Section 2.4.3), and Omphalotus (see Section 2.4.4). Other sesquiterpenes are important mycotoxins; among them are trichothecenes of Fusarium species (see Section 3.2.1). The toxins of the very dangerous macrofungus *Trichoderma cornu-damae* are also trichothecenes (see Section 2.4.5).

2.4.2 Sesquiterpenes as Pungents of Lactarius, Lactifluus, and Russula Species

The sesquiterpenes of the genera Lactarius and Lactifluus (Russulaceae) possess a lactarane, secolactarane, marasmane, protoilludane, and, more rarely, a drimane, guaiane, or humulane base body [6, 42]. In almost all cases, the intact fungi contain sesquiterpene alcohol fatty acid esters that are enzymatically cleaved when the hyphae are injured. The released sesquiterpene alcohols cause the pungent taste of the fungi and have antimicrobial, in some cases also cytotoxic and mutagenic effects. The azulene formed from colorless azulenogenic guaiane derivatives after hydrolysis of the esters under the influence of atmospheric oxygen stains the wound surfaces blue.

Of particular importance are the enzymatically and/or spontaneously modified sesquiterpenes released by some pungent-tasting Lactifluus species. Thus, in *Lactifluus vellereus* (Fr.) Kuntze (*Lactarius vellereus* (Fr.) Fr.), fleecy milkcap (Ph.2.8), the sesquiterpene dialdehydes isovelleral (marasmane-type) and velleral (lactarane-type, Fig. 2.4), formed after cleavage of the velutinal stearyl ester, are responsible for the pungent taste of this fungus. Both or one of the aldehydes can also be isolated from *Lactarius rufus* (Scop.) Fr., rufous milkcap, *Lactarius torminosus* (Schaeff.) Pers., woolly milkcap, *Lactifluus piperatus* (L.) Roussel (*Lactarius piperatus* (L.) Pers.), peppery milkcap, and other pungent-tasting species. 6′-Ketostearylvelutinal was detected in *Lactarius plumbeus* (Bull.) Gray, ugli milkcap, and in *L. necator* (Bull.) Pers., firricer. Other pungent-tasting aldehydes, e.g., piperdial and piperalol, have also been found in Lactarius and Lactifluus species (Figs. 2.4, [6, 10, 44–46]).

Presumably, such and similar aldehydes are also the pungents of Russula species, e.g., of *Russula emetica* (Schaeff.) Pers., sickener (Ph. 2.9). Further examples are lactaral (Fig. 2.4) from *Russula sardonia* Fr. (*Russula drimeia* Cooke), primrose brittlegill, and piperdial from *R. queletii* Fr., fruity brittlegill [3, 44]. In *Russula lepida* Fr., rosy brittlegill, sesquiterpenes of the nardosinane and aristolane types were found [21, 50]. Cucurbitacins, belonging to triterpenes (see **159 = Vol. 1**), have also been detected in Russula species.

The pungent-tasting compounds are responsible for the inedibility (and toxicity?) of the mentioned Lactarius, Lactifluus, or Russula species. Although the pungent taste of these mushrooms makes them easily distinguishable from edible members of the genera, confusion and poisoning are common. Discarding the cooking water, which is occasionally recommended, does not provide reliable protection.

Stearylvelutinal → → Isovelleral → → Velleral   Piperdial R=CHO, Piperalol R=CH_2OH Lactaral

Fig. 2.4: Sesquiterpenes of Lactarius, Lactifluus, and Russula species.

Ph. 2.8: *Lactifluus vellereus*, fleecy milkcap (source: Burkhard Wysekal).

Ph. 2.9: *Russula emetica* agg., sickener (source: Jesko Kleine).

☠ After a short latency period, symptoms of poisoning after ingestion of pungent Lactarius, Lactifluus, or Russula species are nausea, vomiting, diarrhea, and in severe cases unconsciousness and signs of liver and kidney damage **[19]**.

🚑 The poisonings can only be treated symptomatically **[19]**.

(☹) The strong mutagenic effect of isovelleral (unsaturated dialdehyde!) found in the Ames test probably plays no role in vivo [47], since rapid degradation of the compounds seems to occur. Necatorin, first isolated from *Lactarius necator*, shows in vitro also mutagenic effects. *L. necator* is a common edible mushroom in Finland and Eastern Europe. Because of safety reasons, children and pregnant woman should avoid consumption of these mushrooms [36, 48].

2.4.3 Sesquiterpenes as Toxins of Armillaria Species

Fungi of the genus Armillaria (Physalacriaceae) are destructive forest pathogens. The genus includes, among other species, *Armillaria mellea* (Vahl.) P. Kumm., honey tuft (Ph. 2.10), and *A. ostoyae* (Romagn.) Herink, dark honey fungus. One specimen of *A. ostoyae* that has spread over 37 ha and weighs 440 tons is said to be one of the largest organisms on the Earth. In China, *A. mellea* is used in gastrointestinal problems and to strengthen the bones [30, 39].

Ph. 2.10: *Armillaria mellea*, honey tuft (source: Markus Scholler).

Numerous esters of sesquiterpene alcohols with protoilludane base bodies have been detected in the mentioned species. Among the more than 70 representatives described are melleolide and armillyl orsellinate (Fig. 2.5). The compounds are meroterpenoids and possess additionally to the terpene part a phenolic acid (orsellinic acid) derivative that originates from polyketide pathway [13, 14, 19, 30, 33, 34, 37, 38].

Some melleolides inhibit 5-lipoxygenase [24] and exhibit antibacterial, antifungal, and cytostatic activity [7, 30, 34, 38].

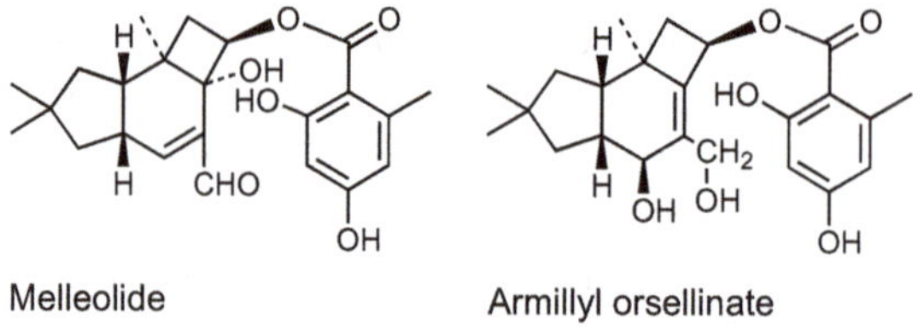

Melleolide

Armillyl orsellinate

Fig. 2.5: Sesquiterpenes of Armillaria species.

(☠) If the honey mushrooms are prepared incorrectly (failure to discard the cooking water, insufficient cooking) and consumed in large quantities, they can cause diarrhea and vomiting [**42**]).

2.4.4 Sesquiterpenes as Toxins of Omphalotus Species

Illudins, especially illudin S (Fig. 2.6), are held to be responsible for the toxicity of *Omphalotus illudens* (Schwein.) Bresinsky & Besl, Eastern jack–o'–lantern, Omphalotaceae. Sesquiterpenes with illudin base bodies could also be detected in the oil tree mushroom species *Omphalotus olearius* (DC.) Singer, jack–o'–lantern, shining oil tree fungus (Ph. 2.11), *Omphalotus olivascens* H.E. Bigelow, O.K. Mill. & Thiers, Western jack–o'–lantern, *Omphalotus guepiniiformis* (Berk.) Neda (*Omphalotus japonicus* (Kawam.) Kirchm. & O.K. Mill.), moonlight mushroom, [31, 32, 35], and *Omphalotus mexicanus* Guzmãn & V. Mora, Mexican blue jack–o'–lantern [15].

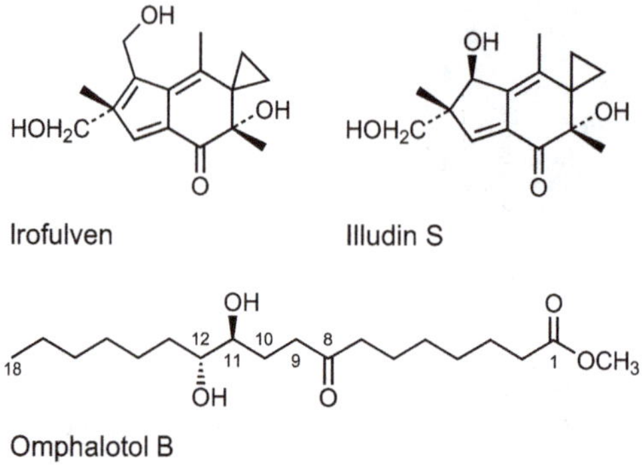

Irofulven

Illudin S

Omphalotol B

Fig. 2.6: Ingredients of Omphalotus species and a semisynthetic illudin derivative (irofulven).

Ph. 2.11: *Omphalotus olearius*, jack–o'–lantern (source: Markus Scholler).

O. illudens and *O. olearius* are distributed in Europe and in North America, *O. olearius* mainly in the Mediterranean region. *O. olivascens* occurs at the Pacific Coast of North America, *O. guepiniiformis* in Eastern Asia, and *O. mexicana* in Central America.

The omphalotols A and B (Fig. 2.6) are fatty acid derivatives from *Omphalotus guepiniiformis* with antimicrobial activity against *Helicobacter pylori* [27]. Additionally, a cytotoxic polyisoprenepolyol, named omphaloprenol A, has been isolated from the mycelium of this species [4].

The illudins inhibit DNA biosynthesis and have a strong cytotoxic effect [31]. In vitro studies show strong mutagenic effects of illudin S that are caused preferentially by alkylating purine residues of DNA and induction of oxidative stress [11].

Poisoning by *Omphalotus olearius* occurs mainly in Southern Europe through confusion with *Cantharellus cibarius* Fr., chanterelle. A case of confusion of *Omphalotus illudens* with *Laetiporus sulphureus* (Bull.) Murrill, chicken of the wood, involving several school children, has been reported from North America [16]. In Japan, confusion of *Omphalotus guepiniiformis* with *Lentinula edodes*, shiitake, led to hemorrhagic enteritis [18].

Salivation, nausea, vomiting, headache, weakness, and sweating or chills set in 1–2 h after ingestion of oil tree mushrooms [**19**].

Treatment must be symptomatic. Within a few days, the symptoms of poisoning are overcome [16, **19**].

Illudin S has antiviral activity [29]. Irofulven (hydroxymethylacylfulvene (HMAF, Fig. 2.6), a semisynthetic derivative of illudin S, was under clinical investigation as an antitumor agent. A clinical study in patients with recurrent ovarian cancer reported dose-limiting retinal toxicity, failing to progress beyond phase II trials [41]. The observation that the compound is particularly effective in tumors with a nucleoside excision repair deficiency led to a revival of interest in the compound [8].

2.4.5 Macrocyclic Trichothecenes as Toxins of *Trichoderma cornu-damae*

The macrocyclic trichothecenes satratoxin G, satratoxin H (Fig. 2.7), satratoxin H 12′,13′-diacetate, satratoxin H 12′-acetate, satratoxin H 13′-acetate, and 12′-episatratoxin H are highly toxic compounds of *Trichoderma cornu-damae* (Pat.) Z.X. Zhu & W.Y. Zhuang (*Podostroma cornu-damae* (Pat.) Boedijn), poison fire coral (Ph. 2.12), Hypocreaceae, Ascomycota. Additionally, roridin D, roridin E (Fig. 2.7, for miotoxin A see Section 3.2.1.2), roridin Q, verrucarin J, and verrucarol have been detected in the fungus [12, 25, 40, 52]. Furthermore, the trichothecenes miophycoten D, roridin F, and satratoxin I have been isolated from the methanolic extract of the fungi grown on culture plates [28]. Roridin F and satratoxin I are not complete macrolides, because the macrolide ester bridge between C-4 and C-15 is not closed. They might be transformed into roridin E and satratoxin H through a one-step esterification. The complete mitochondrial genome of the fungus has been elucidated [26].

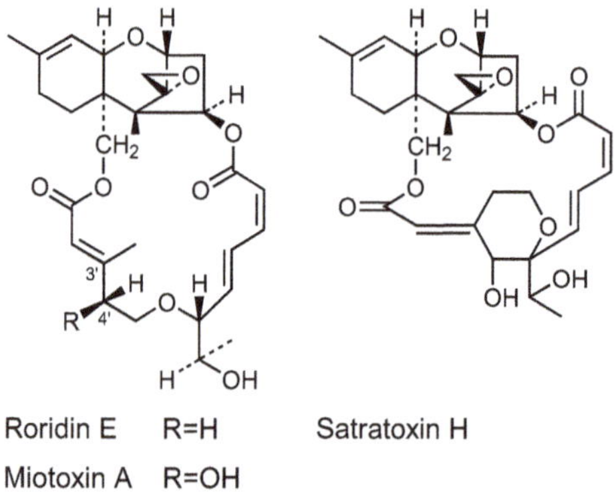

Roridin E R=H

Miotoxin A R=OH

Satratoxin H

Fig. 2.7: Macrocyclic trichothecenes of *Trichoderma cornu-damae* and a transformation product (miotoxin A).

Trichoderma cornu-damae is native to Korea, Japan, China, Papua-New Guinea, and to some parts of Indonesia and Australia. The fruit body of the very rare mushroom is shaped like a deer's red horn. In its immature period it resembles *Ganoderma lucidum* (Curtis) P. Karst, reishi mushroom, and *Cordyceps militaris* (L.) Fr., caterpillar fungus, both well-known medicinal fungi, so that confusion may be possible [2, 22, 26]. Confusion is also possible with *Clavulinopsis miyabeana* (S. Ito) S. Ito, a red coral mushroom [**13**].

The satratoxins inhibit protein synthesis. All the contained trichothecenes expect verrucarin J had a lethal effect on mice by at least 0.5 mg per mouse [40].

Several cases of serious poisonings, some with fatal outcome, after ingestion of the fruit bodies, e.g., boiled, fried, cooked in sake, or as tea, have been reported from Japan and Korea [2, 12, 20, 23, 40, 49]. The species is evaluated to be the most dangerous mushroom species in the world.

Ph. 2.12: *Trichoderma cornu-damae*, poison fire coral (source: Ray Palmer).

A 64-year-old man came to the emergency department with a 10-day history of fever, desquamation on his palms, soles, and scalp. The man complained of chilling sensation, weakness, mild headache, and hair loss. Laboratory tests showed severe pancytopenia. Though intense treatment, the patient died due to multiple organ failure on day 6 in hospital. His wife showed also symptoms of poisoning but survived. The Korean couple had harvested and dried wild mushrooms from a nearby forest, which were mistaken as *Ganoderma lucidum*, and had drunk the boiling water from the mushrooms daily for over a month before admission. A piece of *T. cornu-damae* has been found among the samples of *G. lucidum* [2].

Symptoms of poisoning in the early stages are vomiting, diarrhea, and dehydration. This is followed by changes in perception, leukocytopenia and thrombocytopenia, hair loss, desquamation on the palms and face, neurological problems like speech impediment and movement disorders, sepsis and multiorgan failure including kidney failure, liver necrosis, and disseminated intravascular coagulation [2, 22, 23, 40]. A fatal respiratory failure after ingestion of the mushrooms has been observed [20]. The time till the onset of the symptoms ranges from a few hours after ingestion of the mushrooms itself until a few days after drinking of a mushroom tea [22]. It is controversial whether the fungi are toxic when only touched.

Recommended treatment measures are hemodiafiltration, plasmapheresis, administration of granulocyte colony-stimulating factor, and i.v. application of a large volume of saline (9 L) over a 12 h period [22, 49].

Roridin F and 12′-episatratoxin H possess specific cytotoxicity against mouse breast cancer cells that is stronger than that of doxorubicin [25].

References

For numbers in bold, see cross-chapter literature p. 233.

[1] Abraham WR (2001) Curr Med Chem 8: 583
[2] Ahn JY et al. (2013) Yonsei Med J 54(1): 265
[3] Andina D et al. (1980) Phytochemistry 19: 93
[4] Aoki S et al. (2021) Biosci Biotechnol Biochem 85(6): 1364
[5] Ayer WA, Browne LM (1981) Tetrahedron 37: 2199
[6] Bergendorf O, Sterner O (1988) Phytochemistry 27: 97
[7] Bohnert M et al. (2014) Bioorg Med Chem 22(15): 3856
[8] Börcsök J et al. (2021) Z Clin Cancer Res 27: 2011
[9] Cadelis MM et al. (2020) Antibiotics 9: 928
[10] Camzine S, Lupo AT Jr. (1984) Mycologia 76: 355
[11] Casimir L et al. (2023) DNA Repair 122: 103433
[12] Choe S et al. (2018) Case Forensic Sci Int 2018: 234. Doi: 10.1016/forsciint.2018.0
[13] Donelly DMX et al. (1985) J Nat Prod 48: 10
[14] Dörfer M et al. (2019) Mycol Progr 18: 1027
[15] Eckhardt P et al. (2024) Phytochem Anal 35(3): 469
[16] French AL et al. (1988) J Toxicol Clin Toxicol 26: 81
[17] Gao JM (2006) Curr Org Chem 10: 849
[18] Hori K et al. (2008) Int J Surg Pathol 16: 62
[19] Huang K, Xu B (2023) Food Funct 14(21): 9518
[20] Jang J et al. (2013) Tuberc Respir Dis 75: 264
[21] Jian Wen T et al. (2000) Helv Chim Acta 83(12): 3191
[22] Kim HN et al. (2016) Clin Exp Emerg Med 3/3): 186
[23] Koichi M et al. (2003) Acta Criminol Med Leg Japonica 69(1): 14
[24] König S et al. (2019) Cell Chem Biol 26(1): 60
[25] Lee BS et al. (2024) Separations 11: 65
[26] Lee HY et al. (2022) Mitochondrial DNA Part B 7(11): 1899
[27] Lee S et al. (2022) Pharmaceuticals 15(2): 139
[28] Lee SR et al. (2019) J Nat Prod 82(1): 122
[29] Lehmann VKB et al. (2003) J Nat Prod 66: 1257
[30] Li Z et al. (2016) J Ethnopharmacol 184: 119
[31] McMillan JA et al. (1976) Tetrahedron Lett 4219
[32] McMorris TC et al. (2002) Phytochemistry 61(4): 395
[33] Midland SL et al. (1982) Tetrahedron Lett 23: 2515
[34] Momose I et al. (2000) J Antibiot 53: 137
[35] Nair MSR et al. (1983) Mycologia 75: 920
[36] Nieminen P, Mustonen AM (2020) Toxins 12: 639
[37] Obuchi T et al. (1990) Planta Med 56: 198
[38] Orban A et al. (2024) Chembiochem 13: e202400168
[39] Ren S et al. (2023) Appl Biochem Biotechnol 195(5): 3491
[40] Saikawa Y et al. (2001) Tetrahedron 57(39): 8277
[41] Schilder R et al. (2010) Int J Gynecol Cancer 20: 1137
[42] Schmitt JA (1974) Z Pilzkunde 39: 219
[43] Steglich W (1981) Pure Appl Chem 53: 1233
[44] Sterner O et al. (1982) Mutat Res 101: 269
[45] Sterner O et al. (1985) Tetrahedron Lett 26: 3163

[46] Sterner O et al. (1985) J Nat Prod 48: 279
[47] Sterner O et al. (1987) Mutat Res 188: 169
[48] Suortti T et al. (1983) Phytochemistry 22: 2873
[49] Suzuki M et al. (2002) Chudoku Kenkyu 15(2): 177
[50] Vidari G et al. (1998) Tetrahedron Lett 39: 6073
[51] Wu J et al. (2022) J Fungi 8: 913
[52] Zhu M et al. (2020) Toxins 12: 417

2.5 Triterpenes as Toxins of Macrofungi

2.5.1 General

Fungi contain a high variety of triterpenes and steroids. Triterpenic acids are particularly widespread in wood-destroying fungi, e.g., trametenolic acid, polyporenic acids, and eburicolic acid [30]. Only in a few cases the pharmacological effects of the compounds have been studied. However, it may be assumed that, like the sesquiterpenes of fungi (see Section 2.4), they are part of chemical defense and exert effects on microorganisms and animal organisms.

Of toxicological relevance are triterpenes from Tricholoma species (see Section 2.5.2), the fasciculols from *Hypholoma fasciculare* (see Section 2.5.3), and the hebelomic acids and hebevinosides from Hebeloma species (see Section 2.5.4).

2.5.2 Triterpenes as Possible Toxins of Tricholoma Species

The genus Tricholoma (Tricholomataceae) includes about 200 species that are found in temperate and subtropical zones in the Northern and the Southern Hemisphere. The gilled mushrooms form ectomycorrhiza with trees of the families of Pinaceae, Betulaceae, and Fagaceae. Phylogenetic analyses have led to numerous changes in taxonomical classification during recent time and may explain discrepancies in the metabolites found in mushrooms known with the same name.

Triterpenes that may be responsible for the toxicity of some Tricholoma species are, among others, the saponaceolides A–S, the tricholopardins, the terreolides, and the tricholidic acids.

Saponaceolide A and B (Fig. 2.8) were first isolated in 1991 by De Bernardi et al. [7] from the fruit bodies of *Tricholoma saponaceum* (Fr.) P. Kumm., soapy knight, soap-scented toadstool (Ph. 2.13), together with some related saponaceolides. Saponaceolides have also been found in *T. terreum* (Schaeff.) P. Kumm., grey knight, dirty tricholoma (Ph. 2.14), and in *T. scalpturatum* (Fr.) Quél., yellowing knight [6, 9].

The tricholopardins A–D (Fig. 2.8), together with antibacterial pardinumones [26], have been isolated from *Tricholoma pardinum* (Pers.) Quél., spotted tricholoma, tiger tricholoma [19]. The tricholidic acids (Fig. 2.8), together with some saponaceolides and

Ph. 2.13: *Tricholoma saponaceum*, soapy knight (source: Markus Scholler).

Ph. 2.14: *Tricholoma terreum*, grey knight (source: Geert Schmidt-Stohn).

tricholomenyn C, have been obtained from *T. ustaloides* Romagn., charred knight, and from an unspecified Japanese Tricholoma species, probably *T. albobrunneum* (Pers.) P. Kumm. [13, 18].

The saponaceolides and the tricholopardins are characterized by a central methylenecyclohexane ring linked to a γ-lactone substituent and a pyrane ring that is part of a spiroketal with the adjacent heterocyclic system [19].

Representatives of both groups exhibit cytotoxic effects, especially against tumor cells [23]. The LD_{50} of saponaceolide B is 88.3 mg/kg BW, p.o., mice. It led to a minimal increase in the creatine kinase level in mice (1.52–1.65 times), which was assumed to be a cause of rhabdomyolysis [29].

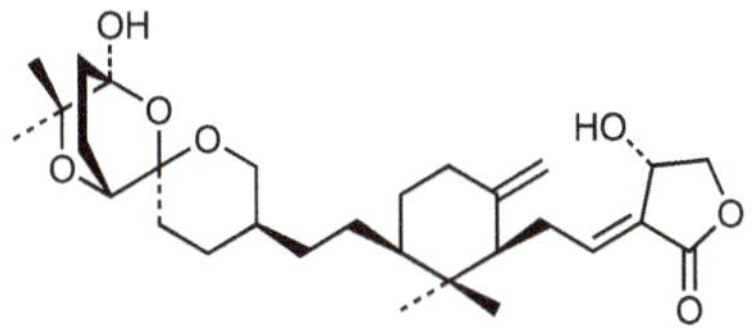

Saponaceolide A

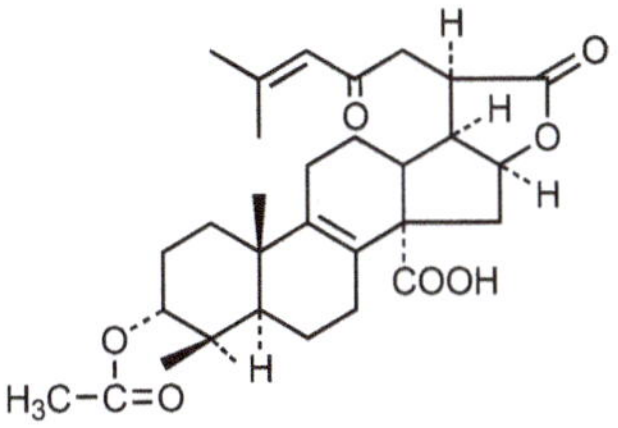

Tricholidic acid

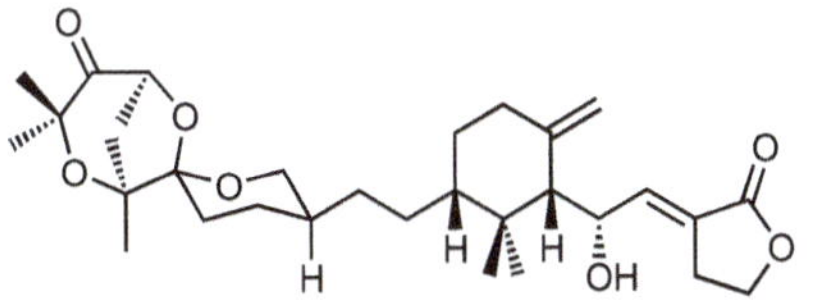

Terreolide A

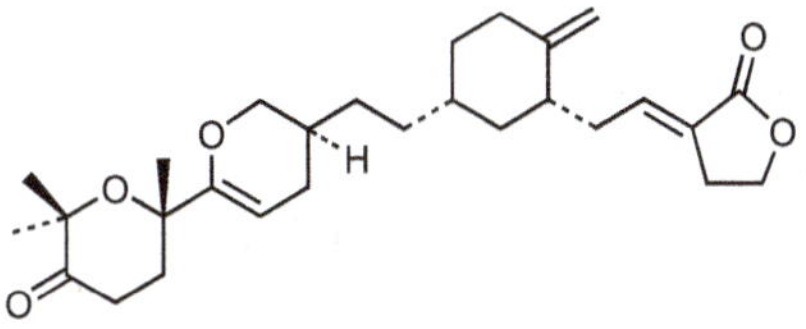

Tricholopardin C

Fig. 2.8: Triterpenes of Tricholoma species.

Tricholoma saponaceum gets its name from the soap-suds-like odor. Its cap is olive-green, rarely blackish in color, the lamellae are whitish with a greenish-yellowish tinge. It is found in deciduous and coniferous forests. It is toxic when eaten raw. Even cooked, it may cause nausea and vomiting when it is ingested in larger quantities **[107]**.

Tricholoma pardinum is widely distributed across Europe, North America, and some parts of Asia. Eating of these mushrooms leads to gastrointestinal symptoms such as nausea, vomiting, and diarrhea.

Tricholoma terreum is a popular edible mushroom (vernacular name 'moretta'), mainly in alpine regions of France and Italy. There are no documented cases of poisoning due to its ingestion. Nevertheless, Chinese scientists found numerous triterpenes (terreolides, terreumols, 16 saponaceolides including saponaceolide B) in *T. terreum*, collected in Europe, and evaluated the mushrooms as possibly poisonous

[29]. Newer investigations confirmed only the presence of saponaceolide B in fruit bodies of this species. Considering the mentioned LD_{50} value of 88.3 mg/kg BW for saponaceolide B this value would approx. correspond to a toxic dose of 6.16 g for a 70 kg person. Assuming a saponaceolide B content in the fruit bodies of around 47 mg/kg FW means that serious intoxication of an adult human would require the consumption of more than 130 kg of fresh fruit bodies. Thus, *T. terreum* can be considered an edible mushroom [6, **13**].

Tricholoma ustaloides, T. albobrunneum, T. focale (Fr.) Ricken, booted knight, *T. fulvum* (DC.) Bigeard & H. Guill., bitch knight, and *T. pessundatum* (Fr.) Quél., tacked knight, are suspected of being poisonous or inedible due to their unpleasant smell and taste.

Poisonings occurring in the United States and Canada after consumption of *Tricholoma magnivelare* (Peck) Redhead, American matsutake, were characterized by renal failure. They were due to confusion of the mushrooms with *Amanita smithiana* Bas [17] that contains 2-amino-4,5-hexadienoic acid as possible toxin (see Section 2.7.1.1).

The compounds that are responsible for the possible toxicity of *Tricholoma equestre* (see Section 2.12.1) have not been identified until now. *T. equestre* was for a long time highly valued as culinary delicacy.

2.5.3 Fasciculols as Toxins of Hypholoma Species

Hypholoma species (Strophariaceae) are distributed worldwide. From a toxicological point of view, *Hypholoma fasciculare* (Huds.) P. Kumm. (Ph. 2.15), sulphur tuft, clustered woodlover, and *Hypholoma lateritium* (Schaeff.) P. Kumm. (*H. sublateritium* (Fr.) Quél.), brick tuft, should be mentioned.

The genus Hypholoma is characterized by mushrooms with small to medium-sized caps, mostly thin-fleshed. The stalk is un-ringed and bears at most remnants of the veil. The lamellae are attached, deeply bulged, rounded, or broadly adnate. The spore dust is dark brown to blackish purple in color. Most of the species are wood dwellers.

Hypholoma fasciculare has a sulfur-yellow cap with a fox-colored apex and first yellow, then greenish lamellae. The mushroom grows up to 10 cm high, the cap up to 7 cm wide. The flesh is sulfur yellow and has a bitter taste. It occurs from May to November on tree stumps and roots, rarely on living wood, and is in Europe very common. *Hypholoma lateritium* has a brick-reddish cap and a yellowish stem with a rusty brown base. In Japan, Korea, and the United States it is considered edible, in Europe it is considered inedible or poisonous.

Fasciculol E and fasciculol F, depsipeptides of a polyhydroxylanostane, were found to be the main active ingredients and responsible for poisonous effects of *H. fasciculare* (Fig. 2.9). The fasciculol type also includes fasciculic acids A to F, from which D–F are of interest as inhibitors of 3α-hydroxysteroid dehydrogenase [16]. The

Ph. 2.15: *Hypholoma fasciculare*, sulphur tuft (source: Burkhard Wysekul).

fasciculols A, B, C, and E and the sublateriols A–C have been detected in *H. lateritium* [4, 27]. Fasciculol E has also been isolated from the slender pholiota, *Pholiota spumosa* (Fr.) Singer (Strophariaceae, [5]). Whether the amounts of these compounds present in this species are toxicologically relevant is unclear. The pholiols are cytotoxic triterpenes in *Pholiota populnea* (Pers.) Kuyper & Tjall.-Beuk., destructive pholiota [28]. *Pholiota aurivella* (Batsch) P. Kumm., golden pholiota (Ph. 2.16), *Ph. lenta* (Pers.) Singer, and *Ph. squarrosa* (Vahl) P. Kumm., shaggy pholiota, are also suspected to be poisonous.

Noteworthy are also the sesquiterpenes fascicularenone E, F, and G that have been obtained from the culture medium of *Hypholoma fasciculare* [20], and naematolin from cultures of *H. lateritium*. Naematolin is a bicyclic caryophyllene sesquiterpene with cytotoxic and antimicrobial activity [2]. A caryophyllene synthase necessary for its biosynthesis could be identified [1].

The fasciculols are calmodulin antagonists. Death in mice is caused by paralysis of the respiratory center. The LD_{50} is 50 mg/kg BW fasciculol E or 168 mg/kg BW fasciculol F, i.p., mice [22]. The LD_{50} of a methanolic extract of the fruit bodies of *Hypholoma fasciculare* is 243 mg/kg BW in mice [3].

Severe poisoning of humans by *Hypholoma fasciculare*, with symptoms similar to those after ingestion of *Amanita phalloides* (see Section 2.10), has been described only in the older literature [14, 24, 25]. We are not aware of cases of poisonings from the Pholiota species mentioned and from *H. lateritium*.

The consumption of wild-collected *Hypholoma fasciculare*, along with *Lepiota brunneoincarnata* (see Section 2.10) and *Coprinopsis atramentaria* (see Section 2.7.2.1), was one of the causes of a poisoning of more than 1,200 patients in Northwest Iran in spring 2018. In this period were optimal conditions for growth of mushrooms [21].

Ph. 2.16: *Pholiota aurivella*, golden pholiota (source: Markus Scholler).

☠ Consumption of *H. fasciculare* causes nausea, vomiting, diarrhea, and abdominal pain. The symptoms may be delayed for 5–10 h after consumption but resolve over a few days. The bitter taste is likely to prevent consumption of larger quantities. In fatal cases, fulminant hepatitis with kidney and myocard damage have been reported [**19, 42, 50**].

🚑 Treatment is symptomatic after toxin removal.

2.5.4 Triterpenes as Toxins of Hebeloma Species

Of the European representatives of the genus Hebeloma, Hymenogastraceae, *Hebeloma crustuliniforme* (Bull.) Quél., poison pie (Ph. 2.17), *H. sinapizans* (Paulet) Gillet, bitter poison pie, scaly-stalked hebeloma, rough-stalked hebeloma, and *H. laterinum* (Batsch) Vesterh. (*H. senescens* Berk. & Broome), browning pleco, are considered toxic. *Hebeloma vinosophyllum* Hongo occurs in Asia (Japan, Vietnam).

Compounds with polyhydroxylanostane parent bodies, the cytotoxic hebelomic acids A, B, E, F, H, and I (Fig. 2.10) and their depsipeptides, the crustilinols, have been isolated from them. They are structurally very similar to fasciculols and presumably also similar in action [8, 12, 15].

Poisoning by the European Hebeloma species occurs only occasionally.

Ph. 2.17: *Hebeloma crustuliniforme*, poison pie (source: Jesko Kleine).

☠ Symptoms of poisoning by Hebeloma are gastrointestinal irritation after a short latency period [**19, 42**].

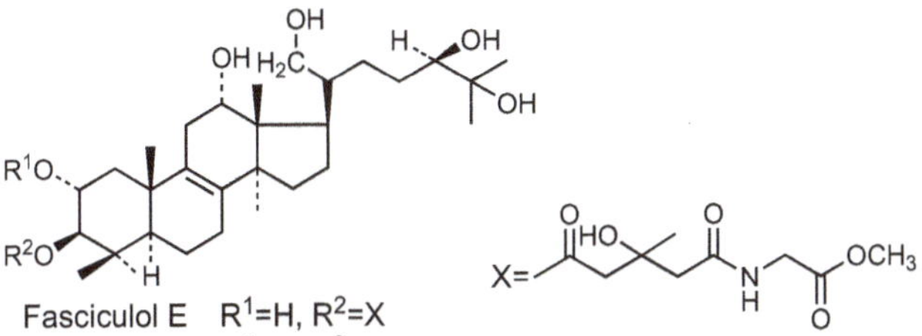

Fasciculol E R^1=H, R^2=X
Fasciculol F R^1=X, R^2=H

Fig. 2.9: Triterpenes of Hypholoma species.

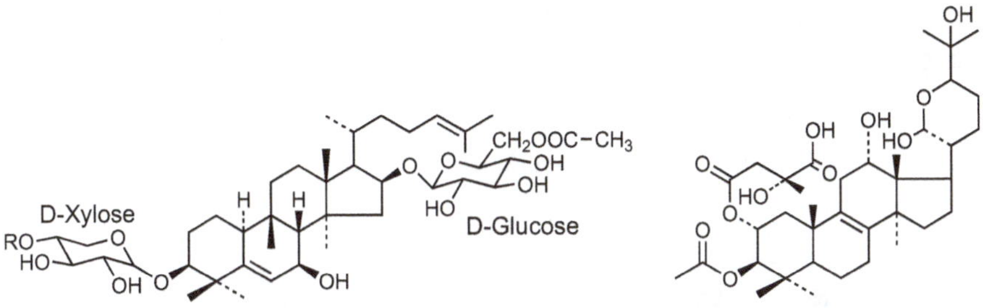

Hebevinoside II R=—CO—CH_3
Hebevinoside III R=H

Hebelomic acid A

Fig. 2.10: Triterpenes of Hebeloma species.

The glycosides hebevinoside II, III, VI, VII, and IX (Fig. 2.10, the other hebevinosides are artifacts) were detected in *Hebeloma vinosophyllum*. The aglycones have a 3,7,16-trihydroxy-cucurbita-5,24-diene parent body (hydroxyhebevinogenin). Monosaccharide components of the mostly bisdesmosidic biosides are D-glucose and D-xylose in positions 3 and 16, rarely only one of both sugars. The sugar residues are partially acetylated [10, 11].

The hebevinosides are neurotoxic to mice and cause paralysis. The LD_{50} of hebevinoside VI in mice is 66 mg/kg BW, i.p. Essential for toxicity is the glucose residue attached to C-16, substitution of the α-positioned hydroxy group present at C-7 by a methoxy group increases the toxicity [11]. Cases of intoxications of humans by *H. vinosophyllum* are not known to us.

References

For numbers in bold, see cross-chapter literature p. 233.

[1] Al-Salihi SAA et al. (2019) Mol Biotechnol 61: 754
[2] Backens S et al. (1984) Justus Liebigs Ann Chem 1984: 1332
[3] Badalyan SM et al. (1997) Mikol Fitopatol 31: 42
[4] Chuluunbaatar B et al. (2024) Molecules 24: 301
[5] Clericuzio M et al. (2004) Croatia Chem Acta 77: 605
[6] Clericuzio M et al. (2024) Molecules 29: 1794
[7] De Bernardi M et al. (1991) Tetrahedron 47: 7109
[8] Dossena A et al. (1996) Tetrahedron Assymmetry 7: 1911
[9] Feng T et al. (2015) Nat Prod Bioprospect 5: 205
[10] Fujimoto H et al. (1986) Chem Pharm Bull (Tokyo) 34: 88
[11] Fujimoto H et al. (1987) Chem Pharm Bull (Tokyo) 35: 2254
[12] Garlaschelli L et al. (1995) J Nat Prod 58: 992
[13] Gilardoni et al. (2023) Molecules 28: 3864
[14] Herbich J et al. (1966) Arch Toxicol 21: 310
[15] Hirotani M, Furuya T (1990) Phytochemistry 29: 3767
[16] Kleinwächter P et al. (1999) J Basic Microbiol 39: 345
[17] Leatham AM et al. (1997) J Toxicol Clin Toxicol 35: 67
[18] Nowe S et al. (1982) Chem Lett 1679
[19] Shi C et al. (2021) Nat Prod Bioprosp 11: 235
[20] Shiono Y et al. (2005) Z Naturforsch B 60: 880
[21] Soltaninejad K et al. (2018) Environ Med 9: 152
[22] Suzuki S et al. (1983) Chem Pharm Bull (Tokyo) 31: 2176
[23] Vidari G et al. (1995) Tetrahedron Asymmetry 6: 2977
[24] Wasiljkov BP (1961) Botan Shurnal XLVI. 581
[25] Wasiljkov BP (1963) Schweiz Z Pilzkunde 41: 117
[26] Yang HX et al. (2021) ACS Omega 2021, doi.org/10.1021/acsomega.1c04418
[27] Yaoita Y et al. (2014) Nat Prod Commun 9: 419
[28] Yazdani M et al. (2023) Pharmaceuticals 16: 104
[29] Yin X et al. (2014) Chem Eur J 20(23): 7001
[30] Yokoyama A et al. (1975) Phytochemistry 14: 487

2.6 Terphenylquinones as Toxins of Macrofungi

Terphenylquinones serve as precursors of pigments of Basidiomycetes. They are phenylpropanoids and formed on the shikimate pathway from amino acids by quinone synthetases [4]. A representative of toxicological relevance is polyporic acid (Fig 2.11). It occurs in high concentration (20–40% of the FW) in *Hapalopilus rutilans* (Pers.) Murrill (*Hapalopilus nidulans* (Fr.) P. Karst, Ph. 2.18), tender nesting polypore, cinnamon bracket, Phanerochaetaceae, and gives this species a violet color. Its structure was elucidated in 1926 by Kögl.

Ph. 2.18: *Hapalopilus rutilans*, tender nesting polypore (source: Markus Scholler).

Polyporic acid is an inhibitor of dihydroorotate dehydrogenase. The enzyme is situated in mitochondria and important for pyrimidine biosynthesis. Rats given polyporic acid (100–800 mg/kg BW, p.o. via probang) developed strongly reduced locomotor activity and depressed visual placing response within 24 h. Laboratory investigations showed hepatorenal failure, metabolic acidosis, hypokalemia, and hypocalcemia [2].

Until now, only few poisoning cases by *H. rutilans* have been described, all from Europe [1, 3, 5]. They were caused mainly by confusion with *Fistulina hepatica*.

O
HO
OH
O
Polyporic acid

Fig. 2.11: A toxin of *Hapalopilus rutilans*.

The poisoning, the polyporic acid syndrome, is characterized by late onset (about 12 h) of gastrointestinal symptoms including jaundice, diarrhea, vomiting, neurological effect, e.g., visual disturbances and vertigo, signs of mild liver and kidney toxicity as proteinuria, leukocyturia, and a conspicuously purple urine. The syndrome resolves over 2–7 days [**180**].

Treatment is symptomatic.

Some inhibitors of dihydroorotate dehydrogenase are used to treat autoimmune diseases like rheumatoid arthritis. Possibly, polyporic acid can be a suitable drug candidate.

References

For numbers in bold, see cross-chapter literature p. 233.

[1] Herrmann M et al. (1989) Mykol Mitteilungsblatt 32(1): 1
[2] Kraft J et al. (1998) Arch Toxicol 72: 711
[3] Langner J (1992) Z Mykol 58(2): 173
[4] Seibold 2023 Fungal Biol Biotechnol 10: 14
[5] Villa AF et al. (2013) Clin Toxicol 5: 798

2.7 Amino Acids as Toxins of Macrofungi

2.7.1 Aliphatic Amino Acids

2.7.1.1 L-2-Amino-4,5-hexadienoic Acid (AHDA) and L-2-Amino-4-pentynoic Acid as Toxins of Amanita Species (*Amanita smithiana*, *Amanita proxima*, and Other)

L-2-Amino-4,5-hexadienoic acid (AHDA, norleucine toxin) and L-2-amino-4-pentynoic acid (Fig. 2.12) are the cause of kidney damages that have been observed after ingestion of *Amanita smithiana* Bas, toxic lepidella, North American lepidella (Ph. 2.19, Amanitaceae) in North America and Taiwan [52, 109, 110, 114] and of *A. pseudoporphyria* Hongo, Hongo's false death cap, in Japan [41]. The compounds also seem to be responsible for poisoning caused by *A. proxima* Dumée (Ph. 2.20) in Mediterranean countries (proximien syndrome: [11, 20, 21, 57]), by *A. cheelii* P.M. Kirk (*A. punctata* Lam., [42], contains according to Vetter also amanitins, see Section 2.10) and *A. neo-ovoidea* Hongo in Korea [53], and by *A. boudieri* Barla, *A. gracilior* Bas & Honrubia and *A. echinocephala* (Vittad.) Quél., solitary amanita, in Portugal and Germany [46].

It is doubtful whether the poisoning of a man on Sicily attributed to *A. ovoidea* (Bull.) Link, bearded amanita [30], was really caused by this species. Because of the similarity to *A. proxima* there was possibly an ambiguous identification [61].

Ph. 2.19: *Amanita smithiana*, toxic lepidella (source: Dick Culbert, Wikimedia, CC-BY-2.0).

Ph. 2.20: *Amanita proxima* (source: Nicos D Karabelas, Wikimedia, CC-BY-SA-4.0).

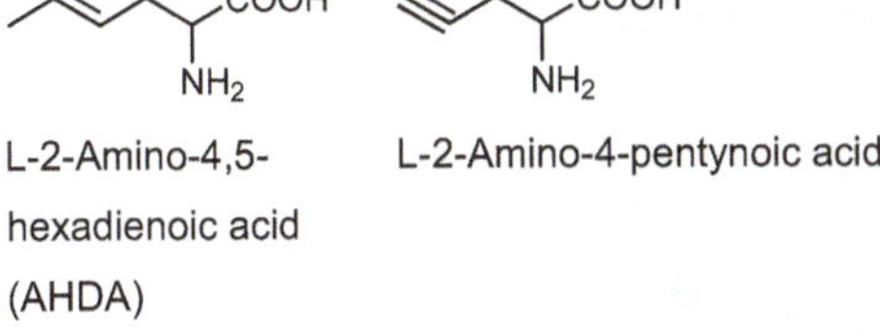

Fig. 2.12: Aliphatic amino acids of Amanita species.

Symptoms occurring 30 min to 12 h after ingestion of the mushrooms are nausea, vomiting, diarrhea, abdominal cramps, headache, weakness, fatigue, myalgias, and rash. They may persist for up to several days. Acute renal failure follows 2–5 days after ingestion and lasts several days. Signs of hepatotoxicity are possible.

Treatment must be symptomatic. Dialysis can become necessary. Final renal recovery is possible [92, **50, 178**].

2.7.1.2 2*R*-Amino-4*S*-hydroxy-5-hexynoic Acid and 2*R*-Amino-5-hexynoic Acid as Toxins of *Trogia venenata*

The unusual amino acids 2*R*-amino-4*S*-hydroxy-5-hexynoic acid and 2*R*-amino-5-hexynoic acid (Fig. 2.13) and γ-guanidinobutyric acid are the causative agents for the Yunnan Sudden Unexplained Death (SUD) characterized by acute cardiac failure. SUD often strikes in time-space clusters during the rainy saison in villages between 1,800 and 2,400 m above sea level in the Yunnan Province in South China. Epidemiological investigations implicated a connection with ingestion of the lignicole mushroom *Trogia venenata* Zhu L. Yang, Yan C. & L.P. Tang, little white mushroom (Ph. 2.21), Marasmiaceae. Identification of this formerly undescribed toxic mushroom and its components in 2012 explains several hundred death cases (since 1978 >400) of apparently healthy people in the mentioned area.

O OH OH NH_2 — 2*R*-Amino-4*S*-hydroxy-5-hexynoic acid

O OH NH_2 — 2*R*-Amino-5-hexynoic acid

Fig. 2.13: Aliphatic amino acids of *Trogia venenata*.

Ph. 2.21: *Trogia venenata*, little white mushroom (source: Ray Palmer).

Administration of an extract of the fungi to mice led to hypoglycemia. The mentioned amino acids caused a strong increase in creatine kinase level in serum of mice. It is assumed that the compounds have a similar biochemical effect as hypoglycin ([95, 96, 116], see **[160]** = **Vol. 2**). However it remains unclear if the compounds have the same targets as hypoglycin (β-oxidation or ATP production **[160]** = **Vol. 2**). γ-Guanidinobutyric acid is a GABA antagonist and acts proconvulsive.

Further components of *T. venenata* are monoterpenes (6,7,8-trihydroxy-2,6-dimethyloctanoic acid), sesquiterpenes (β-caryophyllene-4*R*,5*R*-oxide), and ergosterol derivatives [113].

The LD_{50} of 2*R*-amino-4*S*-hydroxy-5-hexynoic acid is 71 mg/kg BW, p.o., mice, those of 2*R*-amino-5-hexynoic acid 84 mg/kg BW, p.o., mice. If the LD_{50} values would be applied to human (60 kg BW), this would correspond to the ingestion of about 400 g of dried mushrooms.

☠ The time until onset of symptoms can be >6 h. Initially, patients report nausea, vomiting, diarrhea, abdominal pain, and fatigue. Further symptoms are palpitations, chest discomfort, dizziness, syncopes, and seizures in the hours preceding death. Ventricular tachycardia and fibrillation may precede death. Postmortem examinations showed focal lymphocytic myocarditits with breakage of the muscle fibers, kidney and hepatocyte necrosis, and congestion of the liver, lung, and spleen [95, 96, **31**, **50**].

2.7.2 Amino Acids with a Cyclopropane Ring

2.7.2.1 Coprin as Toxin of Coprinopsis and Coprinellus Species

Some species of the genera Coprinopsis and Coprinellus are poisonous in connection with alcohol consumption (disulfiram-like mushroom poisoning, antabus-like reaction). This phenomenon is known from *Coprinopsis atramentaria* (Bull.) Redhead, Vilgalys & Moncalvo (*Coprinus atramentarius* (Bull.) Fr., Ph. 2.22), common inkcap (160–360 mg coprin/kg FW), *Coprinellus micaceus* (Bull.) Vilgalys, Hopple & Jacq. Johnson (*Coprinus micaceus* (Bull.) Fr.), glistening inkcap (Ph. 2.23, 26 mg coprin/kg FW), and *Coprinopsis alopecia* (Lasch) LaChiusa & Boffelli (*Coprinus alopecia* Lasch), distinguished inkcap [65], all Psathyrellaceae.

Such toxicity can also be assumed for some other ink caps. Coprin (Fig 2.14.), which is responsible for this effect, was also found in *Coprinellus disseminatus* (Pers.) J.E. Lange (*Coprinus disseminatus* (Pers.) Gray), fairy inkcap, and in *Coprinopsis picacea* (Bull.) Redhead, Vilgalys & Moncalvo (*Coprinus picaceus* (Bull.) Gray), magpie inkcap (Ph. 2.24). *Coprinus comatus* (O.F. Müll.) Pers., shaggy mane, contains at most very small amounts of coprin (10–15 mg/kg FW) and is considered safe to eat also in connection with ethanol [65, **107**].

Coprin has also been detected in *Cantharocybe virosa* (Manim. & K.B. Vrinda) T.K.A. Kumar (Hygrophoraceae), responsible for severe cases of gastrointestinal irrita-

Coprin

Fig. 2.14: A possible toxin of Coprinopsis and Coprinellus species.

Ph. 2.22: *Coprinopsis atramentaria*, common inkcap (source: empire 331/iStock/Getty Images Plus).

Ph. 2.23: *Coprinellus micaceus*, glistening inkcap (source: Geert Schmidt-Stohn).

tion in Thailand. In these cases, the poisoning was probably triggered by a stomachic remedy containing 6.65% ethanol [82]. Also suspected of being poisonous by an antabus–like reaction to ethanol is *Ampulloclitocybe clavipes* (Pers.) Redhead, Lutzoni, Moncalvo & Vilgalys (*Clitocybe clavipes* (Pers.) P. Kumm.), club foot (Hygrophoraceae), a gilled mushroom from Europe and North America [19].

Ph. 2.24: *Coprinopsis picacea*, magpie inkcap (source: Markus Scholler).

Imperator torosus (Fr.) Assyov, Bellanger, Bertêa, Courtec., Koller, Loizides, G. Marques, J.A. Muñoz, Oppicelli, D. Puddu, F. Rich. & P.-A. Moreau (*Boletus torosus* Fr.), browny bolete, blood-spotted boletus (Boletaceae), possibly also contains coprin [47]. However, intoxications of coprin-type after ingestion of *I. torosus* and ethanolic beverages have not been observed. Gastrointestinal symptoms after consumption of this species are more likely to be caused by poor digestability [27]. Nevertheless, the observed alcohol intolerance after ingestion of *Suillellus luridus* (Schaeff.) Murrill (*Boletus luridus* Schaeff.), lurid bolete, was explained as possible confusion with *I. torosus* [51].

The cause of the disulfiram-like mushroom poisoning described following ingestion of *Echinoderma asperum* (Pers.) Bon (*Lepiota aspera* (Pers.) Quél.), freckled dapperling, is unclear. Five patients had mistaken the freckled dapperling for edible mushrooms as *Amanita rubescens* Pers. or *Macrolepiota procera* (Scop.) Singer. The mushrooms were tolerated well until an alcoholic beverage was consumed [31].

The ink mushrooms occur worldwide and prefer nutrient-rich soil. They are characterized by a more or less wrinkled, moderately fleshy to skin-like, fragile cap, which often, at least when young, still carries remains of the cover. The lamellae are first white, then become pink, finally dark brown to black. The cap of most types dissolves due to autolysis from the edge to a liquid colored by the dark brown to black spores. *Coprinopsis atramentaria*, which is usually more abundant, and the single standing *C. alopecia* have ash-gray to gray-brown caps. *Coprinellus micaceus* has a yellow-brown cap with a brown center and micaceous glittering granules.

Coprin is N^5-(1-hydroxycyclopropyl)-L-glutamine. It has been isolated and structurally elucidated nearly at the same time in 1975 by Lindberg et al. [59] and Hatfield and Schaumberg [35]. Heating or cooling of the fungi does not significantly change the content [59].

The active metabolite of coprin seems to be 1-aminocyclopropanol. It forms a thiohemiketal with a thiol group in the active center of the acetaldehyde dehydrogenase and thus inhibits the further oxidation of the toxic acetaldehyde produced during ethanol degradation to the nontoxic acetate [112]. That means that coprin only causes symptoms of poisoning in combination with alcohol, just as the synthetic disulfiram (antabus [36, 104]).

Poisoning occurs, for example, when nontoxic ink mushrooms are confused with *C. atramentaria* and when alcoholic beverages are consumed in temporal proximity [64].

A 56-year-old woman had prepared herself a mushroom dish at noon from supposed *Coprinus comatus*, which were, however, *Coprinopsis atramentaria*. After two sips of tea with rum in the evening, she began to suffer from feelings of heat in her head, palpitations, and nausea. To combat this, she drank a gentian schnapps, which massively intensified the symptoms. During the following medical examination, a massive redness of head and trunk was noticed, the patient was slightly tachycardic and hypotonic, she complained of head pressure, dizziness, and abdominal pain. The symptoms disappeared within a few hours. When the patient drank a Campari about 76 h later, she again felt hot [64].

The latency period until the onset of action after ingestion of coprin-containing mushrooms and alcohol in the vicinity of time is usually ½ to 4 h. Due to the very slow elimination of coprin, poisoning is still possible if alcohol is consumed only on the third day after the mushroom meal. Symptoms of poisoning are sensation of heat, redness of face, neck and eyes, dizziness, headache, arrhythmia, tachycardia, shortness of breath, and drop in blood pressure with danger of collapse [14, 37, 69, **42, 107**]. In most cases, the symptoms of poisoning are overcome after 2–4 h, and no health damage remains. Danger exists for asthma and circulatory patients.

Apart from a strict ban on alcohol for about 10 days, no further treatment is usually necessary. The treatment can be supported by administration of fomepizole, an inhibitor of alcohol dehydrogenase [9].

2.7.2.2 Pleurocybellaziridine as Toxin of *Pleurocybella porrigens*

Pleurocybellaziridine (PA, 3,3-dimethylaziridine-2-carboxylic acid, Fig. 2.15) is an unstable amino acid which has been made, among others, responsible for the toxicity of the angel wing mushroom, *Pleurocybella porrigens* (Pers.) Singer (*Phyllotus porrigens* (Pers.) P. Karst) (Ph. 2.25), Tricholomataceae. The mushroom is widespread throughout the world's temperate regions and grows on conifers in humid and shady environment. In Germany, it occurs in higher regions of low mountain ranges. The fruit bodies are initially white and then change to a tinge of yellow. The stipe is very short or absent. Younger specimens are reminiscent of auricles due to their curved caps and bent edges.

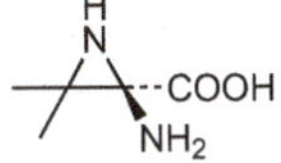

Pleurocybellaziridine (PA)

Fig. 2.15: A toxin of *Pleurocybella porrigens*.

Ph. 2.25: *Pleurocybella porrigens*, angel wing (source: Mangkelin/iStock/Getty Images Plus).

The fruit bodies have long been consumed and were considered a delicacy. In autumn 2004, 59 poisonings were reported following consumption of *P. porrigens* in Japan. Seventeen people died from acute encephalopathy. Most of the poisoned victims suffered before from kidney failure and needed dialysis before the poisoning [43, 108]. The special sensitivity of people with kidney disease can be explained by the fact that the toxin is presumably excreted via the kidneys and that this is not sufficiently possible in such patients.

Because of its instability pleurocybellaziridine could not be isolated in intact form from the mushrooms but related amino acids whose common precursor it is [8, 43, 44]. The compound could be synthesized and was shown to be toxic to rat oligodendrocytes.

In addition to the amino acids, a lectin, named *Pleurocybella porrigens* lectin (PPL), and a glycoprotein, pleurocybelline (PC), have been found. PPL is structurally similar to the ricin B chain (see **[161] = Vol. 3**) and the hemagglutinin component of botulinum toxin (see Section 4.3.8). The complex of PPL and PC shows protease activity.

It is assumed that the three components together cause the high toxicity. The complex of PPL and PC damages the blood brain barrier by its protease activity. Thereafter, PA can pass the barrier, attacks the brain, and causes acute encephalopathy. A very specific demyelinating lesion in the myelin sheath was observed in the brains of patients who died from encephalopathy [43].

☠ Symptoms after consumption of the mushrooms are weakness, confusion, tremor, epileptic seizures, coma, and stroke-like symptoms [**31**]. The latency period is 1–31 days, in average 9 days.

It remains unclear why the mushroom that were previously considered edible caused so serious poisonings. Possibly unusual high concentrations of the toxic constituents (mass fructification following suitable weather conditions in Japan in 2004) or/and the pre-damage of the patients (kidney failure!) contribute. To our knowledge, poisonings have only been reported from Japan so far [**31, 42**].

A cyclic peptide and two lactols isolated from the fungus inhibit the expression of the receptor tyrosine kinase Axl and the programmed death ligands 1 and 2 (PD-L1 and PD-L2). They are of interest as possible anticancer drugs [86].

2.7.3 Amino Acids with a Five-Membered Heterocyclic Ring System (Ibotenic Acid and Related Compounds) as Toxins of Amanita Species (*Amanita muscaria*, *Amanita pantherina*, and Other)

Ibotenic acid is the toxin of several Amanita species (Amanitaceae). The cosmopolitan genus Amanita comprises about 500 described species [115]. Among this genus are some of the most toxic mushroom species, e.g., *Amanita phalloides*, found worldwide (see Section 2.10). *Amanita muscaria* (L.) Lam., fly agaric, fly amanita, *A. pantherina* (DC.) Krombh., panther cap, false blusher, *A. regalis* (Fr.) Michael, Royal fly agaric, and *A. ibotengutake* T. Oda, C. Tanaka & Tsuda, Japanese ringed-bulb [38], are toxic because of their content of ibotenic acid and related compounds. It is not clarified if *Amanita gemmata* (Fr.) Bertill., gemmed amanita, jonquil, contains ibotenic acid or not. Possibly different chemical races exist [**31**].

Amanita species are gill fungi, which have an overall envelope that covers the cap and the stem. The remains of the envelope stay as spots on the cap and often as a partial envelope at the underside of the cap. The lamellae are attached to the stem. The spores are white or greenish colored. Amanita species are soil-dwelling fungi, which are bound to trees by an ectomycorrhiza, so they do not occur in treeless areas.

Amanita muscaria (Ph. 2.26), *A. pantherina* (Ph. 2.27), and *A. regalis* have a membranous manchette on the stem, the cuff. The limb of the cap is striate. In *A. muscaria* the cap is scarlet red, rarely orange-yellow colored and carries white warts. The white flesh of the cap is colored red-yellow under the skin. A bulb with a warty zone round the stem closes the stem at the bottom. The cap of *A. pantherina* is gray-brown to dark brown; it carries concentrically arranged white spots. The flesh is white and smells slightly like radish. The stalk carries a bulbous-edged tuber. *A. regalis* resembles the fly agaric except for the brown cap skin and several flocci on the tuber of the stem. The species occur in deciduous and coniferous forests on acidic and neutral soils in the northern temperate climate zone. The fly agaric and the panther cap are frequently found among birches, the royal fly agaric occurs, only rarely, in Germany in the area of the low mountain ranges. The fly agaric got its name because of the use of mushroom slices pickled in sugared milk to kill flies [62].

Ph. 2.26: *Amanita muscaria*, fly agaric (source: Michael Lalk).

Ph. 2.27: *Amanita pantherina*, panther cap (source: Markus Scholler).

The fungi contain ibotenic acid (Fig. 2.16) as the main active ingredient, namely 0.1–2.8% in *Amanita muscaria* and 0.02–0.53% in *A. pantherina*, besides its decarboxylation product muscimol (pantherin, agarin), <0.01–1.0% in *A. muscaria*, 0.19–1.9% in *A. pantherina*. In *A. regalis* a content of both compounds of 0.10–0.62% was determined. The flesh of the fruit bodies contains higher concentrations of ibotenic acid and muscimol than the cap skin; the content in the stems is lower than in the fruit bodies. Content and composition of the contents of the fly agaric are strongly dependent on location and weather conditions [98, 100, 105]. The preparation method also determines the content of these toxins in 'mushroom meal'. After cooking, preserving with salt or peeling off the skin of the cap, the fly agaric is eaten in some regions (e.g., France, Siberia) without adverse side effects [2, 112].

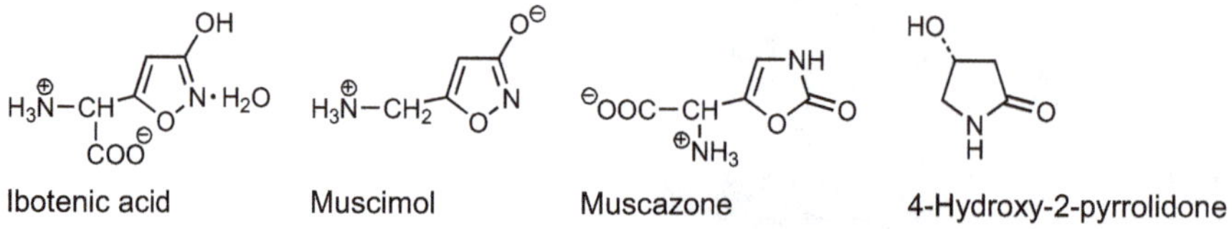

Ibotenic acid | Muscimol | Muscazone | 4-Hydroxy-2-pyrrolidone

Fig. 2.16: Amino acids with a five-membered heterocyclic ring system and derivatives of Amanita species.

Ibotenic acid is an isoxazole derivative that was isolated and structurally elucidated independently by Eugster and colleagues [24] and by Takemoto and colleagues [101] in 1960 and 1964, resp. In *Amanita muscaria*, ibotenic acid and muscimol are accompanied by the oxazole derivative muscazone, which can be formed from ibotenic acid by rearrangement in the light [107], small amounts of muscarine (0.0002–0.0003% FW, Fig. 2.21, see Section 2.8.2), and (*R*)-4-hydroxy-2-pyrrolidone. A recently published work demonstrated muscarine concentrations ranging from 0.004% up to 0.043% in *A. muscaria* thus explaining possible cholinergic effects of the mushrooms [26].

The biogenesis of ibotenic acid is initiated with the hydroxylation of glutamate by the enzyme complex IboH (2-oxoglutarat-dependent dioxygenase). The genes for this complex have recently been identified. The last step, the decarboxylation of ibotenic acid to muscimol, is catalyzed by IboD (tryptophan decarboxylase). The order of the intermediate reactions is somewhat ambiguous [78, 80].

The red and yellow dyes of the cap are betalains and related compounds in which betalamic acid has formed enaminium salts with ibotenic acid, stizilobic acid, α-aminoadipic acid, and other amino acids [22, 75]. Also noteworthy are the pantheric acids A–C4, isolated from *A. pantherina*. These fatty acids promote lipid accumulation in adipocytes [54].

While ibotenic acid as an agonist predominantly targets NMDA receptors and mimics glutaminergic activities [33, 63], muscimol is an agonist at the $GABA_A$ (ionotropic GABA receptor) and $GABA_c$ receptors [7, 50, 80].

The LD_{50} of ibotenic acid is 129 mg/kg BW, p.o., in rats and 38 mg/kg BW, p.o., in mice. The LD_{50} of muscimol for rats is 4.5 mg/kg BW, i.v., or 45 mg/kg BW, p.o., and for rabbits 10 mg/kg BW, p.o. [103, **41**]. Muscazone is hardly pharmacologically active [31]. Human poisoning has been described after ingestion of mushroom amounts corresponding to 6 mg muscimol and 30–60 mg ibotenic acid [17, 23, 34].

The fly agaric and in smaller extent also the panther agaric were used as entheogen (psychoactive substance) by many of the indigenous people in Siberia [5, 16, 39, 68, 79, **41, 107, 135, 147**]. The hypothesis that the ancient Indian miracle drug soma was made from fly agarics has been refuted. This also applies to the assumption that the 'superhuman powers' ('berserk rage') of the berserkers, the bodyguards of Nordic kings in the ninth to twelfth centuries, were based on the consumption of fly agarics. According to a more recent hypothesis, it is more likely that *Hyoscyamus niger*, henbane (see **[161] = Vol. 3**), was responsible for the berserker rage state [25].

The most dangerous of the mentioned Amanita species is the panther cap, *Amanita pantherina*. Poisonings are due to confusion with the blusher, *Amanita rubescens* Pers., the European false blushing amanita, *A. excelsa* (Fr.) Bertill., the gray spotted amanita, *A. spissa* (Fr.) Opiz., or with the parasol mushroom, *Macrolepiota procera* [6, 18, 32, 56, 85, 91, **19, 41**]. Accidental poisonings by *Amanita muscaria* occurred after confusion with the emperor mushroom, *Amanita caesarea* (Scop.) Pers. [13], Caesar's amanita, with *Amanita hemibapha* (Berk. & Broome) Sacc., half-dyed slender Caesar, a species growing in Southeast Asia [102], or with other edible mushrooms [84]. Due to the high degree of popularity of the mushrooms, however, accidental intoxications occur very rarely. For the royal fly agaric, *Amanita regalis*, also only very few cases of poisoning have been described [100]. In a case of poisoning that occurred in Germany (Erzgebirge) in the year 1999, only two persons of a family of five showed symptoms of poisoning [1].

Poisoning by the fly agaric is more likely to be caused by its misuse as an intoxicant or its consumption to reduce pain, anxiety, stress, or insomnia [66, 81]. A survey among young addicts in Germany showed that 25 of 76 subjects under 30 years of age occasionally consume fly agarics or panther caps. They stated the intake of 'half a large or a small mushroom', mostly taken raw or in the form of salad, as their usual dosage [60]. Adverse effects like headache and gastrointestinal symptoms but also severe cases of poisoning are described [40, 66, 67, 81, 90, 97].

✝ A 44-year-old man in the United States presented to the emergency department after cardiopulmonary arrest approx. 10 h after ingesting 4–5 dried caps of the fly agaric, which he used for their mind-altering effects. Despite successful resuscitation, he remained unresponsive and hypotensive and died 9 days later [67].

Warnings must be issued against the consumption of sweets containing muscimol, for example fruit gums, which are offered on the Internet or in several countries via vending machines. Cases of poisoning occurred.

Symptoms that usually occur within 15 min to 4 h after oral intake of mushrooms containing ibotenic acid or muscimol include dizziness, coordination disorders, confusion, sensory delusions, and fits of raving madness. The toxic psychosis with sensory delusions, tantrums, and the urge to move reaches its full effect after 2–3 h. In many cases, the poisoned people think that they are in possession of enormous powers. In addition to the suppression of fear and anger, which is desired during the misuse, gaps in the sense of time and place are described as threatening. When the effect of the poison wears off, the intoxication turns into a deep sleep, from which the victim awakens partially without any memory of the event. Loss of consciousness often occurs. Burst suppression pattern on EEG during comatose state have been observed. Deaths are rare. A danger can result from careless actions performed in a state of intoxication. Severe neuron damage and brain lesions can develop in cases of recurrent consumption [32, 45, 60, 67, 68, 70, 74, 79, 80, 84, **41, 42, 107, 182**]. A small study reported that patients poisoned by *A. muscaria* (32 patients) were more confused and agitated than those poisoned by *A. pantherina* (17 patients), while those were more commonly comatose [106].

The following case of poisoning shows that cholinergic symptoms can also occur:

A 41-year-old man, who described all synthetic drugs as poisonous, prepared a large pan of fly agaric in the evening to find out 'how far he could go'. About 30 min after consuming the entire portion, severe nausea and dizziness set in, after which he lost consciousness. After about 14 h the unconscious person was found by the emergency doctor. The patient showed the picture of a severe cholinergic syndrome with hypersalivation, jaw spasm, sweaty pale skin, and miosis with deep coma. The cholinergic symptoms could be resolved by high doses of atropine. The patient awoke from his coma after 24 h. Optical hallucinations persisted for days [40]. It can be assumed that not only ibotenic acid and muscimol but also muscarine [26] or structurally unknown toxins were involved in the symptoms.

Treatment of poisoning by ibotenic acid containing mushrooms is usually done after emptying the gastrointestinal tract, which is only useful within 1 h after ingestion of the mushrooms, giving charcoal, and by administration of sedatives, e.g., benzodiazepines and barbiturates. Patients occasionally require ventilator support [15, 32, 68, 87]. In mild cases, monitoring and letting the patient 'sleep it off' are often sufficient. In severe cases with cholinergic or anticholinergic symptoms, the administration of the respective antagonists (atropin or physostigmin) can become necessary [10].

Intoxications by *Amanita pantherina* and *A. muscaria* have also been observed in animals, e.g., young dogs. Possibly dogs are attracted to the fishy odor of these mushrooms. Symptoms of poisoning were anorexia, vomiting, convulsions, dizziness, aggressiveness, paralysis, dyspnoea, and in fatal cases death [58, 77, 88, 89, 94].

Ibotenic acid is used as a 'brain-lesioning agent' through cranial injections for neurological research [**182**]. The compound was a lead compound for pharmacological research. Its derivative AMPA (α-**a**mino-3-hydroxy-5-**m**ethyl-4-isoxazolyl**p**ropionic **a**cid), for example, had given its name for a subtype of glutamate receptors in human brain. Muscimol is a widely used experimental tool in neuroscience research and a candidate for the treatment of pain [83]. Muscimol analogs are of interest as possible GABA agonists to treat neurodegenerative diseases [80].

2.7.4 Amino Acids with a Six-Membered Heterocyclic Ring System (Acromelic Acids) as Toxins of Paralepistopsis Species

The isomeric acromelic acids A and B (ACRO A and B, for B the oxo and carboxyl groups are reversed, Fig. 2.17) contain both a pyridine and a pyrrolidine ring. They were first isolated from *Paralepistopsis acromelalga* (Ichimura) Vizzini (*Clitocybe acromelalga* Ichimura), poison dwarf bamboo mushroom (Tricholomataceae), a Japanese poisonous mushroom. In the fungus they are accompanied by acromelic acids C–E and by the toxic pyridine nucleoside clitidine (Fig. 2.17), the weakly neurotoxic 4-amino-quinolinic acid, the sulfur-containing amino acid clithioneine (Fig. 2.17, yield 0.00023%), and β-cyano-L-alanine (see **[160]** = **Vol. 2**) [28, 29, 48, 49]. The acromelic acids have also been found in *Paralepistopsis amoenolens* (Malençon) Vizzini (*Clitocybe amoenolens* Malençon, Ph. 2.28), paralysis funnel, a species native to Morocco and also found in Southern Europe and Switzerland [73]. Acromelic acid A content was determined at 325 ng/mg DW in *Paralepistopsis amoenolens* and 283 ng/mg DW in *P. acromelalga* [12].

Acromelic acid A | Clitidine | Clithioneine

Fig. 2.17: An amino acid with a six-membered heterocyclic ring system and further ingredients of Paralepistopsis species.

Ph. 2.28: *Paralepistopsis amoenolens*, paralysis funnel (source: Mario Ianotti).

Acromelic acids are strong glutamate agonists and attack the same receptor as kainic acid (see **[160] = Vol. 2**). Thereby they exert depolarizing and neurotoxic effects. Peripheral glutamate receptors of cutaneous nerve endings may be involved. Acromelic acid A has a 100 times stronger depolarizing effect than kainic acid [71] and a much stronger effect than acromelic acid B [**182**].

Poisoning by *Paralepistopsis acromelalga* has been described from Japan and South Korea [72, 76], poisoning by *P. amoenolens* from France, Italy, and Morocco [55, 93, **42**]. However, some of the poisonings allegedly caused by *P. amoenolens* in France are said to be caused by *Paralepista flaccida* (Sowerby) Vizzini (*Lepista flaccida* (Sowerby) Pat.), tawny funnel. Although the mushrooms of this species are considered edible by most authors [99], they have already caused intolerance reactions and are difficult to digest [4]. Because of danger of confusion *P. flaccida* was canceled from the list of edible mushroom species, at least in Switzerland [3].

☠ Poisoning by fungi containing acromelic acids manifests itself in very strong intermittent pain, burning, reddening, and swelling of extremities (acromelagia, erythromelagia), which occur after 3–7 days and may persist for weeks to months. Cyanosis or erythema may worsen during pain paroxysms. More severe cases are associated with local diaphoresis and trophic changes of the digits. The symptoms may spontaneously resolve over subsequent weeks [71, 76, **50**].

A causal therapy is not known. The pain cannot be relieved by analgesics, but only by holding the hands and feet under running water. Heat and movement intensify the pain, cold alleviates it. A warning is given against the use of ice because of the danger of cold-related secondary damage to the skin nerves [**42**]. High dose i.v. administration of nicotinic acid may be efficacious because of its vasodilating effect [76].

Acromelic acids are used as tools for neuropharmacological research [**182**].

References

For numbers in bold, see cross-chapter literature p. 233.

[1] Ahnert E (1999) Z Mykol 65: 36
[2] Amelang N (2003) Tintling 8: 15
[3] Anonym (2014) Tintling 19(6): 26
[4] Anonym (2019) Tintling 24(3): 66
[5] Bauer W et al. (2002) Der Fliegenpilz: Traumkultur, Märchenzauber, Mythenrausch. ATV-Verlag, Aarau
[6] Benjamin DR (1992) Clin Toxicol 30: 13
[7] Benkherouf AY et al. (2019) J Neurochem 149: 41
[8] Berndt S (2012) Tintling 17(2): 86
[9] Berndt S (2017) Tintling 22(5): 36
[10] Berndt S (2021) Dtsch Ärzteblatt Int 118: 197
[11] Besancon A et al. (2012) Ann Fr Anesth Reanim 31: 466
[12] Bessard J et al. (2004) J Chromatogr A 1055: 99

[13] Brvar M et al. (2006) Wien Klin Wochenschr 118: 294
[14] Buck RW (1967) N Engl J Med 276: 391
[15] Caffrey CR, Lang PM (2018) Open Access Emerg Med 10: 9
[16] Carboué Q, Lopez M (2021) Encyclopedia 1: 905
[17] Chilton WS (1994) in: Handbook of Mushroom Poisoning. Diagnosis and Treatment. Spoerke DG, Rumack BH (eds), CRC Press, Boca Raton, p. 165
[18] Cirronis M et al. (2025) Toxicon 254: 108208
[19] Cochran KW, Cochran MW (1978) Mycologia 70(5): 1124
[20] Courtin P et al. (2009) Clin Toxicol (Phil.) 47(9): 906
[21] De Haro L et al. (1998) Nephrologie 19: 2
[22] Doepp H et al. (1982) Liebigs Ann Chem 254
[23] Eugster CM (1968) Naturwissenschaften 55: 305
[24] Eugster CH et al. (1965) Tetrahedron Lett 1813
[25] Fatur K (2019) J Ethnopharmacol 244: 112151
[26] Feeney K et al. (2025) Int J Med Mushrooms 27(7): 1
[27] Flammer R (2008) Schweiz Z Pilzk 4: 146
[28] Fushiya S et al. (1990) Tetrahedron Lett 31: 3901
[29] Fushiya S et al. (1992) Heterocycles 34: 1277
[30] Gioacchino LC et al. (2019) Indian J Nephrol 29: 73
[31] Haberl B et al. (2011) Clin Toxicol (Phil.) 49: 113
[32] Handl L et al. (2023) Neuro Endocrinol Lett 44(8): 500
[33] Hastings MH et al. (1985) Brain Res 360: 248
[34] Hatfield GM, Brady LR (1975) J Nat Prod 38: 36
[35] Hatfield GM, Schaumberg JP (1975) J Nat Prod 38: 489
[36] Hatfield GM, Schaumberg JP (1978) in: Ü 114, p. 181
[37] Herrmann M (1966) Mykol Mitteilungsblatt 10: 39
[38] Hiroshima Y et al. (2010) Ther Apher Dial 14: 483
[39] Hobbs C (1995) Medicinal Mushrooms: An Exploration of Tradition, Healing & Culture, Botanica Press, Summertown, Tenn (USA)
[40] Hohn H, Schoenemann J (2000) Dtsch Med Wochenschr 125: 1366
[41] Iwafuchi Y et al. (2003) Int Med 42: 78
[42] Kang E et al. (2015) Kid Res Clin Pract 34: 233
[43] Kawagishi H (2023) Proc Jpn Acad Ser B 99(7): 191
[44] Kawaguchi T et al. (2010) Tetrahedron 66(2): 504
[45] Keller SA et al. (2018) Int J Environ Res Public Health 15: 2855
[46] Kirchmair M et al. (2012) Nephrol Dial Transplant 27: 1380
[47] Kiwitt U, Laatsch H (1994) Z Mykol 60: 423
[48] Konno K et al. (1984) Phytochemistry 23: 1003
[49] Konno K et al. (1988) J Am Chem Soc 110: 6926
[50] Krogsgaard-Larsen P et al. (1979) J Neurochem 32: 1717
[51] Laatsch H (1999) Tintling 4(6): 7
[52] Leathem AM et al. (1997) J Toxicol Clin Toxicol 35: 67
[53] Lee JH et al. (2018) J Korean Med Sci 33: e230
[54] Lee SR et al. (2019) J Nat Prod 82(12): 3489
[55] Leonardi M et al. (2002) Micol Vegetazione Mediterr 17: 133
[56] Leonhardt W (1948) Integration 2&3, Nachdruck in: Der Tintling 7: 54, 2002
[57] Leray H et al. (1994) Nephrologie 15: 197
[58] Lindberg H, Holmgren A (2012) Clin Toxicol 50: 289
[59] Lindberg P et al. (1977) J Chem Soc Perkin Trans 1: 684

[60] Löhrer F, Kaiser R (1999) Nervenarzt 70: 1029
[61] Loizides M et al. (2023) Indian J Nephrol 33(2): 155
[62] Lumpert M, Kreft S (2016) J Ethnopharmacol 187: 1
[63] Marmo E (1988) Med Res Rev 8: 441
[64] Marty H (1998) Schweiz Med Wochenschr 128: 598
[65] Matthies und Laatsch (1992) Pharm in uns Zeit 21(1): 14
[66] Maung AC et al. (2023) Clin Med (Lond) 23(4): 417
[67] Meisel EM et al. (2022) Wilderness Environ Med 33(4): 412
[68] Michelot D (1992) Nat Toxins 1: 73
[69] Michelot D et al. (2003) Mycol Res 107: 131
[70] Mikaszewska-Sokolewicz MA et al. (2016), Acta Biochi Pol 63: 181
[71] Minami T et al. (2004) Brit J Pharmacol 142: 679
[72] Mizusawa K, Shimizu T (2021) Internal Med 60: 1637
[73] Moreau PA et al. (2001) Mycol 22: 1
[74] Moss MJ, Hendrickson RG (2019) Clin Toxicol (Phila) 57(2): 99
[75] Musso H (1979) Tetrahedron 35: 2843
[76] Nakajima N et al. (2013) Clin Toxicol (Phila) 51: 451
[77] Naudé TW, WL Berry (1997) Tydskr S Afr Veter Vet 68, 154
[78] Obermaier S, Müller M (2020) Angew Chem Int Ed 59: 12432
[79] Ogawa Y et al. (2015) Epilepsy Behavior Case Rep 4: 82
[80] Okhovat A et al. (2023) Int J Med Mushrooms 25(9): 1
[81] Ordak M et al. (2023) Toxics 11: 383
[82] Parnmen S et al. (2019) Toxicol Res 36(3): 239
[83] Ramawad HA et al. (2023) Korean J Pain 36(4): 425
[84] Rampolli FI (2021) Eur J Case Rep Intern Med 8(3): 002212
[85] Rass H (2022) Tintling 138(4): 85
[86] Ridwan AY et al. (2020) J Antibiot 73: 733
[87] Riethmüller J et al. (2004) Monatsschr Kinderheilk 152: 892
[88] Romano MC et al. (2019) J Vet Diag Inv 31: 485
[89] Rossmeisl JH et al. (2006) J Vet Emerg Crit Care 16: 208
[90] Satora L et al. (2005) Toxicon 45: 941
[91] Satora L et al. (2006) Toxicon 47: 605
[92] Saviuc P, Danel V (2006) Toxicol Rev 25: 199
[93] Saviuc PF et al. (2002) Rev Med Interne 23: 394
[94] Seljetun KO, Kragstad HR (2023) Vet Rec Open 10: e60
[95] Shi GQ et al. (2012b) PLoS One 7: e35894
[96] Shi GQ et al. (2012a) PLoS One 7: e38712
[97] Soyka M (2023) Nervenarzt 94: 849
[98] Stijve T (1982) Coolia 25: 94
[99] Stijve T (2001) Field Mycol 2: 77
[100] Stijve T (2004) Eleusis 8: 55
[101] Takemoto T et al. (1964) J Pharm Soc Jpn 84: 1186
[102] Taylor J et al. (2019) MMWR (Morbidity and Mortality Weekly Report) 68: 483
[103] Theobald W et al. (1968) Arzneim Forsch 18: 311
[104] Tottmar O, Lindberg P (1977) Acta Pharm Toxicol 40: 476
[105] Tsujikawa K et al. (2006) Forensic Sci Int 164: 172
[106] Vendramin A, Brvar M (2014) Toxicon 90C: 269
[107] Wagner I, Musso H (1983) Angew Chem 22: 816
[108] Wakimoto T et al. (2011) Angew Chem Int Ed 50: 1168

[109] Warden CR, Benjamin DR (1998) Emerg Med 5: 808
[110] West PL et al. (2009) J Med Toxicol 5(1): 32
[111] Wiseman JS, Abeles RH (1979) Biochemistry 18(3): 427
[112] Wurster M et al. (2004) Z Mykol 70: 161
[113] Xu et al. (2018) Nat Prod Res 32(21): 2547
[114] Yang WS et al. (2006) J Formos Med Assoc 105: 263
[115] Zhang P et al. (2015) Mycology 6: 86
[116] Zhou ZY et al. (2012) Angew Chem Int Ed Engl 51: 2368

2.8 Amines as Toxins of Macrofungi

2.8.1 Aliphatic Amines and Azocompounds

2.8.1.1 Hydrazine Derivatives as Toxins of Gyromitra, Paragyromitra, and Discina Species

Hydrazines, among them gyromitrin, are chemically highly reactive compounds. Gyromitrin and related compounds (Fig. 2.18) are the poisonous compounds in some Gyromitra species (Discinaceae, Ascomycota). According to new taxonomical investigations the scope of this genus has been changed. Some species are transferred to other genera, e.g., to Paragyromitra or Discina [177].

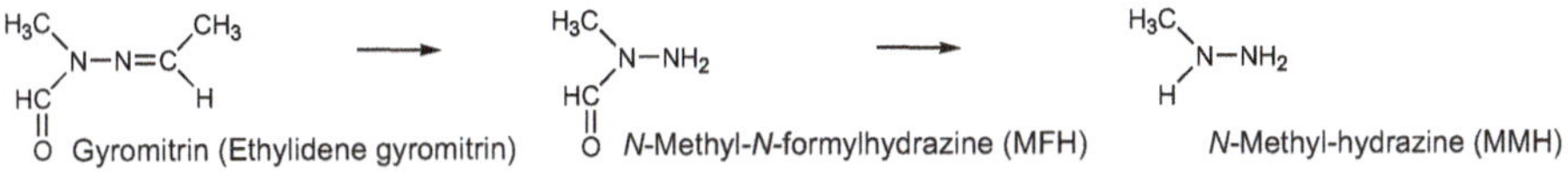

Gyromitrin (Ethylidene gyromitrin) → *N*-Methyl-*N*-formylhydrazine (MFH) → *N*-Methyl-hydrazine (MMH)

Fig. 2.18: Gyromitrin and its degradation products.

A recent study of representatives of the genus Gyromitra and related species from across North America found gyromitrin in *G. esculenta* (Pers.) Fr., false morel (Ph. 2.29), *G. antarctica* Rehm, *G. splendida* Raitv., *G. venenata* Hai J. Li, Z.H., Chen & Zhu L. Yang (Ph. 2.30), and *G. leucoxantha* (Bres.) Harmaja (*Discina leucoxantha* Bres.), fishtail cup. Despite indications of toxicity, gyromitrin was not detected in *Paragyromitra ambigua* (P. Karst.) X.C. Wang & W.Y. Zhuang (*Gyromitra ambigua* (P. Karst) Harmaja), changeable false morel, and *Paragyromitra infula* (Schaeff.) X.C. Wang & W.Y. Zhuang (*Gyromitra infula* (Schaeff.) Quél.), pouched false morel [38]. Hydrazine derivatives have also been found in very small amounts in *Discina gigas* (Krombh.) Eckblad (*Gyromitra gigas* (Krombh.) Cooke), snow morel [91], and *Discina fastigiata* (Krombh.) Svrček & J. Moravec (*Gyromitra fastigiata* (Krombh.) Rehm), brown false morel, brown gyromitra [163, 172]. Of special toxicological interest are *Gyromitra esculenta* and *G. venenata.*

Most Gyromitra, Paragyromitra, and Discina species are widely distributed in the Northern Hemisphere. *G. venenata* has first been described from China. *G. antarctica* occurs in the South of Argentina and Chile. *G. esculenta* is a spring mushroom with a wrinkled brain-like folded, mostly reddish-brown cap. The stem is whitish or brownish. The fungus is found in pine forests especially on sandy soils. The other named species have a similarly shaped cap. Their ascospores are marked by appendages at both ends.

Ph. 2.29: *Gyromitra esculenta*, false morel (source: Geert Schmidt-Stohn).

Ph. 2.30: *Gyromitra venenata* (source: Dirk Alden).

Gyromitra esculenta may be confused with Morchella species, especially with *Morchella esculenta*, true morel (Ph. 2.31). Both species occur in spring at similar locations. Morchella species are conditionally edible but can cause neurological symptoms if poorly cooked (see Section 2.12.3). Confusion is also possible with *Verpa bohemica*

Ph. 2.31: *Morchella esculenta*, true morel (source: Markus Scholler).

Krombh.) J. Schröt. (*Morchella bohemica* Krombh., *Ptychoverpa bohemica* (Krombh.) Boud.), early morel, another false morel, wrinkled thimble-cap, Morchellaceae, Ascomycota. *V. bohemica* can produce severe gastrointestinal reactions and cerebellar syndrome in susceptible persons (see Section 2.12.3). Because of some similarities in poisoning symptoms caused by *Gyromitra esculenta* or *Verpa bohemica* it is assumed that both species possess gyromitrin as responsible toxin. For the FDA, Gyromitra species, Helvella species, and *Verpa bohemica* are lumped together as toxic false morels [90]. The symptoms of poisonings with *Sarcosphaera coronaria* (Jacq.) J. Schröt., violet crown cup (Pezizaceae, Ascomycota), remember also those of poisonings with false morels but until now there was no detection of gyromitrin in this species **[69]**.

In the year 1967, List and Luft identified the hydrazine derivative gyromitrin (ethylidene gyromitrin, acetaldehyde *N*-methyl-*N*-formylhydrazone, Fig. 2.18) as the toxin of *Gyromitra esculenta* [102]. It is accompanied by eight homologues in which the ethyl group is replaced by longer alkyl groups. The information on the content of gyromitrin in *G. esculenta* varies strongly (40–732 mg/kg FW, 41–58 mg/kg DW [7, 103, 128, 172]). It may be dependent on environmental factors, possible chemotypes, and the age of the fungi. In the study of Dirks [38] the gyromitrin levels decreased with age of the ascocarp.

Gyromitrin is a colorless, volatile, chocolate-like smelling liquid (Mp. 19.5 °C). At high temperature or under specific physiological conditions, it rapidly decomposes to *N*-methyl-*N*-formyl-hydrazine (MFH) and then to *N*-methyl-hydrazine (monomethylhydrazine, MMH) (Fig. 2.18 [117, 129, 173]). It is assumed that there are two distinctive phases of metabolic pathway in the human body: a gastric phase with slow hydrolysis of MFH into MMH and a hepatic phase in which MMH and perhaps MFH are oxidized to toxic and possibly carcinogenic derivatives **[182]**.

Gyromitrin, MFH, and MMH are volatile with the vapors during cooking. There is a danger of poisoning by cooking vapors! Some of the toxins also evaporate during drying. In freshly dried mushrooms 0.1–0.2%, in 6 months of dry storage only 0.03% MMH was detected [3]. The presence of small amounts of hydrazine derivatives, which may not cause acute but chronic poisoning, cannot be excluded in cooked mushrooms, especially if they were heated in a covered vessel, even after pouring off the cooking water [136, 150].

In addition to ethylidene gyromitrin, *Gyromitra esculenta* contains small amounts of the *N*-methyl-*N*-formyl-hydrazones of propanal, butanal, 3-methyl-butanal, pentanal, hexanal, octanal, *trans*-oct-2-enal, and *cis*-oct-2-enal as homologous compounds [118, 128]. A compound related to ethylidene gyromitrin, 1-(2-hydroxyacetyl)-pyrazole (13 mg/kg FW), has been detected in *Discina fastigiata* [77].

Other Ascomycetes also contain very small amounts of *N*-methylhydrazine (and its precursors). A relatively high concentration (0.015% of FW, 150 mg/kg DW), which corresponds approx. to that in *Gyromitra esculenta*, has been detected in *Cudonia circinans* (Pers.) Fr., redleg jellybaby (Ph. 2.32, Cudoniaceae, Ascomycota, [3]). *C. circinans* is a fungus with a size of up to 2 cm, which occasionally occurs in coniferous forests in humid locations, especially in mountainous areas. It is occasionally confused with *Craterellus tubaeformis* (Fr.) Quél., yellowfoot, winter mushroom, Hydnaceae.

Ph. 2.32: *Cudonia circinans*, redleg jellybaby (source: Geert Schmidt-Stohn).

Small amounts of MMH have been detected in *Otidea onotica* (Pers.) Fuckel (1.5 mg/kg FW), *Craterellus undulatus* (Pers.) E. Campo & Papetti (*Pseudocraterellus undulatus* (Pers.) Rauschert, 5 mg/kg FW), sinuous chanterelle, *Spathularia flavida* Pers., yellow earth tongue, fairy fan (4 mg/kg FW), and *Leotia lubrica* (Scop.) Pers., jelly baby (5 mg/kg DW), all Ascomycota. Consumption of the mushrooms in normal quantities is said to be harmless to health [14].

Hydrazine derivatives like MMH react with pyridoxal, the co-factor of pyridoxal-dependent enzymes, and inhibit the activity of such enzymes, e.g., of glutamic acid decarboxylase. Among other things, this impairs the synthesis of γ-aminobutyric acid (GABA antagonism, [107]), inhibits the breakdown of biogenic amines, and causes glutathione depression. Iron present in hemoglobin is oxidized [72].

The LD_{50} of gyromitrin is 320 mg/kg BW, p.o., in rats (109) and 344 mg/kg BW, p.o., in mice [173]. For an adult human it is estimated to be 30–50 mg/kg BW. The LD_{50} of the biotransformation product MMH is lower. It is 33–70 mg/kg BW for rats, 57 mg/kg BW for mice, and 7 mg/kg BW for monkeys, in each case p.o. For children it is estimated at 1.6–4.8 mg/kg BW, for adults at 4.8–8 mg/kg BW [3].

In many countries, e.g., in Germany, false morels are not approved as commercial mushrooms. In Finland, in France, and in some other countries, they are sold as fresh mushrooms, canned food, and in restaurants. Insufficient processing (poor cooking, cooking water not discarded, etc. [3]) or confusion between the false and the true morels are main causes for acute and/or chronic poisonings [6, 18, 55, 57, 72, 95, 109, 121, 126, 165, **41**]. Poisoning by *G. venenata* has been reported from China [100].

The sensitivity to the toxins and the characteristics of occurring symptoms vary greatly from person to person [9, 173]. One reason is the different metabolic capacity, mainly the acetylator status of the patient. 'Fast acetylators' (high level of acetylhydrazine) develop hepatotoxicity with high bilirubin levels and less prominent neurotoxicity; in patients with normal acetylator status dominates neurotoxicity [163, **178**]. Elderly people are more susceptible as younger ones. Furthermore, the susceptibility may be enhanced by concomitant use of some medical drugs, e.g. isoniazid for treatment of tuberculosis [37, 90].

Graeme classifies gyromitrin poisoning as causing epileptogenic syndrome [**50**]. Franke et al. observed a lethality rate of 14.5% [47] when evaluating >500 cases of poisoning caused by *Gyromitra esculenta* (in general approx. 10% [113]).

Poisoning symptoms such as nausea, vomiting, bloody diarrhea, abdominal pain, and dehydration occur 4–12 h after ingestion of false morels. This may be the limit of effects in less severe poisonings. In more severe poisonings epileptogenic neurotoxicity (dizziness, seizures, ataxia, delirium, stupor, and coma), hemolysis, methemoglobinemia, icterus, and liver failure may develop [6, 55, 72, 78, 95, 107, 132, 165, **41**].

Treatment can only be supportive. In addition to primary toxin removal measures such as gastric lavage (as long as spontaneous vomiting or threatening hypotension have not occurred), administration of activated charcoal and laxatives, replacement of water and electrolyte loss, and liver protection must be provided. Administration of pyridoxine hydrochloride (oriented at empiric dosage for poisoning by isoniazid, INH: 5 g i.v. in an adult, 70 mg/kg BW in a child [29]) is recommended, which as a coenzyme promotes GABA formation. Methylene blue or toluidine blue can be given in cases of severe methemoglobinemia [99, 107, 132].

The ability of hydrazine derivatives to alkylate nucleic acids is also toxicologically significant. Metabolites of gyromitrin (MFH) undergo cytochrome P450 regulated oxidative metabolism and, via reactive nitrosamide intermediates, lead to the formation of DNA-damaging methyl free radicals. In animal experiments, they induce tumors of the lung, liver, and vessels when applied chronically or in short-term high doses [3, 137, 159, 160]. MMH shows embryotoxic effects in rats (i.v.). The concentrations of MMH measured in the serum of the animals correspond to those detectable in humans after a fungal meal [144].

It is hypothesized that exposure to methyl-free-radical generating hydrazine derivatives might trigger long latency neurodegeneration culminating in ALS (amyotrophic lateral sclerosis) or other neurodegenerative diseases. Preliminary epidemiological investigations in France and Finland support this hypothesis [90, 91, 145].

Because of the risk of acute and chronic effects, the consumption of false morels should be avoided, even if they have been boiled. A statement published in 2000 by a WHO expert group advises against the consumption of false morels [**31, 35**]. The Finish Food Authority stated in 2019 that *Gyromitra esculenta* is not to be eaten by pregnant and breastfeeding women and children [44].

2.8.1.2 Hydrazine Derivatives as Toxins of Agaricus Species

Hydrazine derivatives have also been detected in Agaricus species (Agaricaceae). The genus Agaricus displays a large species diversity. Some Agaricus species are commercially cultivated and serve as food. *Agaricus bisporus* (J.E. Lange) Imbach, cultivated mushroom, portobello, button mushroom (Ph. 2.33), accounts for about 38% of world production of edible mushrooms [62] *Agaricus subrufescens* Peck (*Agaricus brasiliensis* S. Wasser et al., *A. blazei* Murrill), almond mushroom, is used as medicinal mushroom because of its immunomodulating properties [45].

Agaricus species are easily recognizable by brown or white caps with fibrillose scales on the surface, free lamellae, brown spore print, and annulate stipe. Most common habitats are forests and grasslands.

Ph. 2.33: *Agaricus bisporus*, cultivated mushroom (source: Markus Scholler).

The first naturally detected hydrazine derivative, agaritine (β-*N*-(γ-L-glutamyl)-4-hydroxymethyl-phenylhydrazine, Fig. 2.19), was isolated in 1960 by Levenberg from *Agaricus bisporus* [98]. *A. bisporus* contains, depending on the substrate, 0.04–0.7% (up to 1.3%) of DW and 0.02–0.06% of FW agaritine [46, 84, 116, 147]. At 0.2%, the agaritine content in white *A. bisporus* samples was slightly higher than in the brown variety at 0.15% [112]. The average content in spores is 0.3% [75]. Due to the heat lability of agaritine, the content in preserved mushrooms in tins is lower (0.0004–0.0123%, drained 0.0024–0.0052% [46]. No agaritine was detected in the liquid of the tins [112].

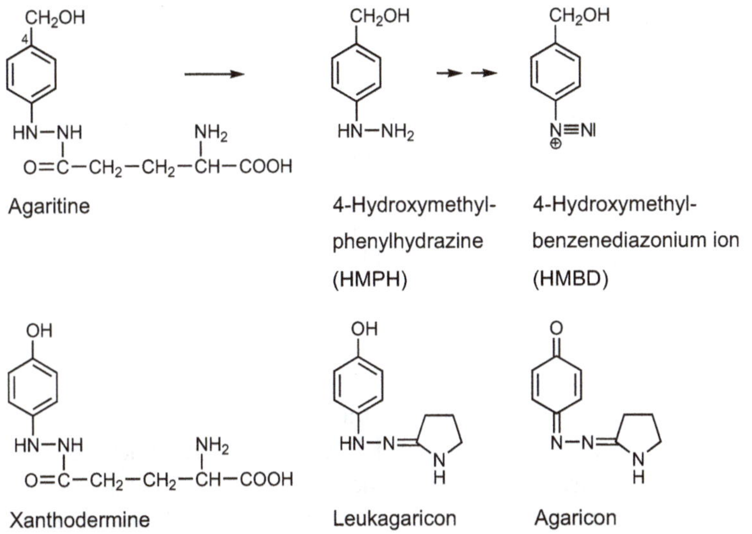

Agaritine

4-Hydroxymethyl-phenylhydrazine (HMPH)

4-Hydroxymethyl-benzenediazonium ion (HMBD)

Xanthodermine

Leukagaricon

Agaricon

Fig. 2.19: Hydrazine derivatives of Agaricus species and their transformation products.

Agaritine is also found in *Agaricus campestris* L., meadow mushroom, field mushroom (0.02–1.0%), *A. bitorquis* (Quél.) Sacc., pavement mushroom (0.05–2.0%), *A. arvensis* Schaeff., horse mushroom (0.02–1.85%), *A. augustus* Fr., the prince (0.1–2.2%), *A. urinascens* (Jul. Schäff. & F.H. Møller) Singer, macro mushroom (0.07–2.5%), and in *A. subrufescens* (0.02–0.2%). In other genera (43 were examined) it was not found [85, 153]. Other investigations determined agaritine contents between 0.0015% and 0.4% in dried fruit bodies of *A. subrufescens* [82, 104, 116]. Another study found in 24 of 53 Agaricus species from wild collection agaritine levels above 0.1% FW [138].

Agaritine can be converted into L-glutamic acid and 4-hydroxymethyl-phenylhydrazine (HMPH) by a γ-glutamyl-transferase of the fungus or by acids. HMPH is easily oxidized, e.g., by an enzyme like laccase, to 4-hydroxymethyl-benzenediazonium ions (HMBD, Fig. 2.19). The resulting radicals are potentially mutagenic. The occurrence of 4-hydrazino-benzoic acid in the fungus is also probable [56, 85, 118, 153, 158]. β-*N*-(γ-L-

Glutamyl)-4-formyl-phenylhydrazine and *N*-γ-L(+)-glutamyl-4-hydroxy-aniline are further ingredients of Agaricus species [31, 74].

LC/MS/MS investigations of mice plasma after oral administration of agaritine (40 mg/kg BW) showed that agaritine is quickly metabolized [86]. Some agaritine metabolites are supposed to bind covalently to DNA, to form adducts, and to damage DNA [68, 123, 141, 175]. DNA damage has been shown by measurement of the concentration of 8-hydroxy-2′-deoxyguanosine as marker for oxidative stress. In opposite to the quick decomposition of agaritine this effect lasts up to 11 days after a single administration to mice [86, 90].

The interaction with DNA explains the development of tumors of liver, lung, blood vessels, and other organs after application of the mushrooms or of hydroxymethyl-phenylhydrazine into animals, especially mice [118, 161, 162]. Lifelong feeding of uncooked mushrooms to mice (from 6 weeks of age, 3 days/week) resulted in a significant increase in tumor incidence compared to control animals. Agaritine itself was not carcinogenic in these experiments [158]. Feeding dried fungi to rats had no effect [110]. The mutagenicity of agaritine is lower than that of its biotransformation products [49], but it is doubled by incubation with kidney homogenate (containing γ-glutamyl peptidase) [175].

☹ Based on results from animal assays it was concluded that agaritine is a mild carcinogen. Extrapolation for humans leads to the conclusion that an intake of about 4 g of *Agaricus bisporus* per day could contribute to a lifetime low cumulative cancer risk in humans of about 20 cases per 10^6 individuals [142]. However, no studies have demonstrated direct implication for increase of carcinogenic risk in humans. Agaritine is classified in the IARC group 3 ('not classifiable as to its carcinogenicity to humans'). However, it was stated that a carcinogenic risk for humans cannot be excluded [60, 90].

The recommendation is not to eat *Agaricus bisporus* in large amounts and to use suitable processing methods. It should not be a daily food, especially when eaten raw. An intake of 100 g fresh mushrooms (or fresh equivalent if dried mushrooms are used) per week is estimated as tolerable. Air-dried mushrooms must be cooked after rehydration and cooking water should be removed. Because of safety reasons a threshold value for agaritine content in Agaricus products including preparations from the medicinal mushroom *Agaricus subrufescens* should be established [60, 90, 112].

Agaricus bingensis Heinem. with an unknown content of agaritine is widely consumed in several East African populations. It is assumed that it may be related, together with malnutrition, to the nodding syndrome, a childhood brain disease [146].

Agaritine exerts activity against leukemic cells, induces apoptosis in leukemia cell lines [122] and should be a potential inhibitor of HIV protease [51]. Whether these positive effects are usable needs further clarification.

In ***Agaricus xanthodermus*** Genev., yellow-staining mushroom, yellow stainer (Ph. 2.34), the compounds xanthodermine, leukoagaricon, 4-diaza-cyclohexa-2,5-dien-1-one, and 4-hydroxybenzene-diazonium salts (Fig. 2.19), which are structurally very similar to agaritine, have been found. Leukoagaricon turns into the yellow agaricon when the fungus is injured in air, xanthodermine into the anion of an intensely yellow-colored acyl-azo compound. The fungus thus contains two chromogens. The detectable phenols (phenol, hydroquinone [56]), which do not occur in the intact fungus, are probably cleavage products of the diazonium compounds [67].

Ph. 2.34: *Agaricus xanthodermus*, yellow-staining mushroom (source: Markus Scholler).

These compound transformations are responsible for characteristic features of *Agaricus xanthodermus*: a strong yellow coloration at the base of the stem when cut and a very unpleasant smell remembering carbolic soap during cooking.

A. xanthodermus is a very common mushroom and widely distributed on all continents. It occurs in woods, lawns, and gardens in autumn. Poisoning is caused by confusion with other Agaricus species, e.g., *A. campestris*, even though unpleasant smell and yellow color should detain people from eating the mushrooms.

Since the mushroom is not suitable for consumption, possible carcinogenic effects of its ingredients do not play a role.

☠ Symptoms of poisoning by *Agaricus xanthodermus* are flushing, sweating, and gastrointestinal irritation symptoms like stomach cramps [**19**]. Some people can eat the mushrooms without ill effects.

Gastrointestinal disturbances can also occur after consumption of other Agaricus species, e.g., *A. freirei* Blanco-Dios, *A. moelleri* Wasser, and *A. pseudopratensis* (Bohus) Wasser [**31**].

2.8.1.3 Dimethyl-methylazoxycarboxamide as Possible Toxin of *Leucocybe connata*

Leucocybe connata (Schumach.) Vizzini, P. Alvarado, G. Moreno & Consiglio (*Lyophyllum connatum* (Schumach.) Singer), white domecap (Ph. 2.35), Tricholomataceae, is a 3–7 cm high, pure white fungus with a thin-fleshed cap. It grows, mostly in clusters, in deciduous and coniferous forests, in parks, along brooks and meadows. It is widely distributed on the Northern Hemisphere.

Ph. 2.35: *Leucocybe connata*, white domecap (source: Markus Scholler).

L. connata was considered edible for a long time. The detection of the compounds lyophyllin (*N,N*-dimethyl-methylazoxycarboxamide, 0.04% in FW), connatin (N^5-hydroxy-N^0,N^0-dimethyl-citrulline, 0.2%), and N^1-hydroxy-*N,N*-dimethylurea (0.01–0.02%, [50], Fig. 2.20) in this fungus led to the conclusion that the mushroom is now considered potentially toxic. In addition, there is a risk of confusion with muscarine containing white funnels (Clitocybe species, see Section 2.8.2) or with the deadly poisonous destroying angel *Amanita virosa* (see Section 2.10).

$(H_3C)_2N-C(=O)-N=N^{\oplus}(O^{\ominus})-CH_3$

Lyophyllin

$(H_3C)_2N-C(=O)-N(OH)-CH_2-CH_2-CH_2-CH(NH_2)-COOH$

Connatin

Fig. 2.20: Ingredients of *Leucocybe connata*.

Based on analogies to other azoxy compounds or *N*-hydroxy ureas, interactions with nucleic acids and resulting mutagenic and carcinogenic effects have been assumed. Whether these are relevant under in vivo conditions in the human organism is questionable [**19, 42**].

☠ Consumption of insufficiently cooked mushrooms can cause gastrointestinal problems. In addition, intolerance reactions have been reported when alcohol is consumed at the same time [80].

The treatment of poisonings is symptomatic.

A ribosome-inactivating protein isolated from *Lyophyllum shimeji* (Kawam.) Hongo by Chinese scientists was also called lyophyllin [92]. It exhibits teratogenic effects in mice [28]. Confusion must be avoided.

2.8.2 Muscarine as Toxin of Inocybe, Clitocybe, and Further Species

Muscarine (Fig. 2.21) is responsible for the toxicity of some Inocybe, Inosperma, Pseudosperma, Clitocybe, and Collybia species.

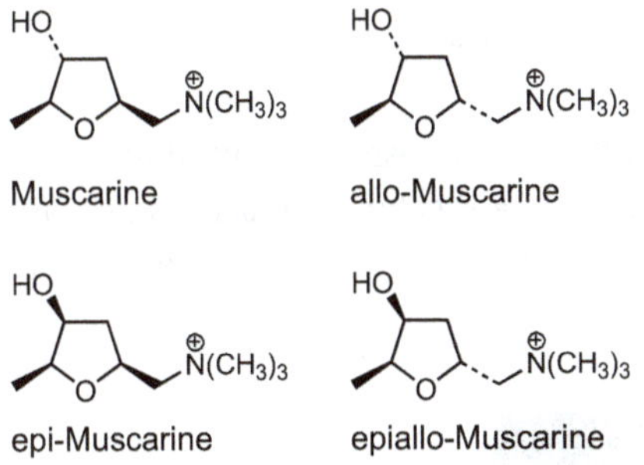

Fig. 2.21: Stereoisomers of muscarin, the toxin of Inocybe and Clitocybe species.

The worldwide occurring genus Inocybe, Inocybaceae, is a very species-rich genus. Most of the species are suspected to be poisonous. Strong fluctuations in muscarine content depending on climate, location, and other factors as well as difficult taxonomic determination make it difficult to classify the Inocybe species reliably into toxic and nontoxic species [**42**]. Based on phylogenetic studies, the genus Inocybe was split and the genera Inosperma and Pseudosperma, among others, were separated.

Some Central European species with known muscarine content are shown in Tab. 2.1. Other Inocybe species contain psilocybin (see Section 2.8.4). In recent years, several other Inocybe, Inosperma, and Pseudosperma species have been identified as containing muscarine in China and India, e.g., *Inocybe serotina* Peck [181], *I. squarrosolutea* (Corner & E. Horak) Garrido, *I. squarrodofulva* [101], *Inosperma virosum* (C.K. Pradeep, K.B. Vrinda & Matheny) Matheny & Esteve-Rav. [94], *Inosperma zonativeliferum* Y.G. Fan., H.J. Li, F. Xu, L.S. Deng & W.J. Yu (muscarine content in the pileus and the stipe 2.08 and 6.53 g/kg resp. [36]), *Inosperma muscarium* Y.G. Fan, L.S. Deng, W.J. Yu & N.K. Zeng, *Inosperma hainanense* Y.G. Fan, L.S. Deng, W.J. Yu & N.K. Zeng [35], and *Pseudosperma arenarium* Y.G. Fan, Fei Xu, Hai J. Li & Vauras [181].

The genus Clitocybe, Clitocybaceae, is also very species-rich and widespread. Known to be poisonous are the pierced funnel, *C. diatreta* (Fr.) P. Kumm., the fragrant funnel, *C. fragrans* (With.) P. Kumm. (content 0.11% DW), the bitter funnel, *C. phaeophthalma* (Pers.) Kuyper, and *C. festiva* J. Favre (content 0.15% DW). A number of other species are suspected to be toxic, e.g., *C. tornata* (Fr.) P. Kumm. and *C. augeana* (Mont.) Sacc., floury funnel [151, **107**]. Poisonous are also *Collybia rivulosa* (Pers.) Z.M. He & Zhu L. Yang (*Clitocybe rivulosa* (Pers.) P. Kumm., *Clitocybe dealbata* var. *rivulosa* (Pers.) P. Kumm.), fool's funnel, false champignon (Ph. 2.36), *Collybia dealbata* (Berk. & M.A. Curtis) Dennis, and *Collybia phyllophila* (Pers.) Z.M. He & Zhu L. Yang (*Clitocybe phyllophila* (Pers.) P. Kumm.), frosty funnel.

Inocybe species are relatively small, thin-fleshed mushrooms with ocher to brown, cracked or scaly, often spherical, more rarely convex caps. The lamellae are gray-brown, deeply indented to attached, only rarely broadly grown. The spore dust is gray-brown to tobacco brown. Inosperma is morphologically distinguished from Inocybe by the absence of pleurocystidia and the shape of the spores. The genus Inosperma is characterized by rimose or scaly pileus, often reddening context, elliptic basidiospores, and thin-walled cheilocystidia [36].

Clitocybe species are mushrooms with thin-fleshed to fleshy cap and thin, white to gray-brown lamellae, which run down the stem to a greater or lesser extent. Veil and ring are missing. The pileus is often funnel-shaped pressed down in the age. The spore dust is white. They are saprophytic ground dwellers.

Ph. 2.36: *Collybia rivulosa*, fool's funnel (source: Geert Schmidt-Stohn).

The structure of muscarine, isolated from *Amanita muscaria*, fly agaric, was elucidated in 1957 by Kögl and colleagues [83]. Muscarine is to be understood as 2,5-epoxide of an aliphatic hexylamine derivative. Due to the presence of three asymmetric carbon atoms, eight stereoisomers of muscarine are conceivable. In nature, (+)-(2*S*,3*R*,5*S*)-muscarine, (–)-(2*S*,3*R*,5*R*)-allo-muscarine, (+)-(2*S*,3*S*,5*S*)-epi-muscarine,

Ph. 2.37: *Inosperma erubescens*, deadly fibrecap (source:Oliver Duty).

Tab. 2.1: Muscarine content of some Inocybe, Inosperma, and Pseudosperma species [41, 140].

Species	English name	Muscarine content (%)	
		DW	FW
Inocybe caesariata (Fr.) P. Karst.		0.52	
I.cincinnata (Fr.) Quél.	Collared fibrecap	0.42–0.53	
I. geophylla (Bull.) P. Kumm.	White fibrecap	0.16–0.26	
I. griseolilacina J.E. Lange	Lilac leg fibrecap	0.17–0.84	
I. lacera (Fr.) P. Kumm.	Torn fibrecap	0.8–1.00	
I. mixtilis (Britzelm.) Sacc.	Yellow fibrecap, yellow-brown fibrecap	0.10–1.33	
I. napipes J.E. Lange	Bulbous fibrecap	0.23–3.15	
I.whitei (Berk. & Broome) Sacc.	Blushing inocybe	0.12–0.17	
Inosperma erubescens (A. Blytt) Matheny & Esteve-Rav. (*Inocybe erubescens* A. Blytt, *Inocybe patouillardii* Bres., Ph. 2.37)	Deadly fibrecap	0.33–0.55	0.04
Pseudosperma rimosum (Bull.) Matheny & Esteve-Rav. (*Inocybe rimosa* (Bull.) Kalchbr.)		0.22–0.50	0.01

and (–)-(2*S*,3*S*,5*R*)-epiallo-muscarine [22, 42] occur. The proportion of the individual stereoisomers in the muscarinic mixture is very different, e.g., in *Inocybe geophylla* 54.2% (+)-muscarine, 45.6% epi-muscarine, and traces of allo- and epiallo-muscarine were found (26). Recently it could be shown that muscarine is released, at least in some species (*Collybia rivulosa, Pseudosperma spectrale* Bandini & B. Oertel, *Inocybe nitidiuscula* (Britzelm.) Lapl.), from its phosphorylated precursor 4′-phosphomuscarine upon cellular injury by unspecific enzymatic ester cleavage. The precursor possesses only weak affinity to the muscarinic acetylcholine receptor M_3 [40].

Muscarine (0.04%) was found in addition to choline (0.14%) and muscaridine ((–)-erythro-(4,5-dihydroxyhexyl)-trimethylammonium, 0.05%) also in the wood pinkgill, *Entoloma rhodopolium* (Fr.) P. Kumm. (*Rhodophyllus rhodopolius* (Fr.) Quél., Ph. 2.38), Rhodophyllaceae. The three quaternary amines are said to be responsible for the emetic effect of the fungus. The watery-gray, poisonous fungus with a cap up to 10 cm wide occurs in autumn in humid deciduous forests. It is easily confused with the edible species *Entoloma crassipes* Petch (*Rhodophyllus crassipes* Imazeki & Hongo) and a common cause of fungal poisoning in Japan. Intoxications manifest themselves mainly in nausea, vomiting, and diarrhea [108]. About the ingredients of the livid pinkgill, *Entoloma sinuatum* (Bull.) P. Kumm. (*Rhodophyllus sinuatus* (Bull.) Quél.) that are responsible for partly severe intoxications with gastrointestinal disturbances mainly in Southern Europe nothing is known to us [**19**]. The mentioned species occur in Europe, North America, and Northern Asia.

Ph. 2.38: *Entoloma rhodopolium*, wood pinkgill (source: Geert Schmidt-Stohn).

In very low concentrations (less than 0.002%) muscarine has also been found in some other fungi, e.g., in the lurid bolete, *Suillellus luridus* (Aschaeff.) Murrill (*Boletus luridus* Schaeff.), in the rufous milkcap, *Lactarius rufus* (Scop.) Fr., in the sulfur knight or stinker, *Tricholoma sulphureum* (Bull.) P. Kumm., and in the wood woolly-foot, *Collybiopsis peronata* (Bolton) R.H. Petersen (*Collybia peronata* (Bolton) P. Kumm., *Gymnopus peronatus* (Bolton) Gray), [148]. Recently, poisoning by *Gymnopus dryophiloides* Antonin, Ryoo & Ka and by *Gymnopus densilamellatus* Antonin, Ryoo & Ka has been reported from China. Intoxication symptoms were gastrointestinal including drowsiness of the mouth [106]. It is unclear whether these species also contain muscarine.

Muscarine acts, like acetylcholine, as an agonist at the muscarinergic acetylcholine receptor. In contrast to acetylcholine, its effect is long-lasting, as it cannot be inactivated by acetylcholinesterase like acetylcholine. It causes, among other things, dilation of blood vessels, drop in blood pressure, stimulation of glandular secretion, mild miosis, and bradycardia. The lethal dose of muscarine for humans is about 0.5 g [125].

Poisonings by muscarine containing mushrooms are relatively common [32, 66, 105, 149, 178, 181, **41**]. They are due, e.g., to confusion of *Inosperma erubescens* or related species with *Calocybe gambosa* (Fr.) Donk (*Tricholoma georgii* (L.) Quél.), St. George's mushroom, or with button mushrooms. Misidentification of Clitocybe species [54] and confusion of these species with *Mycetinis scorodonius* (Fr.) A.W. Wilson & Desjardin (*Marasmius scorodonius* (Fr.) Fr.), common garlic mushroom, and similar-looking small flour mushrooms are also reported.

Four members (three females, one male) of a family in India presented to the emergency department in the evening with symptoms and signs of muscarine toxicity. That afternoon they consumed mushrooms which were plucked up by them that morning from their farmland. About 2–3 h after eating the mushroom dish, they found themselves to be irritable, exhausted, with abdominal cramps and diarrhea. Further symptoms were, among others, salivation, frothing from mouth, sweating, lacrimation, blurring of vision, bradycardia, and miosis. Infusion of atropine and fluid resuscitation led to recovery of all patients over the next 8–10 h [54].

In China, 10 people aged 20–94 from 2 families were poisoned by a formerly undescribed Inosperma species in Hainan Province. They showed sweating, salivation, lacrymation, blurred vision, nausea, vomiting, abdominal pain, and tachycardia. All patients recovered with supportive treatment within 24 h. The species was described as a new species and got the name *Inosperma zonativeliferum* Y.G. Fan, H.J. Li, F. Xu, L.S. Deng, & W.J. Yu [36].

After a short latency period of 15 min to 1 h, symptoms of poisoning occurring after ingestion of muscarine containing mushrooms are a feeling of heat, sweating, increased saliva secretion, nausea, vomiting, diarrhea, mild miosis, intestinal colic, bradycardia, drop in blood pressure, and possibly collapse. In most cases the symptoms subside after about 12 h, in severe cases death may occur [30, 93, 105, 125, 149].

If less than 1 h has passed since ingestion of the mushrooms, primary removal of the poison should be carried out. Otherwise, fluids and electrolytes should be balanced, and atropine should be administered as an antidote (1–2 mg, i.m., or a single dose of 0.02 mg/kg BW very slowly i.v.). Additional circulatory stimulation is required in case of collapse [105, 125, 132, 149, **94**].

Poisonings of dogs have been caused by *Collybia rivulosa* [73] and Inocybe species [124].

2.8.3 Indolylalkylamines as Possible Toxins of Amanita Species (*Amanita citrina* and *Amanita porphyria*)

Bufotenin (Fig. 2.22, 0.1–7.5 mg/g DW, [2, 17, 184]) and small amounts of serotonin are probably responsible for the mild toxicity of *Amanita citrina* Pers., false death cap (Ph. 2.39), *A. porphyria* Alb. & Schwein., gray veiled amanita, and *A. sinocitrina* Zhu L. Yang, Zuo H. Chen & Z.G. Zhang, Amanitaceae. *A. citrina* and *A. porphyria* are widespread on all continents with the exception of the arctic regions, mainly in coniferous forests, *A. sinocitrina* occurs in Asia.

Bufotenin

Fig. 2.22: An indolylalkylamine of Amanita species.

Ph. 2.39: *Amanita citrina*, false death cap (source: Markus Scholler).

Bufotenin occupies 5-HT_{2A} and 5-HT_{2C} receptors. However, it only crosses the blood-brain barrier to a small extent. Its intravenous administration (8 mg) triggers brief hallucinations and visual disturbances in humans. When taken orally, these effects and peripheral effects are minimal or absent [43, 111]. Nevertheless, people with existing hypertension should avoid these mushrooms.

2.8.4 Indolylalkylamines as Active Compounds of Psilocybe Species (Teonanácatl)

Teonanácatl ('flesh of the gods') was already used by the Aztecs 3,500 years ago as an intoxicating drug in ritual acts and is still used as such today in South and Central America [**135, 145, 147**]. It was not until the 1930s that it was recognized by several anthropologists that this name concealed foliose mushrooms of the genus Psilocybe (Hymenogastraceae). In June 1955, the US American banker and hobby mycologist Gordon Wasson together with the photographer Alan Richardson participated as first Western persons at a velada, a shamanic ritual with *Psilocybe caerulescens* Murrill, landslide mushroom. In 1956, the mycologist Roger Heim also participated at a velada and could identify further mushroom species [63]. Publications of G. Wasson and his wife Valentina Wasson in popular journals disseminated the experience with the mushrooms in North America, later also in other parts of the world. As a result, the psilocybin-containing mushrooms gained great popularity, for example in the hippie scene, due to their psychodelic efficacy (overview history: [10, 24, 169]). The term psychodelics (Greek: psyche = soul, delos = evident, manifest) was coined in 1956 as part of an exchange of letters between the psychiater Humphrey Osmond and the writer Aldous Huxley and means psychoactive compounds that can change the mental state of humans [10].

In 1959 Hofmann and coworkers isolated psilocybin (0.2–0.4%) as the main constituent from *Psilocybe mexicana* R. Heim, Mexican liberty cap, teonanácatl, *P. cubensis* (Earle) Singer, golden brownie, and other Psilocybe species, accompanied by small amounts of psilocin (Fig. 2.23), and elucidated the structures of these compounds [10, 69]. Remarkable is the occurrence of a hydroxy group in position 4 of the indole ring system. It is hypothesized that psilocybin and psilocin fulfill a protective function in the mushrooms by altering the behavior of invertebrate predators [131]. When the mushrooms are damaged, a characteristic 'bluing reaction' occurs. It is caused by the enzymatic oligomerization of psilocin [24].

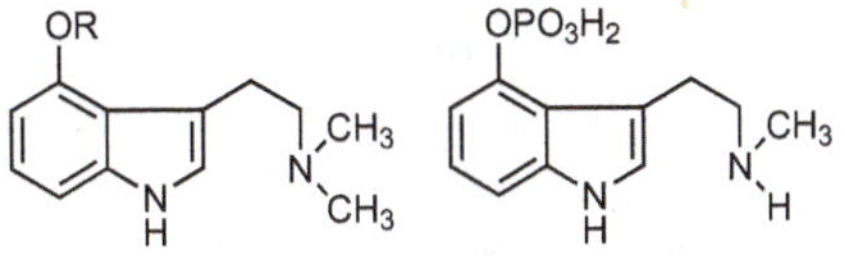

OR — CH_3 — N — CH_3 — N H

OPO_3H_2 — CH_3 — N — H — N H

Psilocybin $R=PO_3H_2$ Baeocystin

Psilocin R=H

Fig. 2.23: Indolylalkylamines of Psilocybe species.

The genus Psilocybe includes about 300 known species, of which 144 exhibit the bluing coloration as a sign of presence of psilocybin and psilocin. Psilocybin has been identified in about 45 accepted Psilocybe species [24]. The most important and most widespread species worldwide is *Psilocybe semilanceata* (Fr.) P. Kumm., liberty cap (Ph. 2.40). The psilocybin content of these fungi is between 0.01% and 2.4% [23, 52, 115,

152, 180]. High concentrations were also detected in *Psilocybe cyanescens* Wakef., blue leg brownie (0.02–1.3% [152]), *P. cubensis* (up to 1.3% [115, 164]), and *P. arcana* Borov. & Hlaváček [155]. The psilocybin content in cultivated *P. cubensis* was 6.3 µg/mg DW in mycelium, 10.5–20 µg/mg DW in caps, and 15.44–18.44 µg/mg DW in stipes. The psilocin content in all samples was between 0.44 and 2.0 µg/mg DW [24]. At least some clans of *P. coronilla* (Bull.) Noordel. (*Stropharia coronilla* (Bull.) W. Saunders & W.G. Sm.), garland roundhead, also contain psilocybin [13].

Ph. 2.40: *Psilocybe semilanceata*, liberty cap (source: Geert Schmidt-Stohn).

Accompanying substances in several Psilocybe species are phenylethylamine (up to 146 µg/g FW in *P. semilanceata* [11]), baeocystin [130], norbaeocystin, *N*-methyl-4-hydroxytryptamine (norpsilocin, liberated from baeocystin, [96]), *N,N*-dimethyltryptophan, aeruginascin, lumichrome, verpacamide A [39] as well as harmane, harmine, and other β-carbolines with MAO inhibiting activities [19]. Genetic and metabolomic studies show that the mentioned mushrooms have the potential to produce much more substances than have been discovered until now [24, 39].

In addition to the genus Psilocybe, representatives of the genera Panaeolus (Galeriopsidaceae), Conocybe (Ph. 2.41), Conobolbitina, Conocybula (Bolbitiaceae), Pluteus (Pluteaceae), Gymnopilus, Galerina (Hymenogastraceae), and Inocybe (Inocybaceae) also contain psilocybin or psilocin [15, 125, 169, 179, **151**]. The psilocybin content has been proven, for example, for *Panaeolus papilionaceus* (Bull.) Quél. (*Panaeolus campanulatus* (L.) Quél.), petticoat mottlegill, *P. cyanescens* Sacc., blue meanies (up to 1.15%, [115]), *P. subbalteatus* (Berk. & Broome) Sacc.), banded mottlegill (up to 0.7%, [52, 154]), *Conobolbitina aeruginosa* (Romagn.) T. Bau & H.B. Song (*Conocybe aeruginosa* Romagn., *Pholiotina aeruginosa* (Romagn.) M.M. Moser), verdigris conecap, *Conocybula cyanopus* (G.F. Atk.) T. Bau & H.B. Song (*Conocybe cyanopus* (G.F. Atk.) Kühner, *Pholiotina cyanopus* (G.F. Atk.) Singer), bluefoot conecap [**31**], *Pluteus salicinus* (Pers.) P. Kumm., knackers crumpet (0.1–0.8% DW), *Galerina steglichii* Besl, Steglich's skullcap, and *Inocybe aeruginascens* Babos.

Ph. 2.41: Conocybe species, cone head species (source: Markus Scholler).

Ph. 2.42: *Gymnopilus junonius*, spectacular rustgill (source: Markus Scholler).

The quaternary ammonium compound aeruginascin (*N,N,N*-trimethyl-4-phosphoryloxytryptamine) was also found in *Inocybe aeruginascens*. It is closely related to bufotenidine, a 5-HT_3 receptor agonist of frog skin [76].

Gymnopilus junonius (Fr.) P.D. Orton (*G. spectabilis* (Weinm.) A.H. Sm., Ph. 2.42), spectacular rustgill, big laughter mushroom, known in Japan as a hallucinogenic mushroom species ('Ohwaraitake'), contains the bitter-tasting neurotoxic gymnopilins (precursor: gymnoprenol, Fig. 2.24). The aliphatic compounds consist of 9–12 isoprene units and inhibit nicotinic acetylcholine receptors. No psilocybin has been detected in Japanese specimens of this species [79, 156].

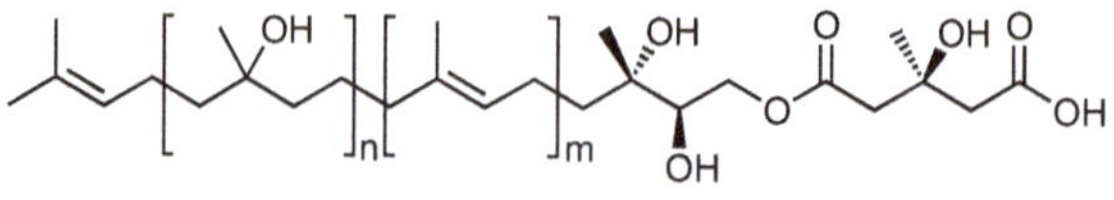

Gymnopilins

Fig. 2.24: Active ingredients (Isoprenoids) of *Gymnopilus junonius.*

The biogenesis of psilocybin starts from L-tryptophan. The amino acid is decarboxylated to form tryptamine which is then hydroxylated on the 4-position of the indole ring by CYP450 monooxygenase to form 4-hydroxytryptamine. Its hydroxyl group is next phosphorylated to form norbaeocystin. Methylation leads to baeocystin and then psilocybin. The enzymes and underlying genes (*Psi* genes, four clusters) have been identified [48, 169]. The gene cluster has been transferred throughout phylogenetically disjunct mushrooms via horizontal gene transfer [131]. L-Tryptophan is also the precursor of the accompanying indole compounds in Psilocybe species [97].

In the human body, the phosphate group of psilocybin is rapidly enzymatically degraded, so that psilocin is the active metabolite. Psilocin is eliminated both renally through O-glucuronylation and by formation of 4-hydroxyindol-3-yl-acetaldehyde. The latter process is catalyzed by MAO A [19].

Due to their structural similarity to serotonin, psilocybin and psilocin are capable to target serotonin receptors. They also alter the serotonin concentration in the brain by influencing the release and metabolism of serotonin. Their effect is like that of LSD (see [**161**] = **Vol. 3**), but a dose around 200 times higher is required for an equally strong effect [61, 115, 119]. The doses of 30 mg psilocybin and 100 and 200 μg LSD produced comparable subjective effects in healthy subjects. LSD had clearly longer effect durations than psilocybin. Psilocybin increased blood pressure more than LSD, whereas LSD increased heart rate more [70].

Psilocybin leads to a reset of neuroplasticity in the human brain. Negative psychic functions are uncoupled and the formation of new networks in the brain becomes possible. An analysis of MRT scans of psilocybin treated depressive patients showed that the increase of the functional connectivity correlates with the clinical improvement of depression. Changes on a molecular level could be detected, e.g., the fast expression of so-called immediate early genes and an increase in the production of the synaptic vesicle protein 2A [34]. In another study, healthy adults were tracked before, during and for 3 weeks after high dose psilocybin (25 mg) and brought back for an additional dose 6–12 months later. The disruption of functional connectivity in cortex and subcortex by psilocybin lasted for weeks. The greatest changes were detected in the Default Mode Network (DMN) [143]. Probably the β-carbolines contained in the fungi contribute to the psychotropic effects because they inhibit MAO and thus the degradation of psilocin [19].

Accidental poisoning with the rather inconspicuous and small psilocybin-containing mushrooms is very rare. In Central Europe, poisoning by psilocybin-containing Panaeolus [12, 174] and Inocybe species (*Inocybe aeruginascens* confused with *Marasmius oreades* (Bolton) Fr., fairy ring champignon [53]) has been observed.

Of much greater significance is the abuse of psilocybin-containing mushrooms as intoxicants ('magic mushrooms'). Since the 1970s, this use has increasingly spread from America to Europe. Today it is mainly propagated on the internet and is particularly widespread among young people. *Psilocybe mexicana, P. cubensis,* and *P. semilanceata* are cultivated illegally in many countries for intoxication purposes. Instructions for use and material for cultivation at home are easily accessible. The dried and powdered mushrooms are consumed suspended in juice, tea or cocoa, eaten in soup or omelets, mixed with honey or made into sweets with chocolate [16, 21, 115, 127, 134, 135, 179].

The main danger of using hallucinogenic mushrooms lies in the possibility of confusion with stronger poisonous mushrooms. Four young people from Bavaria collected *Cortinarius rubellus* instead of *P. semilanceata* and suffered severe poisoning with kidney failure ([5], see Section 2.9.1).

Due to the rapidly growing interest in the medical use of psilocybin (see below) and the associated rise in popularity of psilocybin-containing mushrooms, the use of these fungi for recreational purposes, e.g., to get positive effects on creativity and cognitive function, is increasing rapidly worldwide. This increases the risk of adverse effects. In so-called retreats these mushrooms (often dried pieces), psilocybin, or its derivatives are applied under guidance and help the users to get a state of euphoria. Microdosing means the repeated intake of small amounts of dried mushrooms over a longer period for everyday life optimization without visual or other hallucinogenic effects. In a randomized clinical study low doses of psilocybin led to subjective effects in some participants and altered EEG rhythms, but without evidence to support well-being, creativity, and cognitive function [27].

The lethal dose for humans is assumed to be 20 g of the mushrooms [1, 89, 139, **42, 143**]. The LD_{50} of the whole *P. cubensis* in mice is >2,000 mg/kg BW, p.o. [65]. There were no toxic effects observed in mice for doses of 250–280 mg/kg BW of psilocybin and 180–250 mg/kg BW of psilocin. The lethal effect occurred with a high amount of 400 mg/kg BW [62, 185]. The averaged dose of psilocybin that induces hallucinogenic effects in humans is 4–10 mg (50–300 µg/kg BW). The minimum dose of mushrooms needed to get the desired effect is about 1 g of dried mushrooms [168].

19 out of 9,233 magic mushroom users (0.2%) needed emergency medical treatment because of panic and anxiety. With the exception of one case, the symptoms decreased within 24 h [87]. A comparison of the 'mean-harm score' yielded the lowest value of 19 different illicit drugs for hallucinogenic mushrooms [167, 168].

Whether the so-called 'wood lover's paralysis' that has been observed especially in New Zealand and Australia after consumption of *P. subaeruginosa* Cleland or *P. cyanescens* is related to the mushrooms or their ingredients alone is unclear. The muscle weakness occurs 4–6 h after ingestion and lasts not more than 24 h [169].

The psychosomatic symptoms following the consumption of mushrooms containing psilocybin depend on expectations, constitution, dose, and drug experience. In humans, 4–10 mg psilocybin lead to a trance-like state after 20–30 min, in which a feeling of lightness and detachment often prevails. Hallucinations, in which especially the visual perception is altered ('you hear colors'), are possible. From around 10 mg, there are changes in the sense of time and space. Anxiety, restlessness, and psychoses can occur. Accompanying symptoms are headaches, tingling, feeling cold, balance problems, drop in blood pressure with dizziness. The symptoms last between 2 and 6 h. Mild adverse effects like sleep problems can remain for about 12 h. The disorientation in time and space can trigger panic reactions ('horror trips', 'bad trips') like jumping from high altitudes. The combined use with other psychoactive drugs, including alcohol, increases the risk. The risk of developing dependence is low [71, 166, 168, **42**, **135**]. Suicidal thoughts or 'successful' suicides [114], cases of arrhythmias, cardiomyopathy [88], and acute renal injury [8] are reported. Probably, accompanying substances like phenylethylamine contribute to the side effects.

The most important treatment measure is careful monitoring of the patient to avoid panic reactions. If necessary, tranquilizers can be given [53, 139, **42**].

Increased aggressiveness, salivation, hyperthermia, and howling have been observed in dogs, and excitation, tremors, fever, tenesmus, teeth grinding, and tachycardia in horses.

Because of some severe psychiatric reactions, tragic coincidental events, and probably further reasons the possession and use of psilocybin was prohibited in the late 1960s. A UN convention categorized it as Schedule I substance in Narcotic law. Cultivation of mushrooms containing psilocybin was prohibited in most countries. Nevertheless, cultivation and distribution of magic mushrooms sustained and there is a strongly renewed interest since the 1990s. It is now legal in some countries, e.g., Canada, or some Federal states of the United States, e.g., Oregon.

In 2018, the FDA granted the use of psilocybin the status of 'Breakthrough Therapy' for patients with therapy-resistant depression, later also for unipolar depression. Several clinical studies have demonstrated the benefits of psilocybin in patients with depression [4, 25, 33, 58, 64, 89, 120], addiction disorders [20, 170], to reduce anxiety and to improve quality of life in palliative care patients [59, 133, 171, 183], and in females with Anorexia nervosa [81]. Personal support of the patients before, during, and after a therapeutic session is essential. Psilocybin was already approved as a medicinal product in Australia in summer 2023. Approval in other countries is expected. Standards for taxonomic identification of psilocybin-containing mushrooms and chemical analysis are necessary [24, 169].

References

For numbers in bold, see cross-chapter literature p. 233.

[1] Aboul-Enein HY (1974) Am J Pharm 146: 91
[2] Andary C et al. (1978) Trav Soc Pharm Montpellier 38: 247
[3] Andary C et al. (1985) Mycologia 77: 259
[4] Anderson KAA et al. (2021) Acta Psychiatr Scand 143: 101
[5] Anonym (2002) Tintling 7(3): 46
[6] Arłukowicz-Grabowska M et al. (2019) Ann Hepatol 18(3): 514
[7] Arshadi M et al. (2006) J Chromatogr A 1125: 229
[8] Austin E et al. (2018) Med Mycol Case Rep 23: 55
[9] Azema RC (1979) Documents Mycologiques 10: 1
[10] Bächi B (2020) LSD auf dem Land. Produktion und kollektive Wirkung psychotroper Stoffe. Konstanz University Press, Konstanz
[11] Beck O et al. (1998) J Anal Toxicol 22: 45
[12] Bergner H, Oettel R (1971) Mykol Mitteilungsblatt 15: 61
[13] Berndt S (2017) Tintling 22(5) 36
[14] Berndt S (2022) Tintling 27(5): 96
[15] Besl H (1993) Z Mykol 59(2): 215
[16] Beug MW, Bigwood J (1982) J Ethnopharmacol 6: 271
[17] Beutler Ja, Der Marderosian AH (1981) J Nat Prod 44: 422
[18] Bianchi A et al. (1999) Minerva Anestesiol 65: 811
[19] Blei F et al. (2020) Chemistry Europe 26(3): 729
[20] Bogenschutz MP et al. (2022) JAMA Psychiatry 79: 953
[21] Bogusz MJ et al. (1998) Int J Legal Med 111: 147
[22] Bollinger H, Eugster CH (1971) Helv Chim Acta 54: 2704
[23] Borner S, Brenneisen R (1987) J Chromatogr 408: 402
[24] Bradshaw AJ et al. (2022) Appl Environ Microbiol 88(24): e0149822
[25] Carhart-Harris RL et al. (2021) New Engl J Med 384: 1401
[26] Catalfomo P, Eugster CH (1970) Helv Chim Acta 53: 848
[27] Cavanna F et al. (2022) Transl Psychiatry 12(1): 307
[28] Chan WY et al. (2010) Appl Microbiol Biotechnol 85: 985
[29] Chen HY et al. (2015) Br J Clin Pharmacol 81: 412
[30] Chilton WS (1978) in: **143**, p. 87
[31] Chulia AJ et al. (1988) Phytochemistry 27: 929
[32] Comelli I et al. (2014) Acta Biomed 84(3): 229
[33] Coppola M et al. (2022) J Xenobiot 12: 41
[34] Daws RE et al. (2022) Nature Medicine 28: 844
[35] Deng LS et al. (2021) MycoKeys 85: 87
[36] Deng LS et al. (2022) Front Microbiol 13: 923435
[37] Diaz JH (2016) Wilderness Environm Med 27: 330
[38] Dirks AC et al. (2023) Mycologia 115(1): 1
[39] Dörner S et al. (2022) ChemBioChem 23: e202200249
[40] Dörner S et al. (2024) Angew Chem Int 63(52): e202417220
[41] Eugster CH (1968) Naturwiss 55: 305
[42 Eugster CH, Schleusener E (1969) Helv Chim Acta 52(3): 708
[43] Fabing HD, Hawkins JR (1956) Science 123: 886

[44] Finish Food Authority. Nordic Recipe Archives 1997-1920. http://www.dlc.fi/~{}marianna/gourmet/morel.htm (accessed 07/19/2024)
[45] Firenzuli F et al. (2008) eCAM 5: 3
[46] Fischer B et al. (1984) Z Lebensm Unters Forsch 179: 218
[47] Franke S et al. (1967) Fr Arch Toxicol 22: 293
[48] Fricke J et al. (2017) Angew Chem Int Ed 125: 12524
[49 Friedrich U et al. (1986) Z Lebensm Unters Forsch 183: 85
[50] Fugmann B, Steglich W (1984) Angew Chem 96: 71
[51] Gao W et al. (2007) Med Chem 3: 221
[52] Gartz J (1986) Planta Med 53: 539
[53] Gartz J (1989) Biochem Physiol Pflanz 181: 171
[54] George P, Hegde N (2013) Toxicol Int 20(1): 113
[55] GGIZ (2013) https://ggiz-erfurt.de/saisonales-detail/fruehjahrslorchel.html
[56] Gill M, Strauch RJ (1984) Z Naturforsch C 39: 1027
[57] Giusti GV, Carnevale A (1974) Arch Toxicol 33: 49
[58] Goodwin GM et al. (2022) New England J Med 387: 1637
[59] Griffith RR et al. (2016) J Psychopharmacol 30: 1181
[60] Gry J, Andersen C (2014) Mushrooms Traded as Food. 2014 Vol II sec. 2: Nordic Risk Assessments and Background on Edible Mushrooms, Suitable for Commercial Marketing and Background Lists for Industry, Trade and Food Inspection. Risk Assessments of Mushrooms on the Four Guidance Lists. Nordic Council of Ministers, Copenhagen. https://www.norden.org/en/publication/mushroomson-traded-food-voll-ii-sec-2 (accessed Aprile 11, 2025)
[61] Hasler F et al. (1997) Pharm Acta Helv 72: 175
[62] He MQ et al. (2018) MycoKeys 40: 53
[63] Heim R, Wasson RG (1958) Edition du Museum National d´Histor Naturelle Paris, p. 129
[64] Herdegen T (2021) Dtsch Apoth Ztg 161: 786
[65] Hernandez-Leon A et al. (2024) J Ethnopharmacol 320: 117415
[66] Herrmann M (1964) Mykol Mitteilungsblatt 8: 42
[67] Hilbig S et al. (1985) Angew Chem 97: 1063
[68] Hiramoto K et al. (1995) Chem Biol Interact 94: 21
[69] Hofmann A (1971) Bull Narcotics 1: 3
[70] Holze F et al. (2022) Neuropsychopharmacol 47: 1180
[71] Honyiglo E et al. (2019) J Forensic Sci 64(4): 1266
[72] Horowitz KM et al. (2023) in: StatPearls [Internet]- Treasure Island (FL): StatPearls Publishing; 2023 https://www.ncbi.nlm.nih.gov/books/NBK470580/ (accessed 17.01.2024)
[73] Irwin AG, Leech AR (2014) Vet Rec 175(5): 122
[74] Jadot I et al. (1960) Biochem Biophys Acta 43: 322
[75] Janak K et al. (2006) Food Chem 99: 521
[76] Jensen N et al. (2006) Planta Med 72: 665
[77] Jurenitsch J et al. (1988) Z Mykol 54: 155
[78] Karlson-Stiber C, Persson H (2003) Toxicon 42: 339
[79] Kayano T et al. (2014) Toxicon 81: 23
[80] Kell V (1989) Mykol Mitteilungsblatt 32: 5
[81] Knatz Peck S et al. (2023) Nat Med 29(8): 1947
[82] Koge T et al. (2011) Food Chem 126: 1172
[83] Kögl F et al. (1957) Experientia 13: 137
[84] Kondo K et al. (2006) Food Addit Contam 23: 1179
[85] Kondo K et al. (2006) J Chromatogr B Analyt Technol Biomed Life Sci 834: 55

[86] Kondo K et al. (2008) Food Chem Toxicol 46, 854
[87] Kopra EI et al. (2022) J Psychopharmacol 36(8): 965
[88] Kotts WJ et al. (2022) BMJ Case Rep 15(5): e245863
[89] Kuhn A, Melzig MF (2022) Z Phyother 43: 89
[90] Lagrange E, Vernoux JP (2020) Toxins 12: 482
[91] Lagrange E et al. (2021) J Neurol Sci 427: 117558
[92] Lam SK, Ng TB (2001) Arch Biochem Biophys 393(2): 271
[93] Lampe FK (1978) in: **143**, p. 125
[94] Latha SS et al. (2020) Scient Rec 10: 13669
[95] Leathem AM, Doran TJ (2007) Can J Emerg Med 9: 127
[96] Lenz C et al. (2017) J Nat Prod 80(10): 2835
[97] Lenz C et al. (2021) Chempluschem 86(1): 28
[98] Levenberg G (1960) J Am Chem Soc 83: 503
[99] Lheureux P et al. (2005) Eur J Emerg Med 12: 78
[100] Li HJ et al. (2020) Mycosystema 39(9): 1706
[101] Li SN et al. (2021) MycoKeys 81: 185
[102] List PH, Luft P (1967) Tetrahedron Lett: 1893
[103] List PH, Luft P (1969) Arch Pharm 302: 143
[104] Liu Y et al. (2008) eCAM 5: 205
[105] Lurie Y et al. (2009) Clin Toxicol (Phila) 47/6): 562
[106] Ma J et al. (2024) Int J Med Mushrooms 26(3): 77
[107] Mack RB (1986) NC Med J 47: 535
[108] Maki T et al. (1985) J Agric Food Chem 33: 1204
[109] Märki F et al. (1962) Biochem Biophys Acta 58: 367
[110] Matsumoto K et al. (1991) Cancer Lett 58: 87
[111] McBride MC (2000) J Psychoact Drugs 32: 321
[112] Merdivan S et al. (2016) Int J Med Mushrooms 18: 13
[113] Michelot D, Toth B (1991) J Appl Toxicol 11: 235
[114] Müller K et al. (2013) Arch Kriminol 231(5–&): 193
[115] Musshoff F et al. (2000) Foresnic Sci Int 113: 389
[116] Nagaoka MH et al. (2006) Chem Pharm Bull 54: 922
[117] Nagel D et al. (1977) Cancer Res 37: 3458
[118] Natori S (1987) Bioact Mol 2: 127
[119] Nichols DE (1981) J Pharm Sci 70: 839
[120] Nichols DE et al. (2020) J Antibiot (Tokyo) 73: 679
[121] Niskanen A et al. (1976) Food Cosmet Toxicol 14: 409
[122] Ogasawara A et al. (2023) Fujita Med J 9(2): 147
[123] Oikawa S et al. (2006) Free Radic Res 40: 31
[124] Opdal Seljetun K, von Krogh A (2017) J Vet Emerg Crit Care (San Antonio) 27(2): 212
[125] Patocka J et al. (2021) Int J Mol Sci 22: 2218
126] Perisetti A et al. (2018) Cureus 10(4): e2436
[127] Plazas E, Faraone N (2023) Biomedicines 11: 461
[128] Pyysalo H, Niskanen A (1977) J Agric Food Chem 25: 644
[129] Pyysalo H et al. (1978/79) J Food Saf 1: 295
[130] Repke DB et al. (1977) J Nat Prod 40: 566
[131] Reynolds HT et al. (2018) Evol Lett 2: 88
[132] Riethmüller J et al. (2004) Monatsschr Kinderheilk 152: 892
[133] Ross S et al. (2016) J Psychopharmacol 30: 1165

[134] Satora L et al. (2005) Przegl Lek 62: 394
[135] Schäfer AT (2000) Arch Kriminol 205: 30
[136 Schmidlin-Meszaros J (1974) Mitt Gebiete Lebensm Hyg 65: 453
[137] Schmidlin-Meszaros J (1979) Chem Rundsch 32: 1
[138] Schulzova V et al. (2009) Food Addit Contam 26: 82
[139] Schwartz RH, Smith DE (1988) Clin Pediatr 27: 70
[140] Seeger R (1993) Dtsch Apoth Ztg 133: 108
[141] Shephard SE, Schlatter C (1998) Food Chem Toxicol 36: 971
[142] Shephard SE et al. (1995) Food Chem Toxicol 33: 257
[143] Siegel JS (2024) Nature. Doi.org/10.1038/s41586-024-07624-5
[144] Slanina P et al. (1993) Food Addit Contam 10: 391
[145] Spencer PS (2019) Frontiers Neurol 10: Art 754
[146] Spencer PS et al. (2022) eNeurologicalSci 27: 100401
[147] Speroni JJ et al. (1983) J Food Protect 46: 506, 513
[148] Stadelmann RJ et al. (1976) Helv Chim Acta 59: 2432
[149] Stallard D, Edes TE (1989) Postgrad Med 85: 341
[150] Stijve T (1978) Mitt Geb Lebensm Hyg 69: 492
[151] Stijve T (1982) Coolia 25: 94
[152] Stijve T, Kuyper TW (1985) Planta Med 51: 385
[153] Stijve T et al. (1986) Dtsch Lebensm Rundschau 82: 243
[154] Strauss D et al. (2023) Heliyon 9(6): e16338
[155] Stribrny J et al. (2003) Soud Lek 48:45
[156] Tanaka M et al. (1993) Phytochemistry 34(3): 661
[157] Toth B (1975) Cancer Res 35: 3693
[158] Toth B, Erickson J (1986) Cancer Res 46: 4007
[159] Toth B, Gannett P (1994) In Vivo 8: 999
[160] Toth B, Patil K (1982) Mycopathologia 78: 11
[161] Toth B et al. (1977) Z Krebsforsch Klin Onkol 89: 245
[162] Toth B et al. (1978) Cancer Res 38: 177
163] Trestrail III J (1994) in **151**, p.279
[164] Tsujikawa K et al. (2003) Forensic Sci Int: 138: 85
[165] Turkoglu A et al. (2009) Int J Med Mushrooms 11: 101
[166] Vallersnes OM et al. (2016) BMC Psychiatry 16: 293
[167] van Amsterdam J et al. (2010) Eur Addict Res 16: 202
[168] van Amsterdam J et al. (2011) Regul Toxicol Pharmacol 59: 423
169] van Court RC et al. (2022) Fungal Biol 126(4): 308
[170] van der Meer PB et al. (2023) Front Psychiatry 14: 1134454
[171] Vargas AS et al. (2020) Biomedicines 8: 331
[172] Viernsrein H et al. (1980) Ernährung 4: 392
[173] von Wright A et al. (1978) Toxicol Lett 2: 261
[174] Walting R (1977) Mycopathologia 61: 187
[175] Walton K et al. (1997) Carcinogenesis 18: 1603
[176] Walton K et al. (2001) Toxicology 161: 165
[177] Wang XC et al. (2023) Microbiol Spectrum 11(3): 10.1128/spectrum.00207-23
[178] Wilson D (1947) Brit Med J: 297
[179] Wurst M et al. (1984) J Chromatogr 286: 229
[180] Wurst M et al. (2002) Folia Microbiol 47: 3
[181] Xu F et al. (2020) Toxicon 179: 72
[182] Yan YY et al. (2022) Myco Keys 92: 79

[183] Yu CL et al. (2021) Psychiatry Invest 18: 958
[184] Zhang YZ et al. (2025) Toxins 17(5): 247
[185] Zhuk O et al. (2015) Toxins 7(4): 1018

2.9 Alkaloids as Toxins of Macrofungi

2.9.1 Pyridine Alkaloids as Toxins of Cortinarius Species

With more than 2,000 species worldwide, the genus Cortinarius, webcap (Cortinariaceae), is one of the largest mushroom genera. Its representatives exhibit high morphological variability. The fungi occur both in Northern and Southern Hemisphere from tropical to arctic-alpine habitats and form ectomycorrhiza with a wide range of hosts [25].

Cortinarius species are vivid red, yellow or green leafed mushrooms with bulging, rarely grown lamellae and pinkish-red spore dust. In most cases, there is a veil (cortina) between the brim of the cap and the stem. In addition, fragments of the entire shell in the form of a slimy coating or woolly to membranous flakes are often found on the surface of the cap and on the stem.

Although there are some good edible mushrooms among the webcap species, the majority are poisonous or suspected of being poisonous. Particularly dangerous are *Cortinarius orellanus* Fr., fool's webcap (Ph. 2.43), and *C. rubellus* Cooke (*C. speciosissimus* Kühner & Romagn., Ph. 2.44), deadly webcap, besides *C. orellanosus* Ammirati & Matheny, possibly *C. armillatus* (Fr.) Fr., red-banded cortinarius, bracelet cortinarius (<1% of the orellanine concentration of *C. rubellus* [38]), and *C. eartoxicus* Gasparini **[41]**.

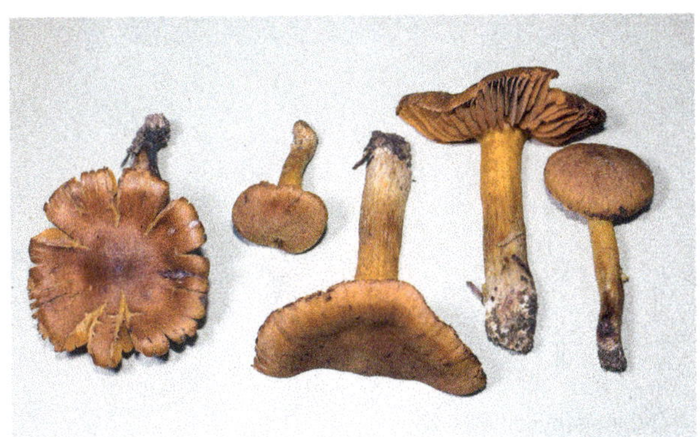

Ph. 2.43: *Cortinarius orellanus*, fool's webcap (source: Geert Schmidt-Stohn).

From *Cortinarius orellanus, C. rubellus, C. rainierensis* A.H. Sm. & D.E. Stuntz, *C. fluorescens* E. Horak, and *C. henricii* Reumaux, the toxic bipyridine alkaloid orella-

Ph. 2.44: *Cortinarius rubellus*, deadly webcap (source: Geert Schmidt-Stohn).

nine, an *N,N'*-dioxide, was isolated. It is photosensitive: once extracted, it is reduced to orellinine, which has the same toxic properties and is rapidly converted into the non-toxic orelline (Fig. 2.25 [3, 22, 26, **41]**). The orellanine content in the mushrooms is 1.1–1.4% in *C. orellanus* and 0.4–0.9% in *C. rubellus* [14]. In some related species, such as *Cortinarius splendens* Rob. Henry (*Calonarius splendens* (Rob. Henry) Niskanen & Liimat), splendid webcap, which cause the same symptoms as *C. orellanus*, the alkaloids could not be detected [10, 28, 40]. Cytotoxic sulfur-containing dipyridine alkaloids including cortamidine oxide and 2,2'-dithiobis(pyridine *N*-oxide) were isolated from an unidentified Cortinarius species in New Zealand (Fig. 2.25 [30]).

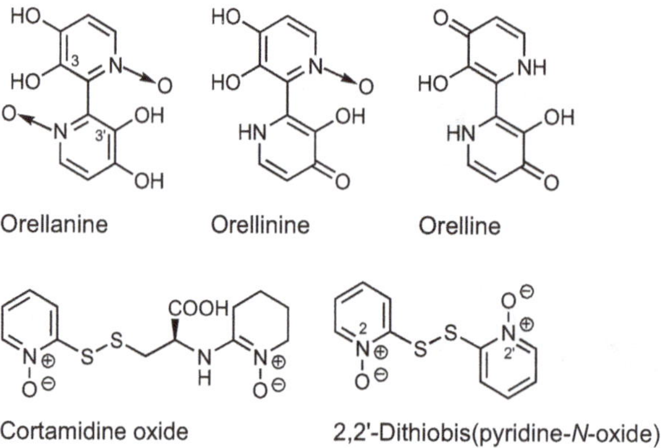

Fig. 2.25: Dipyridine alkaloids of Cortinarius species.

The previously reported presence of toxic bicyclic peptides, the cortinarins A and B, in about 60 Cortinarius species [28] is in doubt [26].

Anthraquinones (emodin, dermocybin, and other, see **[160] = Vol. 2**) are responsible for the intense coloration of the fruit bodies of many Cortinarius species, e.g., of *Cortinarius sanguineus* (Wulfen) Gray (*Dermocybe sanguinea* (Wulfen) Wünsche), blood-red webcap. It cannot be ruled out that they contribute to the toxicity and skin-sensitizing properties of Cortinarius species [45].

Orellanine and orellinine have nephrotoxic effects, probably only after metabolic activation. They induce severe tubulointerstitial nephritis, which leads to necrosis of the proximal tubular epithelia in preserved glomeruli. There are different hypotheses for the mechanism of action. It seems certain that the nephrotoxicity of orellanine is due to the oxidative attack of oxygen radicals on macromolecules. The capacity of orellanine to suppress the expression of antioxidative enzymes increases the susceptibility of cells to the oxidative damage. Effects on the mitochondrial electron transport chain and an inhibition of alkaline phosphatase and DNA and RNA polymerases have been found [6, 11, 15, 26, 28, 33, 34, **41, 178**]. There are still many unanswered questions regarding the pharmacokinetics of orellanine and the specificity of the effect in the kidneys.

Until 1952, *Cortinarius orellanus* was considered harmless. The latency period of up to 20 days before the first symptoms of poisoning appeared made it difficult to establish a link between the symptoms of poisoning and their cause. A mass poisoning of 102 people (11 of whom died) in Poland in the area between Bydgoszcz and Poznán in 1952 [8] drew attention to the toxicity of the fungi. In the meantime, cases of poisoning by *Cortinarius orellanus* and *C. rubellus* have been reported from almost all countries of Europe, North America, and from Australia [12, 13, 16, 17, 19, 28, 31, 36, **19, 41**]. Poisoning by *Cortinarius splendens* has been reported in France [7]; its toxicity appears to be somewhat lower than that of the other species mentioned and caused by other compounds than orellanine [28].

The LD_{50} of orellanine is 12–90 (average 39) mg/kg BW, p.o., mice, and 5 mg/kg BW, i.p., rat. Humans have been shown to be more sensitive than mice to orellanine. The ingestion of two to three mushrooms can make a victim dialysis-dependent for life [14].

Suspected poisoning with Cortinarius species is due to confusion with edible mushrooms such as chanterelles [13, 14] or with hallucinogenic members of the genus Psilocybe [2] (see Section 2.8.4). One poisoning case reports voluntary consumption of *C. orellanus* by a psychiatric patient (overview: [26, **41**]).

☠ The first symptoms of poisoning caused by toxic Cortinarius species appear after 36 h to 20 days. They include tiredness, nausea, vomiting, diarrhea, and somewhat later dryness and burning in the oral cavity, severe thirst, feeling cold, loss of appetite as well as headache and muscle pain. The progressively developing kidney damage manifests itself in oliguria, albuminuria, and finally uremia. Survival of the poisoning results in chronic nephritis. Complete recovery is unlikely. The liver, spleen, and nervous system are only rarely affected. If they are, the consequences are drowsiness, unconsciousness, tremor of the facial muscles, cramps, and possibly hepatitis. In severe cases, death occurs until 2–3 months after consumption of the mushrooms due to kidney failure. Histopathologically, tubular-interstitial nephritis with necrosis of the renal tubules and leukocyte infiltration of the parenchyma are detectable [1, 9, 11, 13, 16, 19, 26, 28, 37, **41, 178**]. A case of myocarditis additionally to nephrotoxicity was recently described [4].

Measures to eliminate the toxins (hemodialysis, hemoperfusion) can still be useful days after ingestion. Symptomatic treatment must be carried out with the usual measures for kidney diseases. In severe cases, repeated dialysis or kidney transplantation may be necessary. The use of high dose antioxidants with or without steroids seems to be not consistently successful [11, 13, 19, 26, **41, 178**].

Cortinarius poisoning has also been reported in sheep [32].

Since 2022, orellanine is in clinical phase I/II trials due to promising data in clear cell renal cell carcinoma [26].

2.9.2 Alkaloids as Possible Toxins of Chlorophyllum Species

Several indole alkaloids, including macrolepiotin (Fig. 2.26) and lepiotins A–C, are possibly responsible for the toxicity of *Chlorophyllum neomastoideum* (Hongo) Vellinga (*Macrolepiota neomastoidea* (Hongo) Hongo, Agaricaceae). The species is widely distributed through Korea and other Asian countries and known there as a poisonous mushroom that causes gastrointestinal symptoms. A case of acute liver failure has been reported [21].

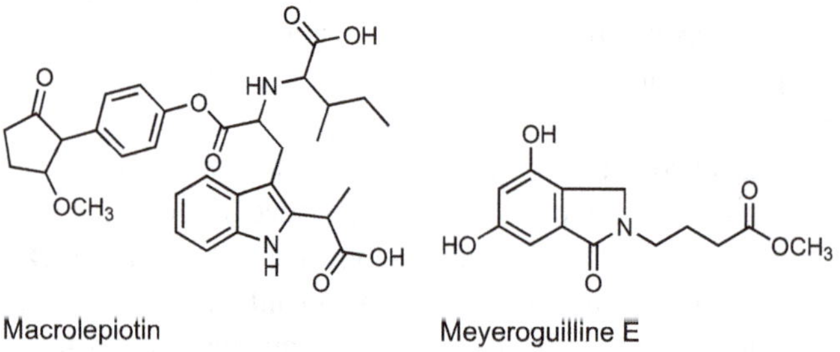

Macrolepiotin

Meyeroguilline E

Fig. 2.26: Alkaloids of Chlorophyllum and Macrolepiota species.

The compounds show in vitro cytotoxic activities [20]. If they are really the cause of the poisoning or possibly other ones (amatoxins?) is unclear.

Chlorophyllum molybdites (G. Mey.) Massee (*Macrolepiota molybdites* (G. Mey.) G. Moreno, Bañares & Heykoop), false parasol, green-spored parasol, vomiter (Ph. 2.45), is another poisonous species of the genus Chlorophyllum. The imposing mushroom with a cap ranging from 8 to 30 cm in diameter is native in North America and is currently spreading worldwide. It usually grows on lawns and parks, mainly in subtropical area, and often forms fairy rings. On the European mainland it has so far only been found occasionally, mostly in greenhouses.

Ph. 2.45: *Chlorophyllum molybdites*, false parasol (source: toa55/iStock/Getty Images Plus).

Ph. 2.46: *Chlorophyllum brunneum*, garden parasol (source: Geert Schmidt-Stohn).

The cap is whitish in color with coarse brownish scales and becomes lead-gray with age. The flesh slowly turns reddish-brown when injured. The stipe can grow up to 25 cm and has a ring. The gills are free and white, turning dark and green with maturity. The spore powder is greenish. The greenish lamellae and green spore powder are the most important distinguishing features to other parasol mushrooms, however they are only found in the older mature fruit bodies [35].

There is an increasing number of reports of poisoning by this species in many countries, among them the USA and Germany [5, 24, 27, 29, 35, 39, 42]. In 2019, there were 55 food poisoning incidents caused by eating this mushroom in China [41].

The poisonings are caused by confusion with other mushrooms, e.g., *Chlorophyllum rhacodes* (Vittad.) Vellinga (*Macrolepiota rhacodes* (Vittad.) Singer), shaggy parasol, *Macrolepiota procera*, parasol mushroom, or *Coprinus comatus* (O.F. Muell.) Pers., shaggy mane.

It is still unclear which ingredients of *C. molybdites* are responsible for the poisoning. Some alkaloids, the protein molybdophyllysin, steroids, and as yet unknown active ingredients come into question. The alkaloids identified include isoindolinone alkaloids, among them meyeroguilline E (Fig. 2.26), and the pyrrolidine alkaloids lepiothin A and B [23]. Molybdophyllisin is a heat-labile metalloendopeptidase [44]. Its high homology to metalloendopeptidases in several edible mushrooms and the fact that poisoning has also occurred after eating heated mushroom meals contradict the responsibility of molybdophyllysin for the intoxication.

The symptoms, which begin within 1–6 h after ingestion, mainly affect the gastrointestinal tract: vomiting, watery and later bloody diarrhea, cramps, accompanied by profuse sweating. In isolated cases, altered perception, blurred vision, hypotension, and tachycardia can occur. The symptoms may persist over 12–24 h [29].

Treatment is supportive and includes gastric lavage, activated charcoal as well as fluid and electrolyte replacement [29].

Meyeroguilline E may reverse the multidrug resistance of cultivated tumor cells [23].

The ingestion of the European species *Chlorophyllum rhacodes* (Vittad.) Vellinga (*Macrolepiota rhacodes* (Vittad.) Singer), shaggy parasol, and *Chlorophyllum brunneum* (Farl. & Burt) Vellinga (*Macrolepiota bohemica* (Wichanský) Krieglst., *Macrolepiota venenata* Bon, *Chlorophyllum venenatum* (Bon) C. Lange & Vellinga), garden parasol (Ph. 2.46), can also lead to gastrointestinal disturbances. Possible toxins are unknown [18, 43].

References

For numbers in bold, see cross-chapter literature p. 233.

[1] Anantharam P et al. (2016) Toxins 8: 158
[2] Anonym (2002) Tintling 7(3): 46
[3] Antkowiak WZ, Gessner WP (1985) Experientia 41: 769
[4] Balice G et al. (2024) Toxins 16(6): 265
[5] Bijeesh C et al. (2017) J Mycopathol Res 54(4): 477
[6] Cantin-Esnault D et al. (1998) Free Rad Res 28(1): 45
[7] Colon S et al. (1982) Kidney Int 21: 121
[8] Erdmann WD et al. (1961) Arzneim Forsch 11: 835
[9] Franck B (1958) Chem Ber 91: 2803
[10] Gehlbach SH, Perry LD (1975) Lancet 1: 478
[11] Grebe SO et al. (2013) Ren Fail 35(10): 1436
[12] Gross D (1970) Fortschr Chem Org Naturst 28: 109
[13] Hedman H et al. (2017) BMC Nephrology 18: 121

[14] Herrmann A et al. (2012) J Nat Prod 75: 1690
[15] Holmdahl J et al. (1987) Toxicon 25: 195
[16] Horn S et al. (1997) Am J Kidney Dis 30(2): 282
[17] Keeler RF, Dell Balls L (1978) Clin Toxicol 12: 49
[18] Keller SA et al. (2018) Int J Environ Res Publ Health 15: 2855
[19] Kerschbaum J et al. (2012) Clin Kidney J 5: 576
[20] Kim KH et al. (2009) J Antibiot 62: 335
[21] Kim SY et al. (2018) Korean J Gastroenterol 71(2): 94
[22] Kurnsteiner H, Moser M (1981) Mycopathilogia 74(2): 65
[23] Lee RS et al. (2022) ACS Omega 7(43): 39456
[24] Lehmann PF, Khazan U (1992) Mykopathologia 118: 3
[25] Liimatainen K et al. (2022) Fungal Diversity 112: 89
[26] Lyons MJ et al. (2023) J Nat Prod 86: 1620
[27] Meijer AA et al. (2017) Braz Arch Biol Techn 50(3): 118
[28] Michelot D, Tebbett I (1990) Mycol Res 94: 289
[29] Negrini VM et al. (2022) Ann Ist Super Sanità 58(3): 213
[30] Nicholas GM et al. (2001) J Nat Prod 64(3): 341
[31] Nolte S et al. (1987) Monatsschr Kinderheilk 135: 280
[32] Overas J et al. (1979) Acta Vet Scand 20: 148
[33] Richard JM et al. (1991) Toxicology 67: 53
[34] Ruedl C et al. (1990) Mycol Helv 4: 99
[35] Scholler M (2018) Z Mykol 32(2): 339 (published by S Berndt)
[36] Schumacher T, Hoiland K (1983) Arch Toxicol 53: 87
[37] Seeger R (1995) Dtsch Apoth Ztg 135(37): 3347
[38] Shao D et al. (2016) Toxicon 114: 65
[39] Stenklyft PH, Augenstein LW (1990) J Toxicol Clin Toxicol 28: 159
[40] Tiecco M et al. (1987) Experientia 43: 462
[41] Wang N et al. (2021) Front Microbiol 12: 638315
[42] Whitaker G, Box J (1985) J Med Soc 82: 220
[43] Winterstein D (2012) Tintling 17(4): 21
[44] Yamada M et al. (2012) Bioorg Med Chem 20: 6583
[45] Yli-Öyrä J et al. (2024) J Fungi 10: 369

2.10 Peptides as Toxins of Macrofungi (Amanita, Galerina, and Lepiota Species)

2.10.1 Occurrence

The toxicity of the world's most important poisonous mushrooms is due to their content of cyclic peptides, especially amatoxins, besides phallotoxins and virotoxins (overview: [100, 103]).

Table 2.2 gives an overview of amatoxin-containing mushrooms based on the overview of Vetter from 2023 [100]. It must be taken into account, that environmental conditions, location, development, and the investigated part of the mushroom influence toxin amounts. It is noticeable that the bioactive peptides occur in phylogenetically distinct mushroom families. This is possibly due to the horizontal transfer of metabolic gene clusters among taxonomically unrelated mushrooms with overlapping habitats [63].

The most important and best studied species containing amatoxins and phallotoxins are *Amanita phalloides, A. virosa*, and *A. verna*.

Amanita phalloides (Ph. 2.47) has a greenish cap 4–12 cm in diameter, white gills, a conspicuous cuff, and, like all Amanita species, a bulbous stem end that is stuck in a sheath-like, lobed sheath, although it often gets stuck in the soil when pulled out. It is found in deciduous forests, especially under oaks and copper beech trees, and in parks on nutrient-rich soils. Characteristic morphological properties are white gills, ring, and volva.

Amanita verna (Ph. 2.48) resembles *A. phalloides*, but with a cap diameter of 5–8 cm it is smaller than that. It occurs preferably in the Mediterranean area, especially under oaks.

Amanita virosa (Ph. 2.49) has a white, often pointed conical cap with a diameter of 4–10 cm, a long fibrous stalk, and a very thin-skinned, ephemeral cuff. It occurs in deciduous forests, especially under copper beech, oak, birch, but also under spruce.

Regarding the distribution of toxins in different parts of the fruit bodies highest concentration was found in gills, lowest in volva and spores. Within the amatoxins, α-amanitin makes up the largest proportion [20, 34, 53, 95, 100, 111, 120, 121, **42**]. The amount of amatoxins in East Asian Amanita species is often higher than in European and North American species so that a higher toxicity can be expected [95]. Among the phallotoxins, phallacin and phallisacin predominate in fungi of European origin [25].

The genus Lepiota comprises 250–400 species. Their fruit bodies are characterized by universal and partial veil, free gills, a species-specific pileus covering, whitish color of the spore powder, dextrinoid basidiospores, and clamp connections. In the Lepiota species summarized in Tab. 2.2 amanitins have been detected. No phallotoxins have been found in Lepiota species so far [83].

Galerina species are small, yellow-orange or yellow-brown mushrooms. The genus includes more than 300 species. According to a combination of chemical tests and DNA barcoding, all known producers of amatoxins in the genus Galerina seem to fall into the subgenus Naucoriopsis and most are in *Galerina marginata* s.l. complex (Ph. 2.51). Among them, *G. castaneipes* and *G. venenata* contain detectable amatoxin quantities. Possibly, *G. sulciceps* that is reported to contain amatoxins belongs taxonomically to this section. The toxin status of *G. badipes* (Pers.) Kühner is uncertain [55].

2.10.2 Chemistry and Biogenesis

The main toxins are the amatoxins (Fig. 2.27). They were first isolated in 1941 by Heinrich Wieland and Rudolf Hallermayer [113]. The toxin mixture was broken down into its components by Theodor Wieland and coworkers [106, 112]. They also elucidated the structures of the individual components.

Tab. 2.2: Occurrence and content of amanitins and phallotoxins in Basidiomycetes (basing on [100] and completed with further data).

Family and species	English name	Occurrence	Contained toxins and toxin content in fruit bodies (mg/g DW, unless otherwise specified)
Amanitaceae			
Amanita amerivirosa Tulloss, L.V. Kudzma & M. Tulloss		North America	
Amanita bisporigera G.F. Atk.	Eastern North American destroying angel	North America	3.03 total amatoxins 0.45 total phallotoxins [95] 1.91 β-amanitin [65]
Amanita brunneitoxicaria Thongbai, Raspé & K.D. Hyde		Asia (Thailand)	
Amanita brunnescens G.F. Atk.	Brown American star-footed amanita	North America	
Amanita cheelii P.M. Kirk		East Asia	
Amanita cokeri E.J. Gilbert & Kühner ex E.J. Gilbert	Coker's amanita	North America	
Amanita decipiens (Trimbach) Jacquet		Europe	
Amanita exitialis Zhu L. Yang & T.H. Li	Guangzhou destroying angel	East Asia	2.05–4.36 total amatoxins 1.22–2.89 total phallotoxins [95]
Amanita fuliginea Hongo	East Asian brown death cap	East Asia	5.50–9.03 total amatoxins 1.47–2.96 phallotoxins [95, 122]

(continued)

Tab. 2.2 (continued)

Family and species	English name	Occurrence	Contained toxins and toxin content in fruit bodies (mg/g DW, unless otherwise specified)
Amanita fuligineoides P. Zhang & Zhu L. Yang		East Asia	0.607 α-amanitin 0.377 β-amanitin 0.069 phalloidin [60] 8.85 total amatoxins 0.25 total phallotoxins [95]
Amanita hygroscopica Coker	Pink-gilled destroying angel	North America	
Amanita ocreata Peck	Western North American destroying angel	North America	
Amanita pallidorosea P. Zhang & Zhu L. Yang	Pale-rose death cap	East Asia	5.40–6.26 total amatoxins 1.62–2.87 total phallotoxins [95]
Amanita phalloides (Vail. ex Fr.) Link (Ph. 2.47)	Death cap, death angel, poison amanita	Europe, North America, North Africa, North Asia, Australia, New Zealand	Switzerland: 5.2–7.4 total amatoxins [88] Turkey: 4.16 total amatoxins [53] Central Europe: 1.24–1.44 total amatoxins 2.81–4.27 total phallotoxins [95]

Amanita rimosa P. Zhang & Zhu L. Yang		East Asia	8.78–12.25 total amatoxins 1.94–2.46 total phallotoxins [95]
Amanita suballiacea (Murill) Murill	Garlic-odored death cap	North America	
Amanita subjunquillea S. Imai	East Asian death cap	East Asia	3.14–8.95 total amatoxins 0.85–3.72 total phallotoxins [95]
Amanita subpallidorosea Hai J. Li		East Asia	
Amanita tenuifolia (Murill) Murill		North America	
Amanita verna Bull. ex Lam. (Ph. 2.48)	Spring amanita, fool's mushroom	Europe	Switzerland: 3.2 total amatoxins 1.55 α-amanitin 1.29 β-amanitin 0.37 γ-amanitin [88] Turkey: 9.96 total amatoxins [53] Mediterranean: 5.47 α-amanitin 10.26 β-amanitin 4.68 phallacidin 0.64 phalloidin [1]
Amanita vidua Gasch, G. Moreno & P.A. Moreau		Southern Europe	4.73–7.18 α-amanitin 5.02–8.50 β-amanitin 6.53–7.78 phallacidin [1]

(continued)

Tab. 2.2 (continued)

Family and species	English name	Occurrence	Contained toxins and toxin content in fruit bodies (mg/g DW, unless otherwise specified)
Amanita virgineoides Bas	False virgin's lepidella	East Asia	
Amanita virosa Bertill. (Ph. 2.49)	European destroying angel	Europe, Northern Asia	1.34–1.60 total amatoxins, no β-amanitin 2.80–4.49 total phallotoxins [95] 0.039 α-amanitin 0.145 β-amanitin [73] 1.17 α-amanitin [65]
Agaricaceae			
Lepiota boudieri Bres.	Girdled dapperling	Europe	Undescribed amanitin derivatives [83]
Lepiota brunneoincarnata Chodat & C. Martin (Ph. 2.50)	Deadly dapperling	Europe Asia	1.29–2.67 total amatoxins 0.72–1.73 α-amanitin 0.57–0.94 β-amanitin [61] α-amanitin, amaninamide [83]
Lepiota elaiophylla Vellinga & Huijser	Lepiota yellow gill	South America	α-amanitin, β-amanitin, undescribed amanitin derivatives [83]
Lepiota farinolens Bon & G. Riousset		North Asia Russia	
Lepiota spiculata Pegler		Central America	
Lepiota subincarnata J.E. Lange (*L. josserandii* Bon & Boiffard)	Fatal dapperling	Europe	3.99–4.24 [90] α-amanitin, γ-amanitin [83]
Lepiota venenata Zhu L. Yang & Z.H. Chen		East Asia	1.97 α-amanitin no β-amanitin [61]

Hymenogastraceae			
Galerina castaneipes A.H. Sm. & Singer		North America	0.99 amanitins [55]
Galerina marginata (Batsch) Kühner (*Galerina autumnalis* (Peck) A.H. Sm. & Singer, Ph. 2.51)	Funeral bell, deadly skullcap, autumn skullcap, deadly galerina	Europe North America Asia	78–243 µg/g FW [22]
Galerina sulciceps (Berk.) Boedijn (*Marasmius sulcipes* Berk.)	Greenhouse skullcap	Tropical and subtropical East and Southeast Asia	
Galerina venenata A.H. Sm.		North America Turkey	1.58 amanitins [55]
Bolbitiaceae			
Conocybe apala (Fr.) Arnolds (*Conocybe lactea* (J.E. Lange) Métrod)	milky conecap	North America, Europe	phallotoxins in small amounts, not amatoxins [42]
Conocybe filaris (Fr.) Kühner	Fool's conecap, ringed cone head	North America, Europe	amatoxins in North American collections, no in European collections [14]

Ph. 2.47: *Amanita phalloides*, death cap (source: Markus Scholler).

Ph. 2.48: *Amanita verna*, spring amanita (source: Geert Schmidt-Stohn).

Amatoxins are bicyclic octapeptides with a ring composed of seven L-amino acids and glycine, which is bridged by a (*R*)-sulfoxide moiety between a cysteine residue and a mostly hydroxylated tryptophan residue. The nine representatives known so far differ in the number of hydroxy groups and in the absence or presence of the amide nitrogen on the aspartic acid residue. Amaninamide differs from α-amanitin in lacking the hydroxyl group of tryptophan.

Amatoxins are readily water-soluble, largely resistant to cooking, freezing, drying, and acid or enzyme degradation. They are not destroyed during food preparation and after ingestion in the gastrointestinal tract. However, it seems that heat treatment procedures at higher temperatures for a longer time can reduce the amount of the toxins

Ph. 2.49: *Amanita virosa*, European destroying angel (source: Markus Scholler).

Ph. 2.50: *Lepiota brunneoincarnata*, deadly dapperling (source: Geert Schmidt-Stohn).

[100]. The so-called Meixner test that is based on the reaction of cyclopeptides and lignin in paper (newsprint) in the presence of concentrated hydrochloric acid and that yields a blue color allows rapid detection or exclusion of amatoxins. However, for an exact analysis of amatoxin-containing materials chromatographic and immunological assays are necessary ([**42**], overview of analytical methods for amatoxin detection: [7, 8, 100, **41**]).

Ph. 2.51: *Galerina marginata*, funeral bell (source: Markus Scholler).

	R^1	R^2	R^3	R^4	LD_{50} [mg/kg BW], (Mice, p.o.)
α-Amanitin	OH	OH	NH_2	OH	0,4 - 0,8
β-Amanitin	OH	OH	OH	OH	0,5 - 1,0
γ-Amanitin	H	OH	NH_2	OH	0,4
Amanin	OH	OH	OH	H	0,5
Amanullin	H	H	NH_2	OH	~ 20

Amatoxins

Fig. 2.27: Amatoxins, toxins of Amanita species.

Phallotoxins (Fig. 2.28) were first isolated in 1937 by Feodor Lynen and Ulrich Wieland [64]. Structural elucidation was performed by Theodor Wieland and coworkers [27, 106, 114, 115]. Phallotoxins are bicyclic heptapeptides. Again, the S atom of a cysteine residue bridges to tryptophan. However, unlike that of amatoxins, the S atom does not carry an oxygen residue. The phallotoxins can be divided into two groups: the neutral phallotoxins phalloin, phalloidin, phallisin, and prophalloin, and the acid phallotoxins phallacin, phallacidin, and phallisacin. The representatives of the former group contain D-threonine, those of the latter in its place D-β-hydroxy aspartic acid.

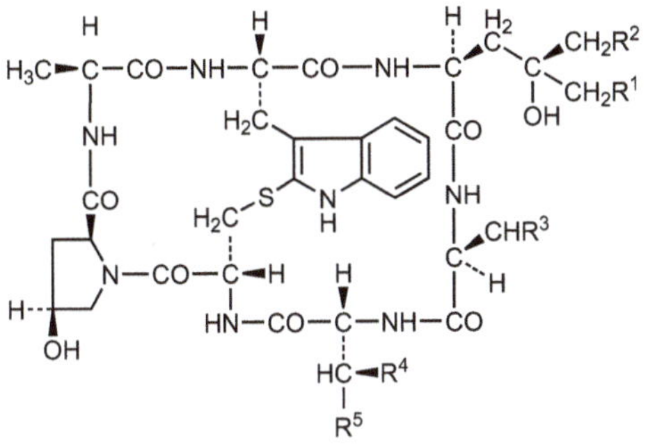

	R^1	R^2	R^3	R^4	R^5	LD_{50} [mg/kg BW], (Mice, p.o.)
Phalloidin	OH	H	H_2	CH_3	OH	2
Phalloin	H	H	H_2	CH_3	OH	1,5
Phallisin	OH	OH	H_2	CH_3	OH	2
Phallisacin	OH	OH	$(CH_3)_2$	OH	COOH	4,5
Phallacidin	OH	H	$(CH_3)_2$	OH	COOH	1,5

Phallotoxins

Fig. 2.28: Phallotoxins, toxins of Amanita species.

In contrast to the amatoxins and phallotoxins, the virotoxins (Fig. 2.29 [28, 29]) are monocyclic peptides. They are composed of seven amino acids, one chromophoric 2-alkylsulfonyl-tryptophan residue, and a 3,4-dihydroxy-proline residue. They presumably arise from phallotoxins by cleavage of the thioether bridge. The dominant virotoxin in the mixture is viroisin (about 50% of the mixture). The virotoxins have been found so far only in *Amanita virosa*, according to some data also in *A. subpallidorosea* [96]. Whether virotoxins are present in all provenances of *A. virosa* and whether they are always accompanied by amatoxins and phallotoxins is still uncertain.

In addition to the representatives of these peptide groups, the phallolysins A, B, and C, proteins with a relative molecular mass of about 34 kDa, have been isolated [87, 116]. When applied parenterally, they act cytolytic by inducing pore formation in cell membranes.

The monocyclic decapeptide antamanide (Fig. 2.30 [15, 114]) found in *Amanita phalloides*, which is not toxic itself, is able to prevent the action of phallotoxins when administered prophylactically or simultaneously with them, presumably by blocking

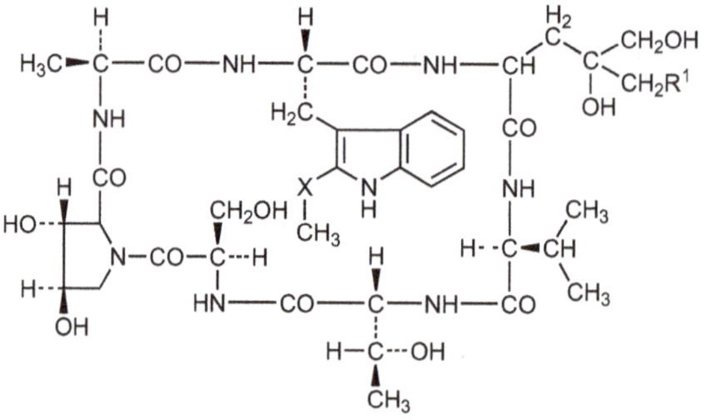

	X	R[1]
Viroisin	SO_2	OH
Desoxyviroisin	SO	OH
Viroidin	SO_2	H

Virotoxins

Fig. 2.29: Virotoxins, toxins of *Amanita virosa.*

their uptake into the cell. It inhibits transporter molecules from the organic anion transporting polypeptide (OATP) family. The cycloamanides [36], monocyclic hexa-, hepta-, or octapeptides, are also nontoxic or antitoxic (overviews of the chemistry of toxins: [19, 26, 99, 109, 110, 111]).

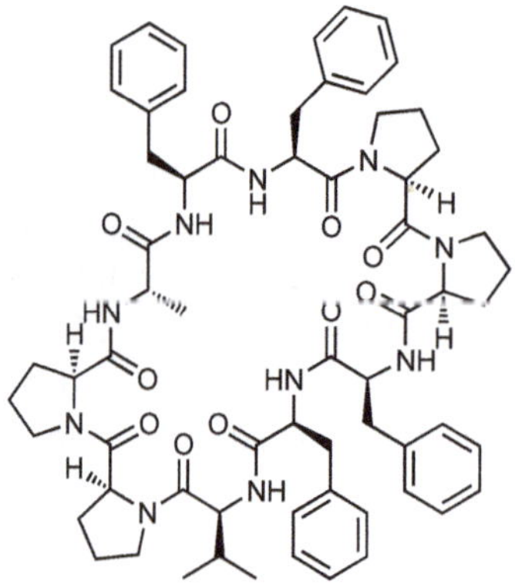

Antamanide

Fig. 2.30: Antamanide, a nontoxic peptide of *Amanita phalloides.*

Amatoxins are ribosomally synthesized and posttranslationally modified peptides (F-RiPPs). The proproteins are encoded by a family of genes called MSDIN family for the first five amino acids in the precursor peptides. The initial posttranslational processing step of the α-amanitin precursor peptide is catalyzed by POPB, a specialized member of the prolyl oligopeptidase family of serine proteases. The distribution of the amanitin biosynthesis pathway in phylogenetically disjunct mushroom genera (Amanita, Lepiota, Galerina) is a result of horizontal gene transfer [63].

2.10.3 Pharmacology and Toxicology

α-Amanitin inhibits RNA polymerase II of eukaryotic cells in a highly specific manner even at concentrations of 10^{-9} M and thus suppresses the formation of mRNA. The target of the toxin is the SB 3 subunit (140,000 Dalton polypeptide) of the enzyme, which forms a 1:1 complex with the toxin [15]. The toxin probably stabilizes a ternary complex of template DNA, enzyme, and nascent ribonucleotide chain. It inhibits elongation at each translocation step. Failure of protein synthesis results in cell death. The organs with intensive protein synthesis, mainly liver and kidney, are affected first [38, 100, 108, 109, 110, 111, 121]. Autophagy and the AMPK-mTOR-ULK pathway that regulates autophagy, oxidative stress induction, and endoplasmatic reticulum stress are involved in the process of α-amanitin-induced liver injury [104, 119]. Toxic effects on hematopoetic cells have been shown and are possibly relevant for in vivo conditions [47].

The other amatoxins have the same mechanism of action. The bicyclic ring system, the sulfoxide bridge, the hydroxylated isoleucine side chain, and the OH group of hydroxyproline are required for biological activity [108, 110].

The amanitins are rapidly absorbed from the gastrointestinal tract and are only bound to a small extent to plasma proteins. They are detectable in plasma up to 48 h after the mushroom meal. Uptake into human liver cells is mediated by transport molecules of the OATP family [58]. The amount taken up into hepatocytes is proportional to the toxin concentration in serum. Immediately after reaching the liver, excretion into the bile begins, so the enterohepatic circulation must be considered in toxin elimination measures. Most of the amatoxins are excreted by the kidney, some are also excreted in the stool. A portion is renally reabsorbed. In some cases, considerable amounts of amatoxins were still detectable in both the liver and kidney on the 9th and 22nd days after mushroom poisoning [51, 54, 100, 109, **42**].

Phallotoxins, particularly studied is phalloidin, bind specifically to the polymeric form of actin. This leads to a stabilization of the F-actin filaments and to a decrease in the concentration of monomeric actin. This results, among other things, in structural changes of actin-containing cell membranes and endomembranes, followed by excessive K^+ loss from the cell, release of degradative enzymes from the lysosomes and finally destruction of the cells [108, 110].

Structural requirements for the biological activity of phallotoxins are the bicyclic ring system, the methyl side chain of the 15-membered ring, the allo-hydroxy group of proline, and the lipophilic indolyl thioether group [29, 108, 110].

In contrast to amatoxins, phallotoxins are hardly absorbed after peroral administration. Apart from their lower toxicity, they do not play a role in fungal poisoning. Hepatic transport proteins of the OATP family, which are physiologically responsible for the transport of bile acids, are responsible for the specific uptake of phallotoxins into hepatocytes [31, 32].

Although the structural requirements necessary for phallotoxin action are not met in virotoxins, they exhibit similar biological activity. They also form complexes with polymeric actin and cause death in animals within 2–5 h by hemorrhages of the liver [62].

LD_{50} values for amatoxins and for phallotoxins for the mouse are shown in Figures 2.27 and 2.28. Because of differences in their absorption from the intestine, reabsorption from the primary urine, and uptake into the enterohepatic circulation, the LD_{50} values for individual mammalian species vary with peroral application. They range from 0.1 mg/kg BW for α-amanitin in the guinea pig to 2.0 mg/kg BW in the rat. Pigs and rabbits are reported to be able to consume green button mushrooms without injury [107]. In mice, the LD_{50} for viroisin is 1.68 mg, for deoxyviroisin 3.35 mg, and for viroidin 1.0 mg/kg BW, i.p. [62]. The LD_{50} of amatoxins for dogs is 0.1 mg/kg BW and for cats 0.1–0.2 mg/kg BW [100].

For humans, the lethal dose of amatoxins is about 0.1 mg/kg BW. This means that a 50 g specimen of *Amanita phalloides* contains the lethal amount of amanitins for an adult [43]. *Galerina marginata* contains the lethal amount of amatoxins in about 100–150 g of fresh mushroom, and *Lepiota brunneoincarnata* also in about 100 g. The lethal dose of phallotoxins to humans is 10–20 times greater than that of amatoxins.

2.10.4 Intoxication Cases, Symptoms, and Treatment of Intoxications

Poisonings by *Amanita phalloides* occur worldwide [1, 2, 5, 17, 23, 41, 50, 54, 57, 68, 75, 76, 85, 87, 101, 118, **31, 41**]. The fungus is blamed for more than 90% of all fatal mushroom poisonings [24].

Poisoning can occur when the mushrooms are mistaken for edible mushrooms, e.g., Agaricus species (to distinguish them: [81, **19, 42, 107**]). There are, for example, reports about mistaken young fruit bodies for *Boletus edulis* Bull., porcini [11], *Volvariella volvacea* (Bull.) Singer, paddy straw mushroom [98], *Macrolepiota procera*, parasol mushroom [91] or possibly *Tricholoma equestre* (L.) P. Kumm. (2 children died in Romania: [66]). Other causes of poisoning are ignorance of the toxicity of these mushrooms or deliberate consumption with suicidal or criminal intent. Even in ancient times, mushrooms were misused as a murder poison [78, 100, **87**]. Recently, the deaths of several people in Australia are said to have been caused by eating a mushroom dish that a family member had served them with murderous intent [9]. In Europe, an increasing number of immigrants and refugees have been affected in recent years, who are familiar with edible mushrooms from their home country that are easily mistaken for the death cap [91]. In case of accidental poisoning, often whole groups or families are affected.

In Northern Germany, the father of a family found green mushrooms with white leaves, cuff, and stalk tuber in an oak stand in August. The man thought they were an edible Russula species and cut the fruit bodies just above the ground. Father, mother, daughter, son, and a guest of the family ate lunch from the dish, which consisted almost entirely of death caps. During the night (about 10 h later), persistent vomiting and diarrhea set in for all individuals. Hospitalization occurred the next morning. The first meal participant died 2 days after consumption, and the others died up to 7 days after eating [86].

The mushrooms retain their toxicity even after several months of storage in the freezer [46].

Poisoning by *Amanita virosa* [16, 18, 59, 77, 93] and *A. verna* [37, 44] occurs particularly in non-European countries, including the United States, Korea, Mexico, Israel, and the Iran [96]. In Germany, a severe poisoning of a family of three persons occurred in the summer of 2005 due to confusion of *A. verna* with button mushrooms [4]. Poisoning by *A. bisporigera* has been reported from North America [72], such by *A. exitialis* and *A. fuliginea* from China [92, 100].

Severe poisoning by Galerina species also occurred [10, 22, 45, 74]. There is a danger of confusion with *Kuehneromyces mutabilis* (Schaeff.: Fr.) Singer & A.H. Sm., changeable pholiota, *Flammulina velutipes* (Curtis) Singer, enoki, velvet foot [3, 52] or Armillaria species [49]. Poisoning from *Galerina sulciceps* has been reported from Japan [71] and China [117]. In Germany, this species has been found in greenhouses [13].

Small Lepiota species can be confused with, among others, the fairy ring champignon, *Marasmius oreades* (Bolton) Fr. [**19**], or parasol mushrooms. Reports of poisoning by amanitin-containing Lepiota species, sometimes with fatal results, are available from the Mediterranean region [17, 40, 69, 70, 80], China [61], and Argentina [82]. The poisoning of an 8-month-old child who had eaten *L. eleiophylla* grown in a flowerpot near a yucca palm is reported. Thanks to timely identification and treatment consequential damage could be avoided [84].

Poisoning by *Amanita phalloides* and other amanitin-containing mushrooms is characterized by a long latency period of 6–24 h, rarely up to 48 h, before the first symptoms appear. If acute gastroenteritis occurs after a mushroom meal with a delay of 6 h or more, always assume serious poisoning and, without waiting for further confirmation of the diagnosis, start therapy. In the first phase of poisoning, vomiting, bloody, watery, cholera-like diarrhea, and colic occur. In this phase death can already occur. However, it can usually still be well controlled by fluid and electrolyte intake. In mild cases, the poisoning is overcome within a few days. In other cases, there is a temporary remission with relative well-being of the patient, which is followed by the third poisoning phase with symptoms of severe liver and usually also kidney damage after 2–5 days. These include extreme elevation of transaminase and other enzyme levels in serum, severe decrease in prothrombin level (a quick value less than 20% is prognostically very unfavorable), liver swelling, and icterus. Signs of renal damage include the appearance of proteins and erythrocytes in the urine and eventually renal failure. Secondary effects involve the heart, CNS, hematopoietic system, and other organs. Death occurs 2–5 days after consumption of the fungi in coma hepaticum or by circulatory failure. Section shows fatty degeneration and necrosis of liver, kidneys, adrenals, cardiac, and skeletal muscles. Poisoning with amanitin-containing mushrooms during pregnancy may result in intrauterine growth retardation of the child but does not necessarily cause abnormalities in the child [48, 54, 57, 100, 102, 107, **41, 42, 173, 178**].

The result is fatal in 10–30% of cases, with the percentage tending to decrease mainly due to liver transplantation [33, **41**].

The poisoned persons and also those suspected of being poisoned must be given intensive medical treatment as soon as possible. Since the severity of poisoning depends decisively on the amount and exposure time of the toxins in the liver, intensive measures must be taken immediately to prevent further toxin absorption into the liver. Gastric lavage is useful only before the appearance of gastrointestinal symptoms. Administration of laxatives as well as administration of activated charcoal, approx. 50 g every 4–6 h by gastric tube, can interrupt the enterohepatic circulation of fungal toxins. An important measure for secondary toxin elimination is plasmapheresis (extracorporeal detoxification using MARS – Molecular Adsorbent Recirculating System).

If gastrointestinal symptoms have already occurred, fluid and electrolyte substitution must be provided under central venous pressure control. Specific measures are the i.v. application of silibinin (5 mg/kg BW in the first hour, then 20 mg/kg BW per day as a continuous infusion) or penicillin G (1 million IU/kg BW per day) until the transaminase values drop. Silibinin, a flavolignan mixture from milk thistle, *Silybum marianum* (L.) Gaertn., blocks toxin uptake into liver cells by competitive receptor binding, as does penicillin G, and is thought to promote the regenerative capacity of the liver. The benefit of administration of *N*-acetylcysteine (NAC) should result from its antioxidative and glutathione-regenerating effects. Therapy with silibinin alone or in combination with penicillin or NAC seems to be advisable.

Supportive measures include the administration of lactulose to remove ammonia, neomycin or paromomycin (p.o.) to reduce the absorption of enterotoxins, the infusion of glucose (20%) with insulin, and, as required, the supply of coagulation factors and heparin for the prophylaxis of consumption coagulopathy.

In case of severe symptoms, e.g., massive increase of transaminase in the blood, quick values below 20%, and renal dysfunction without improvement tendency on the 4th day, only liver transplantation can be lifesaving.

If the poisoning is survived, diet must be maintained for a long time [16, 21, 33, 39, 48, 54, 56, 81, 94, **41, 42, 178**].

In 2023 it was shown that the diagnostic reagent indocyanine green (ICR) blocks amanitin-induced cell death in vitro and in vivo. It inhibits STT3B, an enzyme that is necessary for N-glucan biosynthesis and obviously for α-amanitin toxicity. Further research is needed to determine whether it could become practical importance as antidote for amanitins [100,105]. The concomitant use of cyclosporin A, an inhibitor of OATP, protected mice against the toxic effects of α-amanitin [35].

Animal poisoning by fungi containing amanitins is rare but has been reported in different animal species, for example, in dogs, cats, and cattle [79, 89, 97, 100].

Amanitins, phallotoxins, and their semisynthetic modification products provide useful services in experimental cell research due to their specific effects. For example, α-amanitin serves as an indicator for all processes involving RNA polymerase II. Phallotoxins can be used to detect actin and actin-dependent reactions, e.g., all types of cell motility [30, 110]. Conjugates consisting of α-amanitin or derivatives and a monoclonal antibody, so-called immune toxins or ATACs (antibody targeted amanitin conjugates), are of interest as possible anticancer agents [12, 67]. Of therapeutical interest is also the inhibiting activity of antamanide on the mitochondrial permeability transition pore (mPTP), a protein that is formed in the inner membrane of mitochondria under several pathological conditions, e.g., ischemia [6].

References

For numbers in bold, see cross-chapter literature p. 233.

[1] Alvarado P et al. (2022) Biology 11: 770
[2] Alves A et al. (2001) Eur J Int Med 12: 64
[3] Amelang N (2014) Der Pilz 25: 13 (in German)
[4] Anonym (2005) Ostseezeitung, 02.08.2005 (in German)
[5] Araz C et al. (2006) Transplant Proc 38: 596
[6] Azzolin L et al. (2011) PLoS One 6(1): e16280
[7] Barbosa I et al. (2022) Foods 11: 3929
[8] Barbosa I et al. (2023) J Pharm Biomed Anal 232: 115421
[9] Barkhausen B (2023) Ostseezeitung, 03.11.2023 (in German)
[10] Bauchet JB (1983) Bull Brit Mycol Soc 17: 51
[11] Beaumier M et al. (2019) Kidney Int Rep 4: 877
[12] Berndt S (2015) Deutsches Ärzteblatt 112(17): 782
[13] Besl H (1981) Z Mykol 47: 253
[14] Brady LR et al. (1975) Lloydia 38: 172
[15] Brodner OG, Wieland T (1976) Biochemistry 15: 3480
[16] Broussard CN et al. (2001) Am J Gastroenterol 96: 3195
[17] Cervellin G et al. (2018) Hum Exp Toxicol 37(7): 697
[18] Chaiear K et al. (1999) Asian J Trop Med Public Health 30: 157
[19] Chilton S. (1994) in: Handbook of Mushroom Poisoning. Diagnosis and Treatment (Eds. Spoerke DB, Rumack BH), CRC Press, Boca Raton, p. 165
[20] Enjalbert F et al. (1996) Mycologia 88: 909
[21] Enjalbert F et al. (2002) J Toxicol Clin Toxicol 40: 715
[22] Enjalbert F et al. (2004) Mycologia 96: 720
[23] Fantozzi R et al. (1986) Klin Wochenschr 64: 38
[24] Faulstich H (1979) Klin Wochenschr 57: 1143
[25] Faulstich H et al. (1975) Liebigs Ann Chem 2324
[26] Faulstich H et al. (1980) Biochemistry 19: 3334
[27] Faulstich H et al. (1981) in: Structure and Activity of Natural Peptides, Selected Topics (eds, Voelter W, Weitzel G), de Gruyter, Berlin, p. 189
[28] Faulstich H et al. (1989) J Histochem Cytochem 37: 1035
[29] Faulstich H, Cochet-Meilhac M (1976) FEBS Lett 64: 73
[30] Faulstich H, Zilker TR (1994) in: Handbook of Mushroom Poisoning. Diagnosis and Treatment. Spoerke DB, Rumack BH (eds), CRC Press, Boca Raton, Ann Arbor, London, Tokyo, p: 233
[31] Fehrenbach T et al. (2003) Naunyn-Schmiedeberg's Arch Pharmacol 368: 415
[32] Frimmer M (1987) Toxicol Lett 35: 169
[33] Garcia J et al. (2015) Food Chem Toxicol 86: 41

[34] Garcia J et al. (2015) Mycologia 107(4): 679
[35] Garcia J et al. (2022) Food Chem Toxicol 166: 113198
[36] Gauhe A, Wieland T (1977) Liebigs Ann Chem 859
[37] GGIZ (2024) https://ggiz-erfurt.de/saisonales-detail/knollenblätterpilz.html (accessed 09/06/2024)
[38] Gong XQ et al. (2004) J Biol Chem 279: 27422
[39] Gouvinhas I et al. (2024) EXCLI J 23: 833
[40] Haines JH et al. (1986) Mycopathologia 93: 15
[41] Hallebach M et al. (1985) Z Klin Med 40: 943
[42] Hallen HE et al. (2003) Mycol Res 107(8): 969
[43] Hatfield GM, Brady LR (1975) J Nat Prod 38: 36
[44] Hazani E et al. (2003) Arch Toxicol Suppl 6: 186
[45] He Z et al. (2024) Toxicon 240: 107639
[46] Himmelmann A et al. (2001) Swiss Medical Weekly 131: 616
[47] Hof WFJ et al. (2024) Toxins 16: 61
[48] Horowitz BZ, Moss MJ (2023) StatPearls [Internet] Treasure Island (FL): StatPearls Publishing. https://www.ncbi.nlm.nih.gov/books/NBK431052 (accessed 09/07/2024)
[49] Hulting J et al. (1985) J Toxicol Clin Toxicol 23(4–6): 428
[50] Iskander P et al. (2023) Gastro Hep Advances 2: 544
[51] Jaeger A et al. (1993) Clin Toxicol 31: 63
[52] Kaneko H et al. (2001) J Toxicol Clin Toxicol 39(4): 413
[53] Kaya E et al. (2013) Toxicon 76: 225
[54] Klein AS et al. (1989) Am J Med 86: 187
[55] Landry B et al. (2020) PLoS One 16(2): e0246575
[56] Le Daré B et al. (2021) Toxins 13: 417
[57] Lecot J et al. (2022) Basic Clin Pharmacol Toxicol 132: 528
[58] Letschert K et al. (2006) Toxicol Sci 91: 140
[59] Lim JG et al. (2000) Yonsei Med J 41: 416
[60] Liu LQ et al. (2023) Se Pu 41(11): 976. Doi: 10.3724/SPJ.1123.07013
[61] Long P et al. (2020) Mycobiology 48(3): 204
[62] Loranger A et al. (1985) Fund Appl Toxicol 5 (Pt1): 1144
[63] Luo H et al. (2022) Proc Natl Acad Sci U.S.A. 119(20): e2201113119
[64] Lynen F, Wieland U (1938) Liebigs Ann Chem 533: 93
[65] Malsawmtluanga CD et al. (2024) Toxicon 251: 108134
[66] Mãrgineau CO et al. (2019) Medicine 98: 41
[67] Matinkhoo K et al. (2021) Chemistry 27(49): 10282
[68] McClain JL et al. (1989) J Forensic Sci 34: 83
[69] Mehdi BK et al. (2010) Ann Biol Clin 68(5): 561
[70] Meunier BC et al. (1995) J Toxicol Clin Toxicol 33: 165
[71] Nagayama T et al. (2023) Toxicon 229: 107139
[72] Nici A, Kim S (2011) Case Rep Hepatol 2011: 936867
[73] Ok HJ et al. (2025) J Mass Spectrom 60(6): e5145
[74] Okabe H (1975) Trans Mycol Soc Jpn 16: 204
[75] Opdam FL et al. (2003) Pharm Weekblad 138: 739
[76] Pawlowska J et al. (2006) Wiad Lek 59: 131
[77] Pérez-Moreno J et al. (1994) Mycopathologia 125: 3
[78] Pruß T (2022) Tintling 27(2): 31
[79] Puschner B, Wegenast C (2018) Vet Clin North Am Small Anim Pract 48(6): 1053
[80] Ramirez P et al. (1993) J Hepatol 19: 51
[81] Riethmüller J et al. (2004) Monatsschr Kinderheilkunde 152: 892

[82] Saracco AS, Lechner BE (2024) Medicina (Buenos Aires) 84: 579
[83] Sarawi S et al. (2022) Phytochemistry 195: 113069
[84] Schabel G (2017) Tintling 22(1) 81: 1917
[85] Schleufe P, Seidel C (2003) Anästhesiologie, Intensivmedizin, Notfallmedizin, Schmerztherapie 38: 716
[86] Schmidt I (1977) Mykol Mitteilungsblatt 21: 74
[87] Schute L, Eyer F (2004) Dtsch Med Wochenschr 129: 1277
[88] Seeger R, Stijve T (1979) Z Naturforsch 34c: 330
[89] Seljetun KO, Kragstad HR (2023) Vet Rec 10: e60
[90] Sgambelluri R M et al. (2014) Toxins 6(8): 2336
[91] Smędra A et al. (2022) Scand J Trauma Resusc Emerg Med 30: 20
[92] Sun J et al. (2018) Hum Exp Toxicol 37(7): 665
[93] Svendsen BS et al. (2002) Tidsskrift for den Norske Laegeforening 122: 777
[94] Tan JL et al. (2022) Clin Toxicol 60(1): 1251
[95] Tang SS et al. (2016) Toxicon 120: 78
[96] Tavassoli M et al. (2019) Toxicol Rep 6: 143
[97] Tegzes JH, Puschner B (2002) Vet Hum Toxicol 44: 96
[98] Tobias M et al. (2024) Clin Pract Cases Emerg Med 8(1): 49
[99] Vetter J (1998) Toxicon 36: 13
[100] Vetter J (2023) Molecules 28: 5932
[101] Vinholt Schiødt F (1995) Ugeskr Leger 157: 4350
[102] Visser M et al. (2024) Toxins 16(2): 67
[103] Walton J (2018) The Cyclic Peptide Toxins of Amanita and Other Poisonous Mushrooms. Springer; Berlin, Heidelberg, New York
[104] Wang H et al. (2025) Toxicology 517: 154208
[105] Wang J et al. (2023) Nat Commun 14: 2241
[106] Wieland T (1968) Science 159: 946
[107] Wieland T (1972) Naturwissenschaften 59: 225
[108] Wieland T (1981) in: Structure and Activity of Natural Peptides, Selected Topics Voelter W, Weitzel G (eds), de Gruyter, Berlin, p: 23
[109] Wieland T (1983) Int J Pept Protein Res 22: 257
[110] Wieland T (1986) Peptides of Poisonous Amanita Mushrooms, Springer, Berlin; Heidelberg, New York
[111] Wieland T, Faulstich H (1978) CRC Crit Rev Biochem 5: 185
[112] Wieland T, Gebert U (1966) Liebigs Ann Chem 700: 157
[113] Wieland H, Hallermayer R (1941) Liebigs Ann Chem 548: 1
[114] Wieland T, Schnabel HW (1962) Liebigs Ann Chem 657: 225
[115] Wieland T, Schön W (1955) Liebigs Ann Chem 593: 157
[116] Wilmsen HU et al. (1985) Eur Biophysics J 12: 199
[117] Xiang H et al. (2018) Clin Toxicol (Phila) 56(5): 365
[118] Xu J et al. (2025) J Emerg Med 70: 87
[119] Xu Y et al. (2023) Toxicol Lett 383: 89
[120] Yilmaz J et al. (2014) Toxicon 87: 68
[121] Zhao J et al. (2006) Peptides 27: 3047
[122] Zhou Q et al. (2017) Mycoscience 58: 267

2.11 Lectins as Possible Toxins of Macrofungi

Lectins are glycoproteins, more rarely proteins, that have high affinity for specific monosaccharide, amino sugar, uronic acid, or oligosaccharide residues of glycoproteins or glycolipids and can form side-valent bonds with the substances bearing such residues. The name lectin was derived from legere (Latin, = to select) because of the ability of some of their representatives to specifically agglutinate erythrocytes of a particular blood group (**[161] = Vol. 3**). They are found in all types of organisms including mushrooms.

In 2015, approx. 105 lectins had been identified in different mushroom species ([3], for overview of lectins and ribotoxic proteins from mushrooms, their binding specificity, their structures, and their biological effects: [3, 9, 10, 14, 20]).

Mushrooms use the lectins as storage substances, for defense, and for interactions with other organisms such as in mycorrhizas [3, 10, 16].

The extent to which lectins contribute to the toxicity of mushrooms is unclear. Since the lectins are likely to be largely destroyed during heating, their relevance as source of poisoning should be low. However, just as lectins from some vegetables, mushroom lectins may contribute to symptoms of poisoning after consumption of mushrooms, at least of raw mushrooms. Some examples should be given:

A lectin-containing fraction of *Cyclocybe aegerita* (V. Brig.) Vizzini (*Agrocybe aegerita* (V. Brig.) Singer, Ph. 2.52), Tubariaceae, is in higher concentrations hepatotoxic in mice (25 and 250 mg/kg BW/day, p.o., for 6 days, [5]). Ageritin, obtained from this species, is the first ribotoxin isolated from a basidiomycete. The protein consists of 135 amino acid residues. It inhibits the proliferation of human cancer cell lines (CNS) and shows larvicidal activities against *Aedes aegypti* mosquitoes [8, 19]. Consumption of *C. aegerita*, popular in Asia as nutritional delicacy, in higher amounts should be avoided. The species is probably synonym or at least closely related to the European species *Cyclocybe cylindracea* (DC.) Vizzini & Angelini (*Agrocybe cylindracea* (DC.) Maire), poplar fieldcap, velvet pioppini [2].

A lectin (CNL) of *Clitocybe nebularis* (Batsch) P. Kumm., clouded funnel cap (Ph. 2.53, Clitocybaceae), binds specific to GalNAcβ1-4GlcNAc [16]. It may be partly responsible for the weak toxicity of *C. nebularis* for sensitive people. The nucleoside derivative nebularin (9-(β-D-ribofuranosyl)-purin) may also contribute. It is an antagonist of adenosine and inhibitor of several enzymes, mainly kinases, and exhibits cytotoxic and antimicrobial activities [21]. Sensitive people may experience gastrointestinal symptoms after ingestion of *C. nebularis* that is widespread in the Northern Hemisphere [11]. In Europe, several intoxication cases by this species have been reported [6, 21].

Bolesatine and bolevenine are lectins from *Rubroboletus satanas* (Lenz) Kuan Zhao & Zhu L. Yang (*Boletus satanas* Lenz), devil's bolete, satan's mushroom (Ph. 2.54, [1]), and from *Sutorius venenatus* (Nagas.) G. Wu & Zhu L. Yang (*Boletus venenatus* Nagas., *Neoboletus venenatus* (Nagas.) G. Wu & Zhu L. Yang), Japanese toadstool [12], that are possibly responsible for gastrointestinal symptoms occurring after consumption of these mushrooms. It is unclear if the lectin bolesatine contributes to the hypercalcitoni-

Ph. 2.52: *Cyclocybe aegerita* (source: Wirestock/iStock/Getty Images Plus).

Ph. 2.53: *Clitocybe nebularis*, clouded funnel cap (source: Markus Scholler).

Ph. 2.54: *Rubroboletus satanas*, devil's bolete (source: Markus Scholler).

nemia that has been observed in seven patients in France after consumption of *R. satanas*. The patients developed gastrointestinal effects about 2 h after ingestion. About 12 h later high plasma levels of procalcitonin and CRP were detectable. They were reduced 24 h later. The other symptoms rapidly resolved ([13, **178**], further case [4]).

Several mushroom lectins have antiproliferative, immunomodulating, and antiviral activities and are, similar like plant lectins, of interest as diagnostic and experimental tools and possible drugs, e.g., for cancer therapy [3, 7, 16, 17, 18].

References

For numbers in bold, see cross-chapter literature p. 233.

[1] Ennamany R et al. (1995) Toxicology 100(1–3): 51
[2] Frings RA et al. (2020) Mycol Progr 19: 1001
[3] Hassan MAA et al. (2015) Int J Mol Sci 16: 7802
[4] Heinig W (2002) Tintling 7(5): 39
[5] Jin Y et al. (2014) Toxicon 90: 273
[6] Keller SA et al. (2018) Int J Environ Res Publ Health 15: 2855
[7] Lampitella E et al. (2021) J Biochem 170(4): 473
[8] Landi N et al. (2019) J Biochem 165(5): 415
[9] Landi N et al. (2022) Toxins 14: 403
[10] Lebreton A et al. (2021) J Fungi 7(6): 453
[11] Löfgren N et al. (1954) Acta Chem Scand 8: 670
[12] Matsuura M et al. (2007) Phytochemistry 68(6): 893
[13] Merlet A et al. (2012) Clin Infect Dis 54: 307
[14] MycoLec databank: https://unilectin.unige.ch/mycolecdatabank
[15] Patocka J (2018) Mil Med Sci Lett 87(1): 14
[16] Sabotič J, Kos J (2019) Molecules 24(23): 4204
[17] Singh SS et al. (2015) Molecules 20(1): 446

[18] Sun J et al. (2014) BioMed Res Internat Art 340467
[19] Tayyrov A et al. (2019) Appl Environ Microbiol 85(21): e01549
[20] UniLectin3D databank: https://unilectin.unige/unilectin3D/
[21] Winterstein D (1999) Tintling 4(6): 27 (oder PZ 8)

2.12 Macrofungi with Unknown Toxins

2.12.1 *Tricholoma equestre*

Tricholoma equestre (L.) P. Kumm., yellow knight mushroom, man on horseback (Ph. 2.55), Tricholomataceae, has been a valuated edible mushroom for a long time. Since about the beginning of the 2000s there are several reports from European countries (France [5], Lithunia [10], Poland [1, 6]) about cases of rhabdomyolysis after consumption of these mushrooms. Fatal cases occurred after the repeated consumption of large amounts (triple meal consisting of 100–400 g prepared fresh fruit bodies: [13]). Animal assays (mice, doses equivalent to 0.67–1 kg fresh mushroom per day for 3 days in human nutrition) confirmed the toxicity [5]. In vitro investigations showed pro-inflammatory activities [13]. Otherwise, no adverse effects have been observed in 10 volunteers consuming 300 g fresh specimens fried for 10 min with butter (mean 4.0 g/kg BW, time of observation 1 week after consumption) [8]. Although *T. equestre* is also collected in North America, no reports of toxicity have emerged as of 2000 from there [14].

Ph. 2.55: *Tricholoma equestre*, yellow knight mushroom (source: Geert Schmidt-Stohn).

Possibly the consumption of *T. equestre* in normal quantities by healthy people may be harmless. It can not be ruled out that reported cases were caused by consumption of poisonous mushroom species that are morphologically similar (*T. terreum*? *Russula subnigricans*? see Sections 2.2.1 and 2.2.2) or that it is an unspecified reaction unrelated to a specific mushroom species that may onset at high and repeated consumption [8].

The possibly contained toxic substances and the mechanism of possible poisoning are unexplained. Because of unclear situation and to ensure safety it is recommended to restrict the consumption for pregnant women, children, and people with underlying conditions, and to avoid large amounts and repeated meals [14]. In Germany and other countries, *T. equestre* is currently (2025) classified as a poisonous mushroom.

2.12.2 Ramaria Species

Among the Ramaria species occurring in Europe *Ramaria pallida* (Schaeff.) Ricken, pale coral (Ph. 2.56), *R. formosa* (Pers.) Quél., and *R. flavescens* (Schaeff.) R.H. Petersen (*Corallium formosum* (Pers.) G. Hahn), pinkish coral mushroom, salmon coral, beautiful clavaria, Gomphaceae, are evaluated poisonous. Ingestion of the mushroom species leads to gastrointestinal disturbances [12].

Ramaria flavobrunnescens (G.F. Atk.) Corner, yellow brownish coral, Gomphaceae, grows in Eucalyptus forests in South America. It causes poisoning in cattle, sheep, horses, cats, and pigs that are known as 'Eucalyptus ill'. Occurrence of the mushrooms is also described for Asia and Australia.

🐈 The poisoning is characterized by ulcerative and gangraneous lesions in the skin, tongue, and esophagus, and loss of hairs of the tip of the tail. The causative agents are unknown [18].

Ramaria rufescens (Schaeff.) Corner and *Amanita franchetii* (Boud.) Fayod., yellow veiled amanita, Franchet's amanita, have been made responsible for fatal poisoning of 10 patients (sudden death by cardiac toxicity) in the year 2005 in China. People died within 5–7 days after mushroom ingestion. Early symptoms began 2–15 h after ingestion and included gastrointestinal disturbances, muscle spasms, or tremors and, after a phase of improvement, cardiovascular collapse [7, **50**]. The symptoms remember to those by amatoxins, but information about causative agents is not available. It is in this case unclear if there is really a linkage between the poisoning and these fungi because other mushrooms were also ingested [**178**].

Ph. 2.56: *Ramaria pallida*, pale coral (source: Getty Tim82/iStock/Getty Images Plus).

2.12.3 Morchella Species

Most of the more than 100 known Morchella species, Morchellaceae, Ascomycota, are edible and very delicious.

However, poisonings by edible morels, e.g., *Morchella esculenta* (L.) Pers., sponge morel, true morel (Ph. 2.31), *M. elata* Fr., black morel, *M. semilibera* DC. (*Mitrophora semilibera* (DC.) Lév.), half-free morel, *M. crassipes* (Vent.) Pers., yellow morel, thick-footed morel (Ph. 2.57), and *Verpa bohemica* (Krombh.) J. Schröt., early morel, have been reported from North America and several European countries. The ingestion of crude or poorly cooked mushrooms or of very high amounts (600 g in one or repeated meals) can cause gastrointestinal (latency <6 h) or/and neurological symptoms (tremor, vertigo, visual disturbances, and movement disorders, latency mean 12 h, Morchella syndrome). The responsible toxins, also those of *M. crassipes*, are unknown [2, 3, 4, 15, 16, 17, **13, 42**]. The clinical symptoms resolved within one day in nearly all of the observed cases. There is a great variability of susceptibility in consumers. To avoid poisoning, the morels should be cooked at least 10 min in boiling water and eaten in normal amounts. Drying and rehydration of the fruit bodies before eating should reduce possible toxicity [9].

Ph. 2.57: *Morchella crassipes*, yellow morel (source: Michael Bachmeier).

References

For numbers in bold, see cross-chapter literature p. 233.

[1] Anand JS et al. (2009) Przegl Lek 66(6): 339
[2] Berndt S (2010) DGfM-Mitteilungen 20(1): 7 in Z Mykol 76/1
[3] Berndt S (2021) Presentation on the Congress „100 Jahre DGfM", Blaubeuren
[4] Berndt S (2024) DGfM-Mitteilungen 2024/2 in Z. Mykol 90/1
[5] Berdy R et al. (2001) New Engl J Med 345(11): 798
[6] Chodorowski Z et al. (2002) Przegl Lek 59(4-5): 386
[7] Huang L et al. (2009) BMJ Case Rep 2009: bcr06.2008.0327
[8] Klimaszyk P, Rzymski P (2018) Toxins 10: 468
[9] Lagrange E, Verneux JP (2020) Toxins 12: 482
[10] Laubner G, Mikulevičiene G (2016) Acta Med Litu 23(3): 193
[11] Merlet A et al. (2012) Clin Infect Dis 54: 307,
[12] Montag K (2021) Tintling 26(4): 65
[13] Muszyńska B et al. (2018) Eur Food Res Technol. https://doi.org/10.1007/s00217-018-3134-0
[14] Nieminen P, Mustonen AM (2020) Toxins 12: 639
[15] Pfab R et al. (2008) Clin Toxicol (Phila) 46(3): 259
[16] Piqueras Carrasco J (2003) Lactarius 11: 83
[17] Saviuc P et al. (2010) Clin Toxicol (Phila) 48(4): 365
[18] Scheid HV et al. (2022) Toxicon 205: 91

3 Toxins of Microfungi (Mycotoxins)

3.1 General

3.1.1 Chemistry and Distribution

Mycotoxins are metabolites of molds that are toxic to humans and animals (overviews: [3, 15, 21, 23, 29, 32, 33, 41, 46, 47, 48, **14, 18, 26, 116, 138, 139, 175**].

Although mycotoxins have been known for more than 80 years (citrinin was discovered in 1931, patulin in 1938, and trichothecene in 1948), their intensive study did not begin until 1960 in connection with a mass turkey death in Great Britain caused by peanut meal infested with *Aspergillus flavus* Link and caused by aflatoxins [40, **116, 138**]:

To date, about 500 mycotoxins are known. They belong to a wide variety of substance classes. Of particular importance are:
- polyketides, which include diketides (moniliformin), tetraketides (e.g., patulin, penicillic acid), pentaketides (e.g., citrinin), octaketides (anthracene derivatives), nonaketides (e.g., citreoviridin, zearalenone), and decaketides (e.g., sterigmatocystin, versicolorins, aflatoxins);
- polyketides with amino acid building blocks (e.g., tenuazonic acid, ochratoxins, fumonisins, cytochalasans);
- polyketides with terpene building blocks (e.g., mycophenolic acid);
- terpenes (e.g., trichothecenes, some of them with polyketide moieties, PR toxin);
- polypeptides (e.g., islanditoxin, cyclochlorotin, roseotoxin, enniatins, beauvericin);
- alkaloids (e.g., α-cyclopiazonic acid, ergoline alkaloids, roquefortines, tryptoquivaline), but also some polyketides are per definition alkaloids (e.g., tenuazonic acid and cytochalasanes).

Producers of mycotoxins are a variety of molds of different taxonomic position. It is estimated that the representatives of 30–40% of all species of molds can produce mycotoxins [**138**]. Chemotaxonomic relationships are not apparent. Fungi of very different classification often form mycotoxins of the same or similar structure (e.g., cytochalasans and trichothecenes). On the other hand, different strains often exist within a species that form structurally different mycotoxins. In addition, there exist mycotoxigenic and non-mycotoxigenic strains within the same species [**158**]. Genetics and biochemistry of mycotoxin formation are increasingly well understood.

The order Mucorales of the class Zygomycetes is particularly rich in species. Although many toxigenic representatives occur in it, their mycotoxins are only seldom

https://doi.org/10.1515/9783110728576-003

investigated. Toxin producers include Absidia, Mortierella, Mucor, Piptocephalis, and Rhizopus species.

The majority of mycotoxigenic fungi can be found in the class Ascomycetes. They include, among others, the genera Alternaria, Aspergillus (Ph. 3.1), Chaetomium, Fusarium (Ph. 3.2), Helminthosporium, Myrothecium, Paecilomyces, Penicillium (Ph. 3.3), Stachybotrys, Trichoderma, Trichothecium, Ascochyta, and Phoma. Many mycotoxigenic representatives are anamorphic fungi with an unknown sexual state.

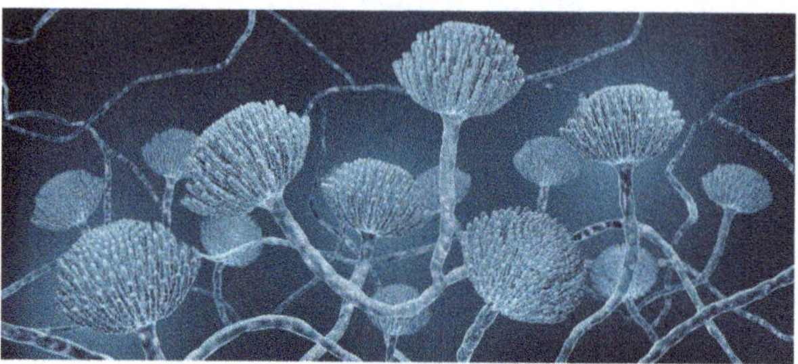

Ph. 3.1: Aspergillus sp. (source: Artur Plawgo/iStock/Getty Images Plus).

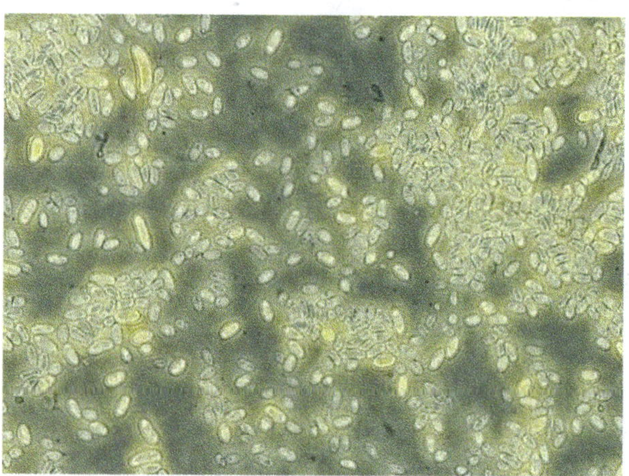

Ph. 3.2: *Fusarium oxysporum* (source: Hannah Beermann & Annett Mikolasch, University Greifswald).

The growth of mycotoxigenic fungi on or in plants, food, feed, spices, or medicinal drugs and the mycotoxin formation depend on genetic and on environmental factors. Most mycotoxin-forming fungi have a growth temperature optimum of 20–25 °C. Many of them still grow at lower temperatures but produce usually no mycotoxins below 10–15 °C. However, this observation cannot be generalized. Some strains produce mycotoxins even at refrigerator temperatures. For good growth, not only a moist substrate is necessary, but usually also a certain humidity (for parasitic fungi about 20–25%, for storage-damaging fungi 12–18%). Fruits become increasingly sus-

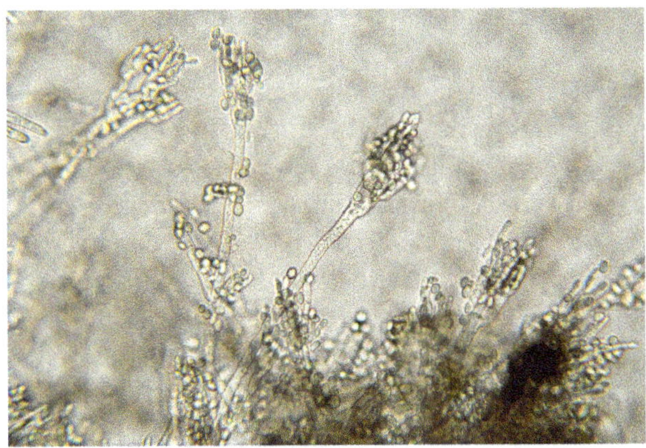

Ph. 3.3: Penicillium sp. (source: Hannah Beermann & Annett Mikolasch, University Greifswald).

ceptible to infestation during the ripening process. The mycotoxin-producing fungi require oxygen for growth, although often only in low concentrations. They thrive on all organic substrates intended for human and animal consumption. In general, plants in tropical and subtropical area are more susceptible to mycotoxin contamination than those in temperate zones due to the presence of high temperature and humidity, which provide optimal conditions for growth and toxin formation [11, 14, 25, 33, 42, **14, 116**].

The mycotoxins in plant foods and animal feeds are either

- produced by fungi living in the plants, on the plants or in the soil,
- produced by saprophytic fungi in the infected raw products or foodstuffs, or
- produced during the use of mycotoxigenic fungi in the production of foodstuffs, e.g., of cheese.

Mycotoxins enter animal products, e.g., meat, milk, cheese, and eggs, either by infestation of the products with mycotoxigenic fungi or by ingestion of mycotoxin-containing feed by the animals (carry over). If different mycotoxins occur in one sample, this can lead to additive or synergistic effects.

The biogenesis of mycotoxins is, at least in the case of polyketides, very complex and often organized in the form of gene clusters. A better understanding of the regulation of mycotoxin's biosynthesis and its connections to environmental signals could be a promising strategy to control contamination [8, 46, 50].

According to estimates by the FAO, around 25% of the crops worldwide are contaminated with appreciable amounts of mycotoxins, which contribute to nearly 1 billion tons of annual losses of feeds/foods [20, 44, 46]. The European Commission has estimated that 5–10% of global crop losses are caused by mycotoxins [21]. The resulting financial losses are due to decreased livestock production, illness or death of animals and humans, increased necessity for medical care and veterinary service, increased costs for regulatory and preventive measures, or detoxification and scrapping of food/feed [46].

Causes for the increasing mycotoxin content of food and feed are the high susceptibility of our high-performance plant breeds to fungal infections on unfavorable sites, the use of agrochemicals, and the mechanical harvesting technologies that damage the crop. Unfavorable storage conditions, air exposure during ensiling, and long storage or transport times promote the spread of infection and mycotoxin formation (overviews on the occurrence of mycotoxins: in fruits, vegetables, and beverages [4, 37]; medicinal herbs [10]; cereals and cereal products [2, 13, 19, 28]; drinking water [24]; non-regulated mycotoxins in different food commodities [35]).

Intoxications of humans and animals by mycotoxins result from the consumption of toxin-containing foodstuffs, food or herbal drugs or products prepared therefrom. Exposure is also possible by inhalation. The hazard of combination of mycotoxins in small quantities each is often underestimated [42, 46].

3.1.2 Pharmacology and Toxicology

The effects of mycotoxins in human and animal organisms can be attributed to the following mechanisms of action:
- influence on the energy metabolism of cells, e.g., by citrinin, citreoviridin, cyclochlorotin, or aflatoxins;
- interaction with nucleic acids or proteins, e.g., by trichothecenes, aflatoxins, ochratoxins, and rubratoxin;
- inhibition of DNA, RNA, and protein biosynthesis, e.g., by trichothecenes, aflatoxins, ochratoxins, patulin, and citrinin;
- induction of cell cycle arrest, apoptosis and pyroptosis, e.g., by citrinin and rubratoxin;
- induction of oxidative stress, leading, e.g., to activation of the NLRP3 inflammasome and the Nf-κB pathway, e.g., by trichothecenes, aflatoxins, ochratoxins, and fumonisins;
- induction of epigenetic changes, e.g., by trichothecenes, patulin, and aflatoxins;
- inhibition of sphingolipid biosynthesis, e.g., by fumonisins;
- inhibition of pyruvate dehydrogenase, e.g., by moniliformin;
- inhibition of kinases, e.g., by wortmannin;
- interaction with receptors for neurotransmitters, e.g., by ergoline alkaloids;
- influence on lipid metabolism, e.g., by citrinin and aflatoxins;
- hormonal effects, e.g., by zearalenones; and
- interaction with the cytoskeleton, e.g., by cytochalasans.

Often these are combined effects, which are also strongly dependent on the metabolites produced in vivo (overviews: [14, 15, 21, 25, 29, 30, 49, **14, 116, 139, 175**]).

The above-mentioned activities result in poisoning symptoms that are mainly hepatotoxic and hepatocarcinogenic (e.g., with aflatoxins and sterigmatocystins), nephrotoxic (e.g., with ochratoxins, citrinin, and fumonisins), neurotoxic (e.g., with citreoviridin and tremorgens), teratogenic (e.g., with aflatoxins and ochratoxins), and immunosuppressive (e.g., with trichothecenes) effects.

3.1.3 Mycotoxicoses in Humans

Acute mycotoxicoses caused by short-term ingestion of larger amounts of material contaminated with mycotoxigenic fungi or their products occur nearly exclusively in farm animals. A much greater role for humans and animals is played by chronic mycotoxicoses, which develop when small amounts of this material are ingested over a long period of time. A particular risk is posed by the carcinogenic activity of many mycotoxins, which is often further supported by the simultaneous immunosuppressive effect (Tab. 3.1).

Tab. 3.1: Carcinogenicity of mycotoxins [38, **65**, **116**].

Mycotoxin	Strength of evidence for carcinogenic activity		
	Human	**Animal**	**Global classification according to IARC**
Aflatoxins (natural mixture)	Sufficient	Sufficient	1
Aflatoxin B1	Sufficient	Sufficient	1
Aflatoxin B2		Limited	1
Aflatoxin G1		Sufficient	1
Aflatoxin G2		Low	1
Aflatoxin M1	Low	Sufficient	2B
Citrinin	No data	Limited	3
Cyclochlorotin (Islanditoxin)	No data	Low	3
Koji acid	Low	Limited	3
Luteoskyrin	No data	Limited	3
Ochratoxin A	Low	Sufficient	2B
Patulin	No data	Low	3
Penicillic acid	No data	Limited	3
Rugulosin	No data		3
Sterigmatocystin	No data	Sufficient	2B
Toxins of *Fusarium gramineum*, *F. culmorum*, *F. crookwellense*	No data		3
Zearalenone		Limited	
Nivalenol		Low	
Fusarenon X		Low	
Deoxynivalenol (DON)		Low	
Toxins of *Fusarium sporotrichioides*	No data		3
T-2 toxin			3

Tab. 3.1 (continued)

Mycotoxin	Strength of evidence for carcinogenic activity		
	Human	**Animal**	**Global classification according to IARC**
Toxins of *Fusarium moniliformis*	Low	Sufficient	2B
Fumonisin B1	Low	Sufficient	2B
Fumonisin B2		Low	
Fusarin C		Limited	

IARC, International Agency for Research on Cancer; 1, for human carcinogen; 2B, for human possibly carcinogen; 3, carcinogenicity for human not classifiable.

Epidemiological studies show that diseases caused by mycotoxins are relatively common, especially in low-income countries in Asia and Africa. The reasons for this are likely to be the climatic conditions in tropical regions, which are favorable for fungal growth, the often inadequate food supply, which forces the consumption of spoiled food, the low level of knowledge of the population about mycotoxins, and inadequate government control of food [11].

Human diseases for which a causal relationship with the ingestion of mycotoxins is assumed or has been proven are:

- ergotism;
- aflatoxicoses, which manifest themselves mainly as hepatitis or liver cancer; whether aflatoxins are involved in the development of 'kwashiorkor', a disease that affects for example African children, is unclear;
- alimentary toxic aleukia (ATA), which caused many deaths in Russia, especially during and after World War II; it is caused by trichothecenes formed in Fusarium-infested overwintered grain and leads, for example, to bone marrow insufficiency;
- Kashin–Beck disease, which is characterized by skeletal developmental disorders in children and is attributed, among others (e.g., deficiency in selenium), to Fusarium toxins;
- yellow rice syndrome, a liver disease occurring in Asia and caused by fungi-infested rice; the mycotoxins responsible are said to be anthracene derivatives, islanditoxin, and cyclochlorotin (not to be confused with fried rice syndrome caused by *Bacillus cereus*);
- onyalai, an acquired form of immune thrombocytopenia endemic to Africa possibly related to mycotoxins such as tenuazonic acid;
- kodo poisoning, a disease caused by α-cyclopiazonic acid that is formed by fungi growing on kodo millet, *Paspalum scrobiculatum* L.

It can be assumed that mycotoxins ingested with food are also involved in the development of many tumors (Tab. 3.1) and in disorders of embryonic development. For example, trichothecenes and fumonisins have been blamed for the high incidence of esophageal cancer and neural tube defects in some regions of China and South Africa (overviews of human mycotoxicoses: [15, 16, **97**, **116**, **139**]).

Children are even more at risk from mycotoxins than adults. Between 2014 and 2021, 13 studies reported the presence of mycotoxins in cereal-based infants'/children's food products in Europe. Most studies verified that trichothecenes and ochratoxin A are the most prevalent. In some studies, DON levels were above the allowed maximal level of 200 µg/kg [39]. The occurrence of aflatoxins in infant's food in several countries has also been shown. For example, 33 of 48 investigated samples in Iran contained aflatoxin B1 with a mean concentration of 2.6 ± 4 µg/kg [1].

Mycotoxins can cross the placenta. A pilot study found different mycotoxins in urine and serum samples of 36 pregnant Dutch women in their first trimester of pregnancy, with DON, zearalenone, and ochratoxin A most frequently detected. However, the risk (except one participant) was evaluated as of low concern [34].

The fetal exposure to mycotoxins is widespread especially in tropical low-income countries indicating frequent mold contamination of maternal diets. A systematic review found evidence for adverse effects of maternal aflatoxins and Fusarium toxins on intrauterine growth and an increased risk for neural tube effects, preterm birth, and late-term miscarriage [26].

3.1.4 Mycotoxicoses in Animals

Mycotoxins negatively affect the health of animals. They lead to clinical symptoms such as gastrointestinal disorders, inflammation, and immunosuppression as well as to breeding problems such as to decreased egg production in poultry or decreased milk and meat production in cattle [18, 27, 29].

The following mycotoxicoses are observed in animals:

- ergotism;
- stachyobotryotoxicosis, a disease of horses and sheep caused by the toxins of *Stachybotrys chartarum* (Ehrenb.) S. Hughes, e.g., macrocyclic trichothecenes, observed mainly in Eastern Europe, characterized by necrosis, hemorrhages, neurological, and respiratory disturbances;
- aflatoxicosis, caused by feed contaminated with aflatoxins, responsible for many deaths of cattle and other animals;
- nephropathy in pigs induced by ochratoxin A and citrinin;
- lupinosis, caused by phomopsins, the toxins of *Diaporthe toxica* P.M. Will., Highet, W. Gams & Sivasith. or *D. leptostromiformis* (J.G. Kühn) Rossman & Udayanga (*Phomopsis leptostromiformis* (J.G. Kühn) Bubák), fungi that live on lupins; the hepatotoxicosis affects mainly sheep in Australia and South Africa;

- slaframine toxicosis, caused by the indolizidine alkaloid slaframine produced by *Slafractonia leguminicola* (Gough & E.S. Elliott) Alhawatema, Baucom, Sanogo & Creamer, living on clover, main symptom is profuse salivation;
- facial eczema (pithomycotoxicosis), observed in sheep in Australia and New Zealand, caused by *Pseudopithomyces chartarum* (Berk. & M.A. Curtis) Jun F. Li, Ariyaw. & K.D. Hyde (*Pithomyces chartarum* (Berk. & M.A. Curtis) M.B. Ellis), responsible is sporidesmin; the poisoning is characterized by hepatogenous photosensitization that causes erythema in nonpigmented skin;
- estrogenic syndrome, especially in pigs, caused by zearalenone;
- disorders of chick maturation in the egg ('avian tibial dyschondroplasia') caused, among others, by wortmannin and fusarochromanone;
- equine leukoencephalomalacia ('moldy corn poisoning'), affecting horses, mules, and donkeys and caused by fumonisins;
- pulmonary edema in pigs ('porcine pulmonary edema'), induced by fumonisins;
- tremorgenic mycotoxicosis in dogs, caused, among other, by penitrems;
- 'Degnala disease', caused by Fusarium and by Aspergillus species, especially in buffaloes in India, Pakistan, and Nepal, characterized by ulceration and necrosis of the extremities, reproductive disorders, and liver damage [12].

In addition, mycotoxins are suspected to be involved in neurological disorders, fertility disorders, and abortions (overviews of mycotoxicoses of animals: [6, 15, 31, **114**]).

3.1.5 Treatment and Mitigation of Mycotoxicoses

The treatment of mycotoxin poisoning can only be symptomatic. Prevention can be achieved through avoiding mold infestation of plants as well as food and feed by good agricultural practice, continuous monitoring for mycotoxins in feed and food, and, if unacceptably high levels are detected, destruction of the correspondinq products. Removal of the funqal mycelium alone is not sufficient.

Complete prevention of mycotoxin contamination of food and feed is practically impossible. But approaches to mitigate mycotoxin contamination have been developed. They include chemical approaches (fungicides, essential oils, detoxification by ozone, ammonia, and other), detoxification by physical methods (dehulling, heating, irradiation, plasma treatment, washing, and other), biological control of toxigenic fungi and biodegradation of mycotoxins (controlling pathogenic microorganisms using microorganisms with antifungal activity or the ability for degradation of mycotoxins, detoxifying enzymes) or deactivation of mycotoxins, e.g., by antioxidative compounds [3, 4, 17, 27, 30, 38, 43].

For the analysis of mycotoxins, primarily chromatographic procedures, often coupled with mass spectrometry, as well as immunological methods are used ([39, 43, 48], for monitoring of mycotoxins in food-producing animals, see [36]). Molecular biological methods are useful, e.g., for species determination and for rapid differentiation of toxigenic and nontoxigenic strains. Masked mycotoxins pose a particular challenge as they may not be detected by routine methods. These are conjugates that are formed when mycotoxins bind to other molecules, such as sugars or amino acids, in the plant material. During digestion, the conjugate can be broken down and release the free mycotoxin [5, 7, 22].

To minimize the risk posed to humans by mycotoxins, European Union, FDA, and other legislators have enacted limit values for some mycotoxins. However, there are many mycotoxins nonregulated until now, the so-called emerging mycotoxins, e.g., nivalenol, the Alternaria mycotoxins, sterigmatocystin, patulin, beauvericin, diacetoxyscirpenol, fusarenon X, moniliformin, and neosolaniol ([35], overviews on regulation in several countries/regions: [3, 9, 45]).

References

For numbers in bold, see cross-chapter literature p. 233.

[1] Açar Y, Akbulut G (2024) Curr Nutr Rep 13: 59
[2] Arce-López B et al. (2024) Environ Toxicol Pharmacol 109: 104489
[3] Awuchi CG et al. (2021) Foods 10: 1279
[4] Azam S et al. (2023) Toxins 13: 323
[5] Bacha SAS et al. (2023) Front Plant Sci 14: 1139757
[6] Bauer J (1988) J Tierärztl Praxis Suppl 3: 40
[7] Brückner L et al. (2024) Mycotoxin Res 40: 709
[8] Caceres I et al. (2020) Toxins 12: 150
[9] Cai YT et al. (2020) J AOAC Int 103(3): 705
[10] Chen L et al. (2020) Toxins 12: 30
[11] Darwish WS et al. (2014) J Vet Med Sci 76(6): 789
[12] De Medeiros Costa Lins A, Salvarani FM (2024) Animals 14(16): 2292
[13] Deligeorgakis C et al. (2023) Foods 12: 4328
[14] Drusch S, Aumann J (2005), Adv Fruit Nutr Res 50: 33
[15] Etzel RA (2005) in: **132**, p. 449
[16] Fink-Gremmels J (1994) Ernährungsumsch 41: 226
[17] Furlong EB et al. (2024) Foods 13: 1112
[18] Gallo A et al. (2015) Toxins 7: 3057
[19] Gozzi M et al. (2024) Mycotoxin Res 40(1): 203
[20] Ismail A et al. (2019) Aflatoxins in Plant-based Foods: Phytochemistry and Molecular Aspects. In: Plant and Human Health, Vol. 2, Ozturk M, Hakeem KR (eds), Springer Nature Switzerland AG, p. 313
[21] Janik E et al. (2020) Int J Mol Sci 21: 8187
[22] Kamle M et al. (2022) Toxins 14: 85
[23] Khan R et al. (2024) Heliyon 10: e28361
[24] Koko DT et al. (2024) RSC Adv 14: 34435
[25] Krogh P (1987) Mycotoxins in Food. Academic Press, London
[26] Kyei NNA et al. (2020) Mycotoxin Res 36: 243
[27] Lach M, Kotarska K (2024) Molecules 29: 4563
[28] Lee HJ, Ryu D (2017) J Agric Food Chem 65(33): 7034

[29] Liao C et al. (2024) Vet Sci 11: 291
[30] Malekinejad H et al. (2024) Front Biosci (Elite Ed) 16(2): 12
[31] Malir F et al. (2023) Mycotoxin Res 39(2): 81
[32] Mafe AN, Büsselberg D (2024) Foods 13(21): 3502
[33] Marroquin-Cardona AG et al. (2014) Food Chem Toxicol 69: 220
[34] McKeon HP et al. (2024) Toxins 16: 278
[35] Mihalache OA et al. (2023) Toxins 15: 583
[36] Munoz-Solano B et al. (2024) Toxins 16: 218
[37] Nan M et al. (2022) Toxins 14: 309
[38] Nazareth TDM et al. (2024) Foods 13(12): 1920
[39] Pereira A et al. (2022) Molecules 27: 8557
[40] Pitt JI, Miller JD (2017) J Agric Food Chem 65(33): 7021
[41] Pohland AE (1993) Food Addit Contam 10: 17
[42] Polak-Śliwińska M, Paszczyk (2021) Molecules 26: 454
[43] Shabeer S et al. (2022) Toxins 14: 307
[44] Smith MC et al. (2016) Toxins 8: 94
[45] Stoev SD (2023) Toxins 15: 464
[46] Stoev SD (2024) Microorganisms 12(3): 580
[47] Teuscher E, Lindequist U (1992) Dtsch Apoth Ztg 132(42): 2231
[48] Trucksess MW (2005) JOAC Int 88: 314
[49] Ülger TG et al. (2020) Toxicon 185: 104
[50] Wang W et al. (2023) J Fungi 9: 21

3.2 Sesquiterpenes as Mycotoxins

3.2.1 Trichothecenes

3.2.1.1 Chemistry and Biogenesis

The trichothecenes (Fig. 3.1), along with the aflatoxins and fumonisins, are among the most toxicologically important mycotoxins. The about 200 representatives are sesquiterpenes with the base body of the tetracyclic 12,13-epoxy-trichothec-9-ene (EPT, the double bond in positions 9,10 may be substituted by a second epoxy group). They are classified into four different groups (types A, B, C, and D) according to structural variations. Type-A trichothecenes contain a functional group other than a ketone or no substituent at C-8 of the base body (e.g., T-2 toxin (T-2), HT-2 toxin (HAT-2), 4,15-diacetoxyscirpenol (DAS), monoacetoxyscirpenol (MAS), neosolaniol (NEO), verrucarin A (VER), NX-2, NX-3). Type B trichothecenes (e.g., deoxynivalenol (DON) and their acetyl derivatives (3-ADON, 15-ADON), nivalenol (NIV), fusarenon X (FUSX)) possess a keto group at C-8. Type C trichothecenes are characterized by a second epoxide ring at C-7,8 or C-9,10. Type D trichothecenes possess a macrocyclic ring system between C-4 and C-15 of EPT, formed by esterification of the hydroxyls at the relevant positions with a 12- or 14-carbon chain. The trichothecenes without a macrolide ring are also called simple trichothecenes. The macrocyclic trichothecenes can be further classified as verrucarins (mainly C_{27} compounds), roridins, and satratoxins (mainly C_{29} compounds: [1, 2, 14, 22, 24, 29, 36, 37, 50, 57, **14, 26, 116**] see also Section 2.4.5).

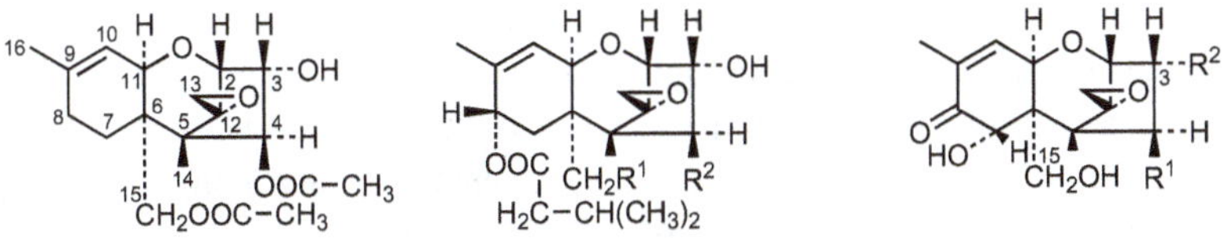

4,15-Diacetoxyscirpenol

T-2 Toxin R^1=OOC−CH_3, R^2=OOC−CH_3
T-2 Triol R^1=OH, R^2=OH
HT-2 Toxin R^1=OOC−CH_3, R^2=OH

Nivalenol R^1=R^2=OH
Deoxynivalenol R^1=H, R^2=OH
Fusarenon-X R^1=OOC−CH_3, R^2=OH

Fig. 3.1: Trichothecenes.

As early as 1948, trichothecene was isolated as the first representative of this group from *Trichothecium roseum* (Pers.) Link during the search for fungistatics. The structure was elucidated in 1965 by Godtfredsen and Vangedal [15].

The biogenesis of simple trichothecenes takes place from farnesol diphosphate by ring closure to form trichodiene, catalyzed by the terpene cyclase trichodiene synthase (TRI5). Following, trichodiene undergoes a series of oxygenations, acylations, and further modifications. Most enzymes and genes (TRIs) responsible for these processes have been identified and form the trichothecene biosynthetic gene cluster [6, 14, 37, 50]. The biosynthetic pathway to the macrocyclic trichothecenes and the responsible genes are not fully understood. It is assumed that the macrocyclic trichothecene roridin E (Fig. 2.7) could be formed by acylation or esterification of the long side chain linked with C-4 to the hydroxyl group at C-15, and vice versa [57]. TRI24 is the first described gene specific to macrocyclic trichothecene biosynthesis. It encodes for an acyltransferase [33]. The trichothecene genotype diversity varies significantly among different hosts and geographic locations [50].

3.2.1.2 Occurrence

Producers of trichothecenes are a variety of fungi from the order Hypocreales, class Ascomycetes, especially of the genus Fusarium. Fungi of this genus are capable of both sexual and asexual reproduction. According to the International Code of Botanical Nomenclature for algae, fungi, and plants, the name of pleomorphic fungi is a case of the 'One Fungus One Name' rule so that the exclusive use of 'Fusarium' is recommended for all species within this genus and the former name Gibberella is only used as synonym [14]. Some of these fungi form trichothecenes already at low temperatures (*F. sporotrichioides* Sherb. already at 8 °C, [14, 50]). Type A trichothecenes are produced by, e.g., *F. sporotrichioides* and *F. poae* (Peck.) Wollenw., type B trichothecenes by *F. graminearum* Schwabe and *F. culmorum* (Wm. G. Sm.) Sacc. Type C and D (macrocyclic) trichothecenes are produced in particular by the genera Myrothecium, Trichoderma (Podostroma, see Section 2.4.5), Stachybotrys, Calcarisporium, Cylindrocarpon, Dendrodochium, Phomopsis, and Verticinimonosporium, not by Fusarium species [36].

Some mycotoxin-producing Fusarium species live parasitic as endophytes and cause diseases of the host plants, e.g., Fusarium head blight [14]. Saprophytic species infect the mature fruits before harvest or during storage. The spectrum of produced mycotoxins and their amount are strongly dependent on fungal species and strain and on environmental conditions. Geographical differences in the distribution of fungal genotypes/chemotypes exist. As an example, 15-ADON types dominate in Southern and Central Europe, whereas 3-ADON is more prevalent in the Northwestern regions [14, 34].

Trichothecenes are detected in cereals (wheat, maize, barley, oak, and rye), cereal products such as bread, pasta, baked goods, fruits, e.g., apples, bananas, grapes, and feeds such as grass, hay, and silage (overview on distribution and content of trichothecenes in food und feed samples around the world: [17, 36]). DON has often the highest share. The occurrence of T-2 toxin in agricultural products is strongly increasing. The contamination of food and feed with T-2 toxin is evaluated by the WHO as a major hazard to the health of humans and animals [51].

Trichothecenes are heat stable and are only partially destroyed during cooking or baking; 60–80% of the mycotoxins in the raw products appear in the processed food. While the DON content decreases with increasing baking time of buns, the baking temperature has no influence on its concentration. Transfer of trichothecenes into milk when contaminated feed is fed to cows occurs only to a minor extent (<1%). Trichothecenes are largely detoxified by the ruminant rumen microorganisms (elimination of epoxy grouping). They also appear only in trace amounts in meat and eggs [32, 55, **14**]. Other food commodities, such as soy, coffee, tea, dried spices, and nuts, can be intermittently contaminated. The detected trichothecene content in water reservoirs is thought to result from runoff from crops, livestock effluent, and human wastewater [38]. DON persists in human breast milk [9, 17].

EFSA investigations found that the DON contamination rate of 26,613 investigated cereal samples from 21 European countries was close to 50% [11]. DON contaminates approx. 60% of wheat, corn, barley, rice, oats, sorghum, and rye worldwide. The percentage of animal feeds contaminated by DON was 42% in the Middle East and North Africa during the years 2012–2020, less than 20% in animal feeds from Italy over 5 years (2018–2022) and 71–76% in animal feeds from Northern Spain (2023). The percentage of animal feeds contaminated by T-2 toxin was 18% in the Middle East and North Africa and 70–100% in animal feeds from Italy during the same time periods [30].

During 2017 and 2018, the average DON content in wheat samples in Europe exceeded 400 µg/kg. It was much higher in countries outside of Europe, e.g., in South America [3].

For humans within the EU, wheat and wheat products are the main sources of trichothecenes [42, 43]. The daily intake of trichothecenes calculated on the basis of the average contamination of cereals and the average amount consumed is below the ADI (1 µg/kg BW per day DON + derivatives; 100 ng/kg BW per day T-2/HAT-2, see below) for adults in Germany. For young children, however, the tolerable amounts are sometimes exceeded [17, **116**].

A particular case of the formation of trichothecenes is the cooperation of the fungus *Albifimbria verrucaria* (Alb. & Schwein.) L. Lombard & Crous. (*Myrothecium verrucaria* (Alb. & Schwein.) Ditmar) with the small shrub *Baccharis coridifolia* DC and other Baccharis species (Asteraceae) that grow in South America. The fungus, which lives on the root surface of the plant, produces roridins that are converted by the plant into the highly toxic miotoxins, especially miotoxin A (Fig. 2.7). They are preferentially stored in the flowering region of female specimens [18, 19]. Gene transfer from the fungus to the plant and independent biosynthesis of the trichothecenes by the plants have also been postulated. Fifty-six trichothecenes, including malonyl-glucose conjugates, were detected in a concentration of 1.2 mg/g plant material.

Poisoning by *Baccharis coridifolia* kills at least 200,000 grazing animals annually in South America. The plant is responsible for spontaneous poisoning of early-weaned beef calves in Uruguay with about 40% lethality [31]. Symptoms of poisoning are like those of trichothecene poisoning and include salivation, anorexia, apathy, dehydration, and diarrhea [18, 19, 31, 41].

3.2.1.3 Pharmacology and Toxicology

The trichothecenes interact with the 60S ribosomes at the A-site of the peptidyl transferase center, indicating that they are peptide elongation (E-type) inhibitors. They inhibit protein, DNA, and RNA synthesis in eukaryotic cells, interfere with mitochondrial and membrane function, induce oxidative stress, and affect pathways associated with MAPK, JAK2/STAT3, and Bcl-w/caspase-3. Cytotoxicity and apoptosis result [14, 22, 36, 37]. The influence on the transcription factor Nfr2 [51] and the activation of the NRPL3 inflammasome [30] seem to play an important role at least in the toxicity of T-2 toxin and DON. Epigenetic changes (DNA methylation, modification of histone, and non-coding RNA) are detectable [28]. As a result, immunotoxicity, hepatotoxicity [45], neurotoxicity [52], reproductive toxicity, and genotoxicity occur [58]. There is insufficient evidence for the carcinogenicity of nivalenol and fusarenon-X (Tab. 3.1). The IARC classifies them as a group 3 carcinogen, which means their carcinogenicity for humans is not classifiable [1, 58].

Structural requirements for the activity are the 12,13-epoxide group and the double bond between C-9 and C-10. If the epoxide group is lost, the trichothecenes lose their efficacy. The quantity and position of hydroxyl groups as well as the type of esterifying acids and other substituents influence the effectiveness. A macrocyclic ring between C-4 and C-15 increases the toxicity [14, 25, 56, **14**].

After dietary ingestion by humans and animals, trichothecenes are absorbed preferentially through the gastrointestinal tract. Resorption can also occur after inhalation or application through the skin, e.g., from mold-contaminated spaces or within working environments like bakeries [7, 17]. Metabolization takes place mainly in the liver [45]. Possible metabolic pathways include hydrolysis by microsomal carboxylesterases and de-epoxidation by the intestinal microbiota. Absorption, conversion to polar metabolites, and elimination, e.g., as glucuronides, probably occur quickly [22, 44, 48].

Information on the acute toxicity of individual trichothecenes is given in Tab. 3.2. Ruminants are less at risk because the conversion to the less toxic de-epoxide occurs in rumens before absorption [46].

Deoxynivalenol (DON) is less toxic than other trichothecenes but nevertheless important because it frequently occurs at levels high enough to cause adverse effects [34].

Tab. 3.2: Acute toxicity of trichothecenes [1, 2, 22, 54, **14, 26**].

Mycotoxin	Mode of application	LD_{50} (mg/kg BW), mice
Deoxynivalenol	i.p.	70
4,15-Diacetoxyscirpenol	i.v.	10
Miotoxin C	i.p.	18
Fusarenon X	i.p.	3.4
Nivalenol	i.p.	4
Roridin A	i.v.	1
Trichothecin	i.v.	About 300
T-2 toxin	i.p.	3
Verrucarin A	i.v.	1.5
Verrucarin A	i.p.	0.5–0.75

☠ Symptoms of acute poisoning by trichothecenes, e.g., by DON and T-2 toxin, include vomiting, diarrhea, mucosal membrane damage in the digestive tract, damage to the bone marrow, decreased leukocyte and platelet counts, and disturbances of the cardiovascular system, motor coordination, and blood coagulation [12, 17, 36]. T-2 toxin [22] and the macrocyclic trichothecenes, in particular verrucarins, also have harmful effects to the skin [26].

From July to September 1987, several thousand persons in India became ill from gastrointestinal disorders caused by bread baked from Fusarium-infested DON-containing grain [4]. A study in Shandong Province, China, showed, that in areas with a high prevalence of DON-contaminated wheat, adverse health effects, e.g., vomiting and diarrhea, were observed in more than 100 of 100,000 residents and that this number was increased with 300 of 100,000 extra cases during heavy precipitation years [17, 27].

☹ Immunosuppression dominates after chronic ingestion of trichothecenes. The high mortality rate (132 per 100,000) due to esophageal and gastric cancer disease in some regions of China (Linxian, Cixiang) compared with that of the white Caucasian population in the United States (less than 5 per 100,000) is attributed to the high level of nivalenol in food (400–800 times higher than in the United States) [21].

Trichothecenes produced by *Fusarium sporotrichioides* and *F. poae*, especially T-2 toxin, are thought to be responsible for alimentary toxic aleukia (ATA). In addition to fever and hemorrhages, the disease is characterized by severe suppression of bone marrow activity and resulting leukopenia. ATA claimed many lives from 1942 to 1948, especially in the former USSR. In some regions, up to 10% of the population was affected, and the lethality rate was about 60%. The cause was contaminated, overwintered grain [14, 22, 24, 35]. Kashin–Beck disease, which occurs in East Asia (Northern China, Northern Korea, Eastern Siberia), has been attributed to consumption of grain contaminated with *F. sporotrichioides*, possibly in combination with selenium deficiency. Samples of brick tea (compressed tea used for the preparation of butter tea in Tibet) contained an extremely high concentration of T-2 toxin (about 400 µg/kg [23]). An association between ingestion of trichothecenes and autism is discussed [8].

Acute poisoning of animals by trichothecenes is characterized by growth retardation, reproductive disorders, and feed refusal in addition to the symptoms mentioned from human poisonings. Affected are, e.g., horses [13], poultry [16, 49], and rabbits [14, 39, 40, 47, 48, **14**]. The stachyobotrytoxicosis occurs particularly in horses. The animals first show skin and mucosal membrane damage as well as edema of the head ('hippopotamus head'). Then there are disturbances in blood formation and, in the final phase, violent diarrhea, rise in body temperature, cardiac insufficiency, and death.

In Europe, the following maximum levels for DON apply according to the Mycotoxin Maximum Level Ordinance and Dietary Ordinance: 1,000 µg/kg in unprocessed cereals, 400 µg/kg in bread and bakery products, and 150 µg/kg in dietary foods for infants and young children [**39**]. According to the FDA, the limits are 1,000 µg/kg in finished wheat products and 500 µg/kg in cereal-based foods for infants and children [17]. The limits for T2/HT-2 toxin are lower. The ADI for T-2 toxin is 100 ng/kg BW [11], according to JECFA 60 ng/kg BW [23]. For regulation in other states and materials, see [14].

3.2.2 PR toxin

PR toxin is a bicyclic sesquiterpene with an eremophilane base body with two epoxide rings and several other functional groups (Fig. 3.2). It is formed by *Penicillium roqueforti* Thom, along with structurally similar compounds, e.g., the eremofortins A–E. The fungus preferably grows in forages/silages under microaerophilic, moderately acidic, and psychrophilic conditions. It has great importance as starter for the preparation of blue cheeses, such as Roquefort or Gorgonzola (see Section 3.4.3). Under natural ripening conditions of blue cheeses, PR toxin converts by spontaneous reaction with amino acids and ammonium salts to PR imine, PR amide, and PR acid [**138**].

PR toxin is formed from farnesyldiphosphate by the aristolochene synthase. The gene cluster for biosynthesis of PR toxin has been identified [5, 10, 20].

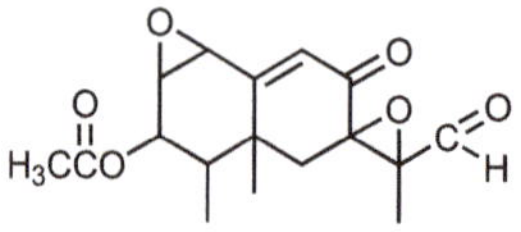

PR toxin

Fig. 3.2: A toxin of *Penicillium roqueforti.*

The mycotoxin inhibits RNA polymerase and protein synthesis, has mutagenic and carcinogenic potential, and damages liver and kidney. The aldehyde group present in PR toxin structure is mainly responsible for its biological effects [10]. Rats that received PR toxin (160 mg/kg BW, p.o.) died of respiratory paralysis within a short time. Histopathologic examination revealed edema and hemorrhages in the liver, lung, kidney, and brain [53]. The LD_{50} in rats is 11.6 mg/kg BW, i.p., and 115 mg/kg BW, p.o. [10].

PR toxin contaminated grass silage, grains, or other food and feedstuffs are a risk for animals. Although PR toxin has been shown to be cytotoxic for human cell lines in vitro there seem to be no reports on potential health hazards for humans. Human health effects from PR toxin in association with cheese were not observed. One reason for this may be that PR toxin in blue cheeses is degraded to the mentioned less toxic metabolites (see Section 3.8.3 [10]).

The level of PR toxin in food and in feed is not regulated.

References

For numbers in bold, see cross-chapter literature p. 233.

[1] Aupanun S et al. (2016) J Vet Med Sci 79(1): 6

[2] Bamburg JR (1976) In: Mykotoxins and other Fungal Related Food Problems. Rodricks JV (ed.) Adv in Chem, Ser. 149, Am Chem Soc, Washington, p. 144

[3] Bertuzzi T et al. (2014) Food Addit Contam Part B 7: 273

[4] Bhat RV et al. (1989) Lancet 1: 35

[5] Chávez R et al. (2023) J Fungi 9: 459

[6] Chen Y et al. (2019) Ann Rev Phytopathol 57: 15

[7] Creasia DA et al. (1990) Fund Appl Toxicol 14: 54

[8] De Santis B et al. (2019) Nutr Neurosci 22: 132

[9] Dinleyici M et al. (2018) Neuro Endocrinol Lett 39: 325

[10] Dubey M et al. (2018) Front Pharmacol 9: 288

[11] EFSA Panel on Contaminants in the Food Chain (CONTAM) (2017) EFSA J 15(9): e04718

[12] Ekwomadu TI et al. (2021) Int J Environ Res Public Health 18(22): 11741

[13] Enyley S, Mostrom M (2024) Vet Clin North Am Equine Pract 40(1): 83

[14] Foroud NA et al. (2019) Toxins 11: 634

[15] Godtfredsen WD, Vangedal S (1965) Acta Chem Scand 19: 1088

[16] Gómez-Osorio LM et al. (2024) Front Vet Sci 11: 1387856

[17] Gonya S et al. (2024) Int J Exp Res Public Health 21: 808
[18] Habermehl G (1989) Dtsch Tierärztl Wochenschr 96: 335
[19] Habermehl G (1989) Pure Appl Chem 61: 377
[20] Hidalgo PI et al. (2017) Appl Microbiol Biotechnol 101(5): 2043
[21] Hsia CC et al. (2004) Oncol Rep 12: 449
[22] Janik E et al. (2021) Molecules 26: 6868
[23] Jiang T et al. (2024) Nutrients 16: 1449
[24] Joffe AZ (1986) Fusarium Species: their Biology and Toxicology. Wiley, New York
[25] Kiesling KH (1986) Pure Appl Chem 58: 327
[26] Kuhn DM, Ghannoum MA (2003) Clin Microbiol Rev 16: 144
[27] Li F et al. (2022) Environ Sci Pollut Res Int 29: 71826
[28] Li L et al. (2023) Environ Toxicol Pharmacol 100: 104154
[29] Li M et al. (2020) Nat Prod Rep 37: 1568
[30] Liao C et al. (2024) Vet Sci 11: 291
[31] Machado M et al. (2023) Toxins 15: 681
[32] Mastanjević K et al. (2023) Life 13: 2211
[33] McCormick SP et al. (2024) Appl Microbiol Biotechnol 108(1): 475
[34] Nesic K et al. (2014) Rev Environ Contam Toxicol 228: 101
[35] Pitt JI, Hocking AD (1986) Proc Nutr Soc Aust 11: 82
[36] Polak-Śliwińska M, Paszczyk B (2021) Molecules 26: 454
[37] Qu Z et al. (2024) mLife 3(2): 176
[38] Ribeiro AR et al. (2016) Arch Environ Contam Toxicol 70: 361
[39] Rocha O et al. (2005) Food Addit Contam 22: 369
[40] Rousseaux CG (1989) Comments Toxicol 2: 37
[41] Schild CO et al. (2020) Toxicon 188: 5
[42] Schollenberger M et al. (2005) Int J Food Microbiol 97: 317
[43] Schothorst RC, van Egmond HP (2004) Toxicol Lett 153: 133
[44] Shintov A et al. (1988) Toxicon 26: 153
[45] Song W et al. (2023) Environ Pollut 330: 121784
[46] Sundstøl Eriksen G et al. (2004) Food Chem Toxicol 42(4): 619
[47] Tsouloufi TK (2024) J Vet Diagn Invest 36(5): 638
[48] Ueno Y (1988) IST Atlas Sci Pharm 2: 121
[49] Vöröshaźi J et al. (2024) Poultry Sci 103: 103471
[50] Wang J et al. (2023) Toxins 15: 446
[51] Wang Y (2023) Toxics 11(4): 393
[52] Wang Y et al. (2024) Mycotoxin Res 40(1): 85
[53] Wei RD et al. (1973) Appl Microbiol 25: 111
[54] WHO Genf (1979) Mycotoxins: Environ Health Criteria 11. www.who.int/publications/i/item/9241540710 (accessed 04/14/2024)
[55] Winkler J et al. (2015) Food Addit Contam Part A 32: 371
[56] Wu Q et al. (2013) Curr Drug Metab 14(6): 641
[57] Zhu M et al. (2020) Toxins 12: 417
[58] Zingales V et al. (2021) Food Chem Toxicol 152: 112182

3.3 Polyketides as Mycotoxins

3.3.1 Patulin, Mycophenolic Acid, and Penicillic Acid

Patulin (expansin, clavatin, and claviform, Fig. 3.3) is an α,β-unsaturated lactone that occurs as a mixture of the *R* and *S* stereoisomers. It is soluble in water and in organic solvents, acid-stable, and alkali-labile (overviews: [7, 73, 100, 128, 133]).

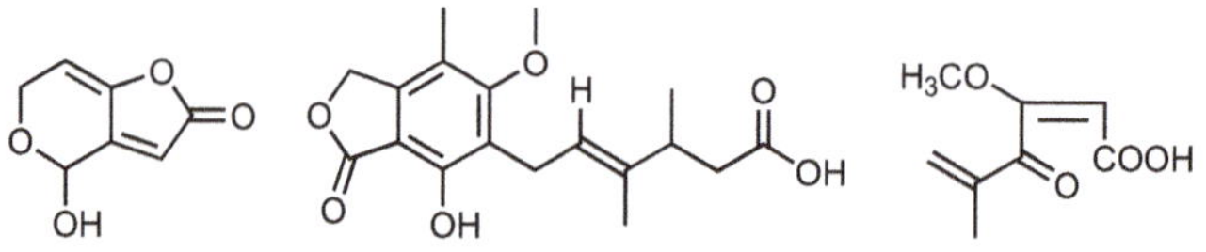

Fig. 3.3: Patulin, mycophenolic acid, and penicillic acid.

It is produced by Penicillium species, e.g., *P. solitum* Westling (*P. patulum* Bainier), *P. expansum* Link, and *P. roqueforti*, by Aspergillus species, e.g., *A. terreus* Thom and *A. giganteus* Wehmer, and by *Paecilomyces niveus* Stolk & Samson and *P. fulvus* Stolk & Samson. It is particularly present in fruits affected by brown rot caused by Penicillium species [140].

The median level of patulin was 10.5–43.1 µg/kg in apple juices from Belgium, 6.3–8.9 µg/kg (maximal level 78 µg/kg) in such from China and 5.1–87.6 µg/kg in dried figs, longan fruits, and hawthorn products from China [131]. The concentration in apples ranged from 8.8–120.4 µg/kg. Its presence in baby food is a major concern. In China, it was detected in 19 out of 30 baby food products analyzed [7]. It has also been detected in bakery products, e.g., bread (up to 0.16 mg/kg). In apples and pears, up to 250 mg patulin per kg of rot were found [90]. The toxin can diffuse from the rotting sites into healthy tissues. It is detoxified by reaction with SH groups of cysteine or glutathione, so it is found in very low concentrations, if at all, in protein-rich products such as raw meat, in raw sausage, and in cheese [**138**].

Biogenetically, patulin is a tetraketide, with 6-methylsalicylic acid and gentisinaldehyde occurring intermediately during its formation. The latter is rearranged into patulin after ring cleavage [49].

The toxicity of patulin is based on its ability to alkylate enzymes and nucleic acids due to its dienone structure and on the induction of oxidative stress. Symptoms of experimental poisoning include pulmonary and cerebral edema, liver, heart, and kidney damage, and immune system dysfunction [69]. A disruption of α1- and α2-adrenergic receptor signaling pathways and epigenetic modifications in the kidneys of patulin-treated mice could be shown [78]. The IARC classifies patulin as Group 3 substance (carcinogenicity for human not classifiable, Tab. 3.1 [7]). The LD_{50} for mice is 35 mg/kg BW, p.o., 10 mg/kg BW, s.c., and 15 mg/kg BW, i.p. The NOAEL is given as 43 µg/g BW/day [27, **14, 26, 116, 139**].

Despite the relatively low risk to humans from patulin, maximum levels have been legally prescribed in several countries. The presence of patulin in fruit juices is regarded as an indication that moldy raw products have been processed and that other mycotoxins may also be present. The maximum permissible level of patulin in fruit juices, prescribed by the FDA, the EU, the Ministry of Health of the People's Republic of China and Health XCanada, is 50 µg/kg, and 10 µg/kg in products for infants. The maximum ADI is 0.4 µg/kg BW [7, **39, 116**]. It is estimated that an average of 3.0 ng patulin per kg BW/day is ingested by adults in EU countries. For apple juice drinkers, which include many young children, the amount is about 21 ng/kg BW/day [129, **116**].

Mycophenolic acid and penicillic acid (Fig. 3.3) are closely related to patulin and formed by some Penicillium species, e.g., *P. roqueforti*. Both are tetraketides. The main human source of mycophenolic acid is cheese (23–38% positive findings, 0.01–15 mg/kg), especially Roquefort cheese (see Section 3.4.3). Penicillic acid has been detected, e.g., in maize and beans (5–230 µg/kg BW [118, **14**]).

Mycophenolic acid inhibits inosine monophosphate dehydrogenase, a key enzyme for de novo purine biosynthesis, which is necessary, for example, in lymphocyte proliferation during the course of an immune response. The compound is acutely nontoxic (LD_{50} mice 500 mg/kg BW, p.o. [16, 28, 34, 35].

Penicillic acid is nephrotoxic, mutagenic, and carcinogenic. The LD_{50} for mice is 5 mg/kg BW, i.v., 2.2 mg/kg BW, s.c. [**14**], and 35–600 mg/kg BW, p.o. [**175**].

A partially synthetic derivative of mycophenolic acid (mycophenolate mofetil) is used as an immunosuppressant, e.g., after transplantation.

3.3.2 Citrinin

Citrinin (antimycin, Fig. 3.4, overview: [38, 52, 106, 138]) is an isochroman derivative. It is nearly insoluble in water but soluble in ethanol, dioxane, and aqueous solutions. It was first isolated by Hetherington and Raistrick from a culture of *Penicillium citrinum* Thom, in 1931 [106].

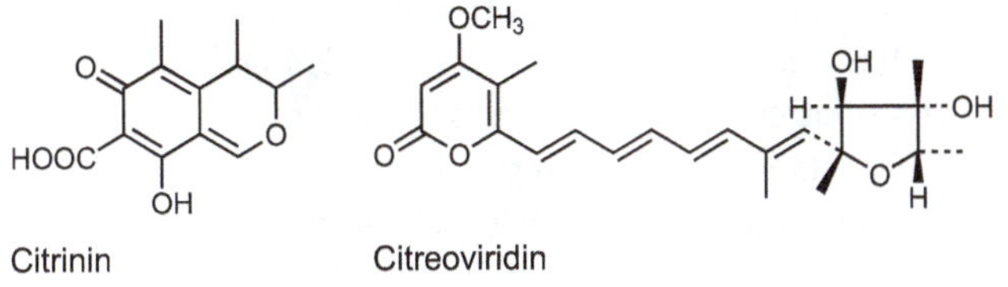

Fig. 3.4: Citrinin and citreoviridin.

It is produced by Penicillium species, including *P. citrinum*, *P. expansum*, *P. roqueforti*, and *P. citreonigrum* Dierckx., by some Aspergillus species, e.g., *A. terreus*, and by *Monascus ruber* Tiegh. and *M. purpureus* Went. Citrinin producers often form patulin, citreoviridin, and ochratoxin A simultaneously.

The biogenesis of citrinin occurs from five acetate units (pentaketide) and three C1 bodies [52, **14, 93**].

Citrinin is found in cereal fruits and in meat products inoculated with molds to promote ripening, on ham, in cheese, in moldy baked goods, and in house dust [116, **14, 26, 138**]. In some 'red rice' products, which are produced by fermentation of rice with the mold *Monascus purpureus* and marketed as dietary supplements to lower cholesterol levels, very high levels of citrinin (>2,000 µg/kg) have been found [25]. However, the analysis of 35 red rice samples purchased from an EU-bound e-commerce platform or online pharmacies detected citrinin in 16 samples in very low amounts below the tolerated maximal level of 100 µg/kg [127]. During thermal food processing citrinin derivatives are generated by reaction with amino acids or proteins so that the concentration in processed cereal products is generally lower than in unprocessed samples [12].

In humans, citrinin is metabolized to dihydrocitrinone. Both compounds may serve as biomarker of exposure to citrinin. The analysis of these biomarkers in urines of German children and adults indicated a widespread exposure to this mycotoxin and does not seem to correlate with the low contamination frequency in Europe. The biomarkers concentrations were higher in children's urine than in urines from adults. The calculated probable median daily intake of citrinin was 0.013 µg/kg BW for adults and 0.05 µg/kg BW for children. This is lower than the value of 0.2 µg/kg BW/day defined as 'level of no concern for nephrotoxicity' by the European Food Safety Authority. However, some individuals exceeded this limit [25].

Citrinin induces oxidative stress and mitochondrial dysfunction. It provokes DNA damage, cell cycle arrest, and apoptosis [116]. It is accumulated in the proximal renal tubules by an active tubular acid transport mechanism and acts predominantly nephrotoxic. Hepatotoxic and teratogenic effects have also been found. Citrinin is classified as 'not classifiable as to its carcinogenicity to humans' (Tab. 3.1 [**65**]). The LD_{50} is 110 mg/kg BW, p.o., and 35 mg/kg BW, i.p., mice.

In addition to ochratoxins, citrinin has been implicated as a cause of kidney disease in pigs and probably in humans [88, **14, 26, 116, 138**]. A human poisoning by red rice contaminated with citrinin has been reported from Japan in 2024 [48].

In Europe, the maximum level for citrinin in Monascus fermented rice products is 100 µg/kg [**39**], in the United States and in Japan 2 mg/kg, and in China 0.05 mg/kg [48]. So far there is no regulation on maximal content in cereals and other food commodities.

3.3.3 Citreoviridin

Citreoviridin (citreoviridin A, yellow rice toxin, Fig. 3.4) has a six-membered, doubly unsaturated lactone ring. It is produced by some Penicillium species, e.g., *P. citreonigrum*, *P. citrinum*, and *P. dierckxii* Biourge, and by *Aspergillus terreus*. It was first detected in contaminated rice ('yellow rice'). It is also found in meat products [**26, 138, 139**].

The biogenesis of citreoviridin occurs from nine acetate units (nonaketide) and five C-bodies [**14**].

Citreoviridin interferes with energy metabolism, induces oxidative stress, and inhibits the absorption of vitamin B_1. It has a predominantly neurotoxic effect and causes severe cardiac dysfunction. Animal studies show that it enhances atherogenesis via upregulating inflammation by activation of NF-κB [43]. The LD_{50} in male mice is 7.2 mg/kg BW, i.p., 11 mg/kg BW, s.c., and 29 mg/kg BW, p.o. [**14, 26, 138, 139, 175**].

The prevalence rates of cardiac diseases in China are high in areas with a high citreoviridin contamination of corn (>5–30 μg/kg [43]). If rice contaminated with citreoviridin can really contribute to cardiac beriberi, a heart disease that is prevalent in Japan and Brazil, is unclear. Beriberi is primarily caused by a deficit of vitamin B_1.

3.3.4 Anthracene Derivatives

More than 30 monomeric and dimeric anthracene derivatives (Fig. 3.5) have been detected as products of a large number of molds. Producers include *Aspergillus aculeatus* Iizuka, *Fulvia fulva* (Cooke) Cif., *Chaetomium elatum* Kunze (all forming emodin), *Penicillium aurantiogriseum* Dierckx., *Talaromyces wortmannii* (Klöcker) C.H. Benj. (*Penicillium wortmannii* Klöcker), *T. rugulosus* (Thom) Samson, N. Yilmaz, Frisvad & Seifert (*Penicillium rugulosum* Thom), *Albifimbria verrucaria* (Alb. & Schwein.) L.Lombard & Crous (all forming rugulosin), *Aspergillus chevalieri* (L. Mangin) Thom & Church, *Didymella pisi* Chilvers, J.D. Rogers & Peever, *Chaetomium elatum* (all forming physcion), and *Talaromyces islandicus* (Sopp) Samson, N. Yilmaz, Frisvad & Seifert (*Penicillium islandicum* Sopp), (forming luteoskyrin, rugulosin, chrysophanol, flavoskyrin, and islandicin).

Anthracene derivatives formed by *T. islandicus*, especially luteoskyrin, have been detected in stored rice and in soybean products. Rugulosin was also found in raw sausages when they had been stored at about 25 °C [122, **14, 26, 138, 139**].

The basic anthracene body of the above-mentioned mycotoxins is built up from eight acetate residues (octaketide [**138, 139**]).

The anthracene derivatives exhibit cytotoxic, hepatotoxic, carcinogenic, and mutagenic effects via inhibition of RNA synthesis, interaction with nucleic acids, interference with respiratory chain and oxidative phosphorylation (partially acting as uncouplers), and formation of oxygen radicals [121].

The LD_{50} of (–)-luteoskyrin for mice is 40.8 mg/kg BW, i.p., and 221 mg/kg BW, p.o. The LD_{50} of (+)-rugulosin for mice is 83.0 mg/kg BW, i.p., and for rats is 44 mg/kg BW,

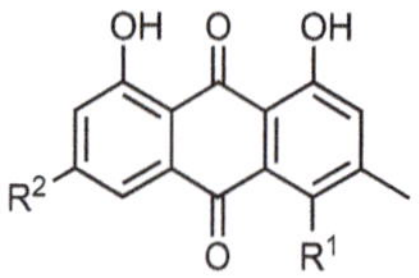

Islandicin R^1=OH R^2=H
Chrysophanol R^1=H R^2=H
Emodin R^1=H R^2=OH

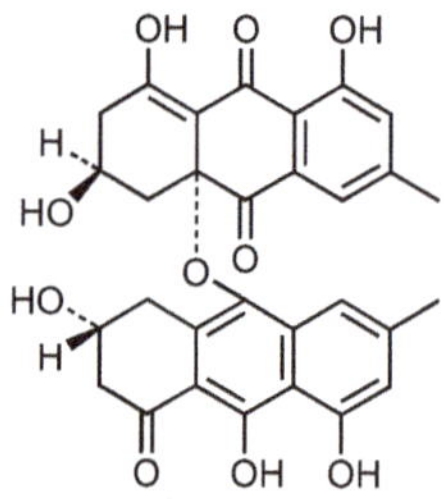

(–)-Flavoskyrin

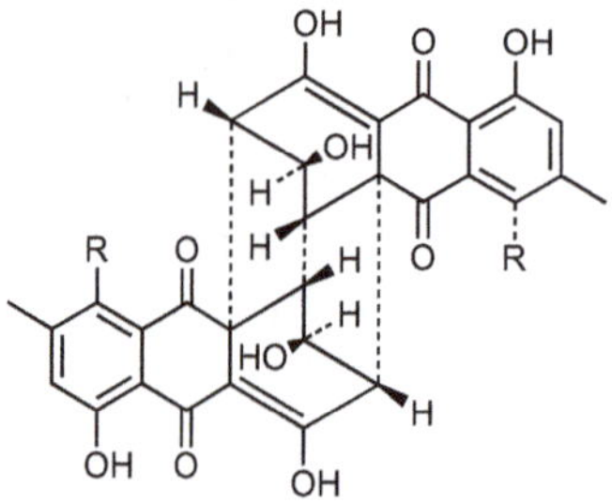

(–)-Luteoskyrin R=OH
(+)-Rugulosin R=H

Fig. 3.5: Anthracene derivatives.

i.p. Of emodin, 3.7 mg/kg BW, p.o., is lethal to day-old chicks [45, **14, 26**]. Luteoskyrin and rugulosin have been implicated in yellow rice syndrome [37, **175**].

3.3.5 Zearalenones

Zearalenones (Fig. 3.6) are β-resorcylic acid lactones with a 14-membered lactone ring. They are poorly soluble in water but readily soluble in alkali solutions and organic solvents. The main representative zearalenone (F-2 toxin) was first isolated in 1962 by Stob and coworkers [110]. Its structure was elucidated by Urry and coworkers in 1966 ([124], overviews: [40, 74, 96, 97]).

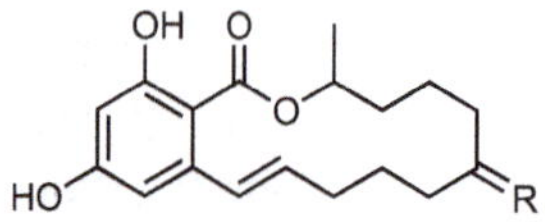

Zearalenone R=O
Zearalenol R=H+α-OH

Fig. 3.6: Zearalenones.

Zearalenones are polyketides composed of nine acetate residues (nonaketides). The responsible gene cluster has been identified in *F. graminearum* [84, 93].

Known producers are about 15 Fusarium species, e.g., *F. graminearum*, *F. sambucinum* Fuckel, *F. oxysporum* Schltdl., *F. sporotrichioides*, and *F. fujikuroi* Nirenberg. Cereal crops, especially maize, are the main substrate of these fungi. They are already infested in the field and are further penetrated by the fungi during storage with moisture contents above 23%. The contamination frequency is worldwide up to 43%. The concentrations of zearalenone were up to 900 µg/kg in corn, up to 8,000 µg/kg in wheat, up to 1,500 µg/kg in barley and malt (also detectable in beer), up to 50,000 µg/kg in grain dust, and up to 450 µg/kg in walnuts. The average zearalenone content in animal feed ranges from 14 to 215 µg/kg, depending on the geographic region and cereal type. Human intake ranges from 100 to 500 ng/kg BW per day. Because of its heat resistance, zearalenone also survives the baking process. When ingested in large amounts by dairy cattle, it can partially transfer into milk [40, 74, 97, 139, **138**].

Zearalenones are among the most potent naturally occurring estrogenic compounds [62, 74, 96, 97, 113]. They attack estrogen receptors and lead to the symptom picture of hyperestrogenism, especially in female juvenile domestic animals. α-Zearalenol, which is formed in the organism from zearalenone together with β-zearalenol, is approx. 10 times more estrogenically active than zearalenone [123]. Studies indicate genotoxic, hepatotoxic, and immunotoxic effects. However, until now the IARC classifies zearalenone as non-carcinogenic for humans (class 3 [40, 74, 97]).

The LD_{50} of zearalenone is with more 2,000 mg/kg BW for mice, rats, and guinea pigs very high. The EFSA Panel on Contaminants in the Food Chain stated a tolerable daily intake of 0.23 µg/kg BW [97].

The levels of zearalenone found in cereal products intended for human consumption are generally not able to endanger humans. However, in some regions of the world, early onset development of female sexual characteristics has been associated with mycotoxin exposure [74, **116**]. The investigation of possible effects of long-term exposition of low doses of the toxin to human, especially to infants, pregnant, and postmenopausal woman, needs further attention [40].

Poisoning by zearalenone has so far only been observed in domestic animals, especially pigs. In female pigs, problems can occur at zearalenone concentrations of 50–200 µg/kg feed [**116**]. Estrogenic activity is manifested in female animals by enlargement of the genitalia and teats, tumor formation in the genital area, ovarian atrophy, and infertility, and in male animals by feminization [32].

In the EU, the permitted level of zearalenone is set at a maximum of 50 µg/kg in bread and other cereal products (in maize 100 µg/kg), and 20 µg/kg in dietary foods for infants and young children. To ensure animal health, an upper limit of 0.05 mg/kg is recommended for prepubertal female breeding pigs, for example [97, **39**, **116**].

3.3.6 Sterigmatocystins and Versicolorins

Sterigmatocystins are difuranoxanthin derivatives (Fig. 3.7, overview: [65, 142]). They are formed by *Aspergillus versicolor* (Vuill.) Tirab.) (*Emericella versicolor* (Vuill.) Pitt. & A.D. Hocking), *A. flavus*, *A. parasiticus* Speare, *A. nidulans* (Eidam) G. Winter (*Emericella nidulans* (Eidam) Vuill.), and other Aspergillus species, in addition to the biogenetically closely related difuronoanthraquinones, the versicolorins. Sterigmatocystins are also produced by Penicillium, Bipolaris, Botryotrichum, and Humicola species [65]. Major representatives are sterigmatocystin and versicolorin A. In contrast to aflatoxins, sterigmatocystin is released only when the fungal hyphae decay [**116**]. In Korea, 1,135 samples of agricultural products and processed foods were analyzed. Sterigmatocystin was found in 4.1% of the samples in the range of 0.08–10.07 ng/g. The risk to cause human health problems was estimated as unlikely [53].

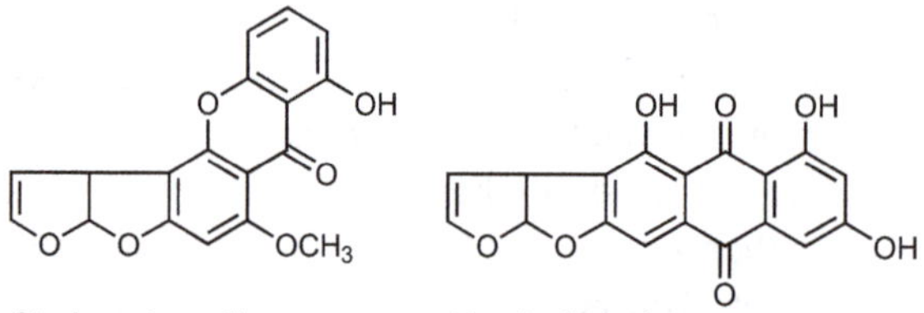

Sterigmatocystin

Versicolorin A

Fig. 3.7: Sterigmatocystin and versicolorin A.

Biogenetically, sterigmatocystins and versicolorins are decaketides. Versicolorins can convert to sterigmatocystins and the latter to aflatoxins [65].

Sterigmatocystin, like aflatoxin B1, is converted to its epoxide and forms then adducts with the guanine of DNA. It has hepato- and nephrotoxic effects. The LD_{50} is 166 mg/kg BW, p.o., 60 mg/kg BW, i.p., in male rats; 120 mg/kg BW, p.o., in female rats; 80 mg/kg BW, p.o., in mice; and 32 mg/kg BW, i.p., in male monkeys [**14, 138, 139, 175**]. Compared with the relatively low acute toxicity, sterigmatocystin has considerable carcinogenic activity (2B, Tab. 3.1). Nevertheless, the health hazard to humans is assumed to be low [90].

3.3.7 Aflatoxins

3.3.7.1 Chemistry, Biogenesis, and Occurrence

Aflatoxins are difuranocyclopentanocoumarins or difuranopentanolidocoumarins (Fig. 3.8, overviews: [3, 65, 85, 94, **138, 139**]). More than 20 different aflatoxins from fungi are known. Numerous others are formed only or also by biotransformation in the human or animal organism (e.g., aflatoxin M1, aflatoxin M2, aflatoxin GM1, aflatoxin GM2) or by action of microorganisms (e.g., aflatoxicol). Major representatives occurring in natural substrates are aflatoxin B1, aflatoxin B2 (B = blue fluorescent), aflatoxin G1, and aflatoxin G2 (G = green fluorescent).

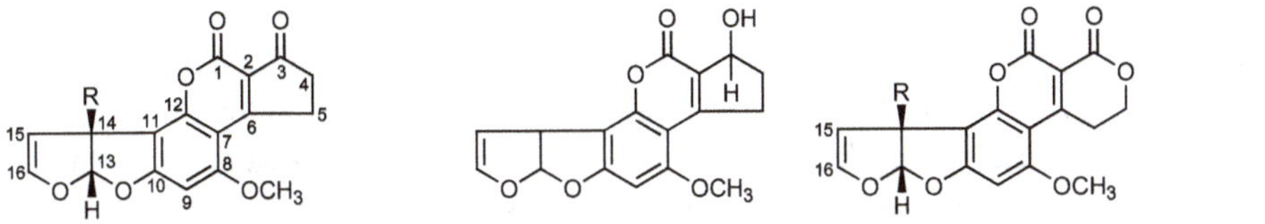

Aflatoxin B_1 R=H
Aflatoxin B_2 15,16-dihydro, R=H
Aflatoxin M_1 R=OH
Aflatoxin M_2 15,16-dihydro, R=OH

Aflatoxicol

Aflatoxin G_1 R=H
Aflatoxin G_2 15,16-dihydro, R=H
Aflatoxin GM_1 R=OH
Aflatoxin GM_2 15,16-dihydro, R=H

Fig. 3.8: Aflatoxins.

Aflatoxins were first isolated 1961/1962 [101, 125]. The structures of aflatoxin B1 and aflatoxin G1 were elucidated by Asao and coworkers in 1965 [6].

Aflatoxins have relatively high melting points (195–320 °C), are poorly soluble in water (10–30 µg/mL), but are readily soluble in methanol or dimethyl sulfoxide. They are heat stable and are destroyed only to a small extent during boiling or baking [**14**].

The biosynthesis occurs from ten acetate units (decaketides) and involves in the case of aflatoxin B1 at least 27 enzymatic reactions. Versicolorins and sterigmatocystins occur as intermediates. In Aspergillus, all 34 genes encoding the enzymes for aflatoxin biosynthesis are located in close vicinity on chromosome III in a gene cluster. Their expression is coordinated by two cluster-specific regulators: *aflR* and *aflS*. Environmental factors such as pH, light, nutrient sources, and oxidative stress response may trigger the toxin formation [14, 65, 115].

Aflatoxins are formed by strains of the closely related Aspergillus species *Aspergillus flavus* (*A. oryzae* (Ahlb.) Cohn, *A. parasiticus*, and *A. nomiae* Kurtzman, B.W. Horn & Hesselt. The fungi thrive at a water content of at least 18% on starchy substrates and of about 10% on oily substrates. They also parasitize on living plants. The fungi produce aflatoxins in a temperature range from 7.5 to 40 °C (optimum between 25 and 30 °C). With ongoing global climate change, aflatoxins are predicted to be an emerging threat in areas where it was not previously present, such in several European regions.

Aflatoxins are found particularly frequently in peanuts, cottonseeds, cereal fruits, hazelnuts, walnuts, Brazil nuts, pecans, pistachio nuts, sunflower seeds, almonds, nutmegs, figs, soybeans, in copra, and in animal feed of plant origin. Infestation with the aflatoxigenic fungi usually occurs before harvest, so that some of the aflatoxins are already present in the seeds or fruits at harvest. In 9 studies from different countries, from a total of different 3,714 samples, 1,246 samples (33.54%) were found positive for aflatoxins (for example until 37.42 µg/kg in wheat in Algeria). Maize and barley were the most contaminated crops among the cereals. Most aflatoxin consumption by humans occurs through rice and wheat. Aflatoxins have also been detected in fruits (33.45% of samples, until 136,000 µg/kg in nuts from India), spices (30.9% of samples, especially red chili, nutmeg, turmeric, black pepper, ginger, for example mean 30.20 µg/kg in red pepper from Iran). From a total of 957 samples of different types of animal feed, 571 samples (59.7%) were found contaminated with aflatoxins (for example until 419 µg/kg in dairy feed in Ethiopia) [46]. Out of 180 maize samples in Ghana, 131 were positive for aflatoxins with levels from 4.27 to 441.02 µg/kg ([1], overviews about the occurrence of aflatoxins in several countries and materials: [3, 27, 85)].

Due to the frequent contamination of animal feed, aflatoxins and their biotransformation products, which are also toxic, can be detected in milk (mainly in the form of aflatoxin M1), dairy products, and, only when large amounts of aflatoxins are fed experimentally, also in eggs and meat [119]. About 0.17–3.0% of the amount fed is excreted in milk, mainly in the form of aflatoxin M1. During cheese making, accumulation and additional mycotoxin formation may occur due to contaminating fungi.

3.3.7.2 Pharmacology and Toxicology

Aflatoxins are rapidly absorbed after peroral administration. Resorption is also possible via the skin or the mucous membrane of the bronchial tract after inhalation of the fungal spores, which also contain aflatoxins. Aflatoxin B1 and its metabolite AFM1 are secreted in breast milk and can impair child growth [44, 137]. Aflatoxin exposure during pregnancy may be associated with increased risk for stillbirth, preterm birth, and low birthweight [4, 22, 58, 108].

In the liver, representatives with a double bond in the outer furan ring, e.g., aflatoxin B1 and aflatoxin M1, undergo oxidation to highly reactive epoxides (toxification) with the aid of CYP450-dependent monooxygenases, especially CYP450 A4. Covalent bonding between the C-8 of aflatoxin B1-8,9-exo-epoxide and the N-7 of a guanine residue of DNA results in an aflatoxin-guanine adduct (8,9-dihydro-8-(7 *N*-guanyl)-9-hydroxy-aflatoxin B1), which is detectable in urine. Secondarily, the formation of further adducts, the cleavage of the alkylated guanine from the DNA, as well as ring openings to formamidopyrimidine derivatives take place. The resulting point mutations particularly affect oncogenes or tumor suppressor genes, which are thus activated (ras oncogene) or impaired in their function (p53). The point mutation (G–T transversion), that is specifically produced by aflatoxin B1 in codon 249 of the p53 tumor suppressor gene [47, **116**], has been particularly well studied.

The representatives without a double bond, e.g., aflatoxin B2, do not form epoxides, but can in part be dehydrogenated to aflatoxin B1. Aflatoxin B1 is reversibly converted to the less toxic aflatoxicol by cytosolic NADPH reductase. Hydroxylated and dealkylated metabolites are less mutagenic than the native compounds. The major detoxification reaction is the conjugation of the epoxide with glutathione catalyzed by glutathione S-transferase [**116**].

Strong carcinogenic, mutagenic, and teratogenic effects of aflatoxins result from the attack on DNA. Epigenetic modifications including DNA methylation, histone modifications, and regulation of non-coding RNA also play a role in aflatoxin-induced carcinogenesis [15, 23].

Aflatoxins also affect enzymes of carbohydrate and lipid metabolism, induce oxidative stress, and disturb mitochondria. Acute toxicity is preferentially due to these attacks. Adverse effects on immune system compromise the ability of both humans and animals to resist infections and cancer [81, 120].

Aflatoxins are primarily hepatotoxic and hepatocarcinogenic. Acute liver injury is characterized by destruction of hepatocytes and proliferation of bile duct epithelial cells. Extrahepatic hemorrhages occur in the lungs, kidneys, and adrenal glands after ingestion of aflatoxin B1. Aflatoxin B1 is the most potent liver carcinogen of natural origin known to date [82]. Even a single administration of approx. 10 µg/kg BW causes hepatocellular carcinomas in rats [**116**].

3.3.7.3 Acute Poisonings by Aflatoxins

Data on the acute toxicity of aflatoxin B1 are summarized in Tab. 3.3. There are significant differences in the effects of aflatoxins in different animals. Mice are relatively insensitive because of their high glutathione S-transferase activity. Male animals tend to be more sensitive than females. In humans, the lethal dose of aflatoxin B1 is estimated to be 1–10 mg/kg BW, with children being particularly sensitive [**116**]. Toxicity decreases in the order B1 > G1 > B2 > G2.

Tab. 3.3: Acute toxicity of aflatoxin B1 [13, 86, 134, **116**].

Animal species	LD_{50} (mg/kg BW), p.o.
Duck chicken	0.3–0.6
Rabbit	0.3
Cat	0.5
Pig	0.6
Dog	0.5–1.0
Sheep	1.0
Guinea pig	1.4
Rat, male	7.2
Rat, female	17.9
Mice, newborn	9.0
Mice, adult	50
Hamster	10.2

Acute human diseases for which a causal relationship with the intake of aflatoxins has been established are listed in Section 3.1.3. In Europe, the ADI of aflatoxin B1 is estimated to be a maximum of 0.25 ng/kg BW [77]. Acute poisoning, which requires aflatoxin concentrations in the range of mg/kg food, is likely to occur in humans only in extreme cases. In India in the mid–1970s, consumption of corn contaminated with aflatoxins (content 0.25–15 mg/kg) resulted in acute toxic hepatitis in several hundred persons, of whom more than 100 died [56, 117]. A severe outbreak of aflatoxicosis in rural Kenya in 2004 resulted in 317 cases and 125 deaths. It was attributed to contaminated maize during a period of food scarcity exacerbated by drought [91]. Locally produced maize was identified as the source of an unfamiliar disease resulting in 68 cases and 20 deaths in Tanzania in 2016 [85].

In animals, however, acute aflatoxin poisoning is possible worldwide. 'Hepatitis X' in dogs observed in the United States in the 1950s was probably caused by dog food contaminated with aflatoxins. In Brazil, an outbreak affected 65 dogs on 9 farms, leading to 60 deaths due to aflatoxin-contaminated mice [85]. Depression, inappetence, and signs of hepatotoxicity were dominant clinical findings in 72 cattle suffered from aflatoxicosis in Egypt [30].

Symptoms of acute aflatoxin poisoning (aflatoxicosis) in animals include growth arrest, weight loss, severe tenesmus, and finally death [29, 37, **26**].

3.3.7.4 Chronic Poisonings by Aflatoxins

More significant are the chronic effects of aflatoxins [72]. In Moçambique and a southern Chinese province, both regions with a high incidence of liver cancer, exposures of 184 ng aflatoxins/kg BW and 2,027 ng aflatoxins/kg BW, resp., were calculated. In about 50% of hepatocellular carcinomas in these regions, the already mentioned point mutation in the p53 gene could be detected [132]. The results of epidemiological studies from African countries [39] also confirm the association between aflatoxin intake and increased incidence of liver cancer in humans.

The total amount of aflatoxins consumed by Chinese people through corn, peanut, and their oil products, observed from 2010 to 2020, was as high as 4.018 ng/kg BW/day leading to 1.53 additional cases of hepatocellular carcinoma per 100,000 people per year [19].

Compared to areas with high exposure to aflatoxins, the risk of tumors in Europe caused by chronic ingestion of these mycotoxins is low. Aflatoxins and hepatitis B viruses are independently but also synergistically involved in the development of liver cancer. The carcinogenic potency of aflatoxin B1 is approx. 30-fold higher in hepatitis B-positive individuals than in hepatitis B-negative individuals [42]. The risk of disease is influenced by individual expression of CYP450 isozymes. Individuals with high activity of CYP450 3A4-dependent enzymes are at greater risk [51].

In the EU [**39**] a maximum of 2 µg/kg aflatoxin B1 and 4 µg/kg of the sum of aflatoxins B1, B2, G1, G2 is permitted for peanut seeds, nuts, dried fruits, and their processed products (for direct consumption or use as a food ingredient) and cereals/cereal products (except maize). For infants and young children's food, a limit of 0.10 µg/kg for aflatoxin B1 is set. The limits are similar in other countries [**40**].

Substances that promote the detoxification of aflatoxins are an option for reducing the carcinogenic risk. One example is the organosulfur compound oltipraz that induces phase II detoxification enzymes, such as glutathione S-transferase. In vitro investigations, animal assays and few clinical trials have shown that oltipraz is able to prevent the conversion of aflatoxin B1 into its carcinogenic epoxide and to accelerate its detoxification [9].

3.3.8 Rubratoxins

Rubratoxin A and rubratoxin B (Fig. 3.9) have a difuranocyclononan parent body. They are produced by *Talaromyces purpureogenus* (Stoll) Samson, N. Yilmaz, Houbraken, Spierenb., Seifert, Peterson, Varga & Frisvad (*Penicillium purpureogenum* Stoll). Rubratoxins have been detected in cereals, especially maize, legumes, and sunflower seeds.

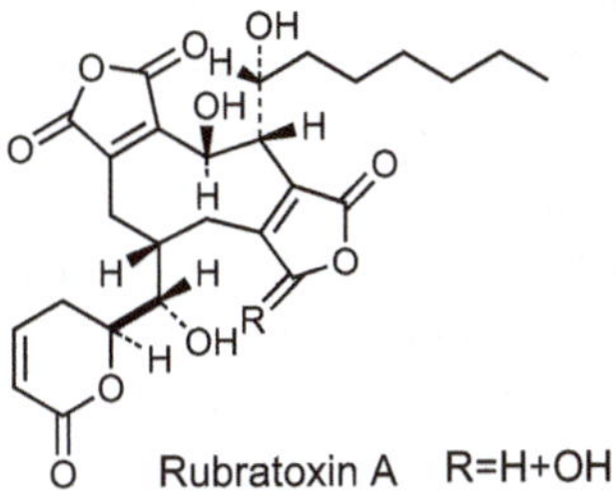

Rubratoxin A R=H+OH
Rubratoxin B R=O

Fig. 3.9: Rubratoxins.

Acetate residues and two oxalic acid residues are involved in the biogenesis of rubratoxins. They are probably dimers of C_{13} monomers [**14**].

Rubratoxins have predominantly a liver-damaging effect [54]. They induce apoptosis of liver cells and the secretion of various cytokines [83]. Prenatal exposure results in developmental delays and behavioral changes in newborn mice [103]. The LD_{50} for rubratoxins in mice is 120 mg/kg BW, p.o., 3.5 mg/kg BW, i.p. (solution in propylene glycol), in the rat 0.35 mg/kg BW, i.p. (solution in DMSO [**14**]). It is not known whether rubratoxins are of toxicological relevance for humans.

3.3.9 Alternaria Toxins

Alternaria species produce a variety of mycotoxins (overviews: [2, 7, 17, 70]). They include tenuazonic acid (TeA, N-containing), tentoxin (TEN), the dibenzo-α-pyrones alternariol (AOH), alternariol monomethyl ether (AME, heptaketide), and altenuene, and the perylene quinones altertoxin I (ATX-I, nonaketide), altertoxin II, altertoxin III, and alterperylenol (ALP) that possess reactive epoxide moieties (Fig. 3.10).

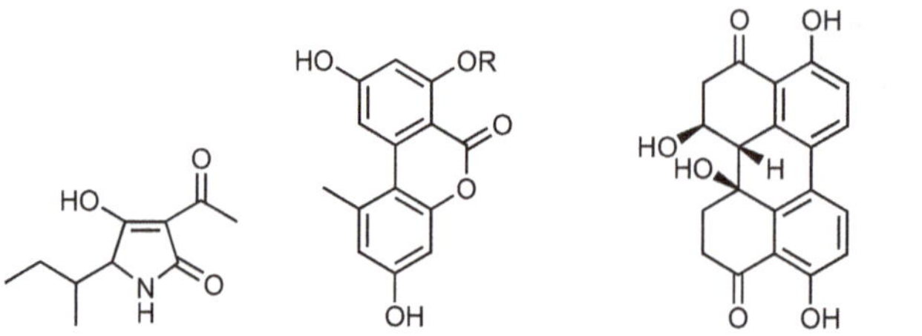

Tenuazonic acid (TeA)

Alternariol (AOH) R=H
Alternariol monomethyl ether (AME) R=CH_3

Altertoxin I (ATX-I)

Fig. 3.10: Alternaria toxins.

The compounds are formed by *Alternaria alternata* (Fr.) Keissl. (*A. tenuis* Nees) and other Alternaria species. The genus Alternaria occurs mainly as a plant parasite, e.g., in tomato fruits. High levels of the toxins have been found in tomatoes, apples, citrus fruits, and fruit juices. Tomato sauces collected from an Austrian retail market contained AOH, AME, TeA, and TEN in concentrations up to 20, 4, 322, and 0.6 ng/g each, while sunflower seed oil and wheat flour samples were contaminated at lower levels [92]. The mentioned compounds were also detectable in 76–100% of dust samples collected at Norwegian industrial grain mills and animal feed mills with maximum concentrations of up to 434 µg/kg. The perylene quinones are rarely found in food [70, 114]. Alternaria toxins, especially TeA (maximal 1,174 µg/kg), were found in all tested samples of hops collected in Croatia. It is unclear if they are transferred to beer [102]. Alternaria toxins occur often as masked mycotoxins (see Section 3.1 [7]).

The biogenesis of alternariol takes place by a polyketide synthase, a ketoreductase, an enoyl reductase, and a cytochrome P450 monooxygenase. The gene cluster for the biogenesis of tenuazonic acid includes genes for a non-ribosomal peptide synthetase, a polyketide synthase, and a thioesterase. Precursor is 1,8-dihydroxynapthalene [7].

The activities of Alternaria toxins and their toxicological significance for humans are insufficiently investigated. Genotoxic, mutagenic, cytotoxic (especially of the perylene derivatives), endocrine disruptive, and immunomodulatory effects of isolated toxins and complex extracts have been shown in vitro and in vivo. Animal assays show an accumulation of short-chain acylcarnitines in the liver and a decreased riboflavin absorption in the gastrointestinal tract leading to decreased β-oxidation and decreased energy supply [2, 89]. Tenuazonic acid generates internal hemorrhages, e.g., in dogs and chicken, and reduces feed efficiency in mice. Alternariol and its methyl ester are reported to possess mutagenic and carcinogenic properties [7].

Because of insufficient data, the EFSA did not establish maximal level until now. But monitoring of the toxins is recommended.

3.3.10 Ochratoxins

3.3.10.1 Chemistry, Biogenesis, and Occurrence

Ochratoxins, of which six are known, possess a 7-carboxy-3,4-dihydro-3*R*-methyl-isocoumarin base body that may be linked via the carboxy group to L-phenylalanine in an amide-like fashion (Fig. 3.11, overviews: [85, 94]). Of this group, ochratoxin A is of particular importance. It is a colorless substance, relatively stable on heating, whose salts are readily soluble in water. In 1965, it was isolated as the first representative of the ochratoxins by Van Der Merve and coworkers [126].

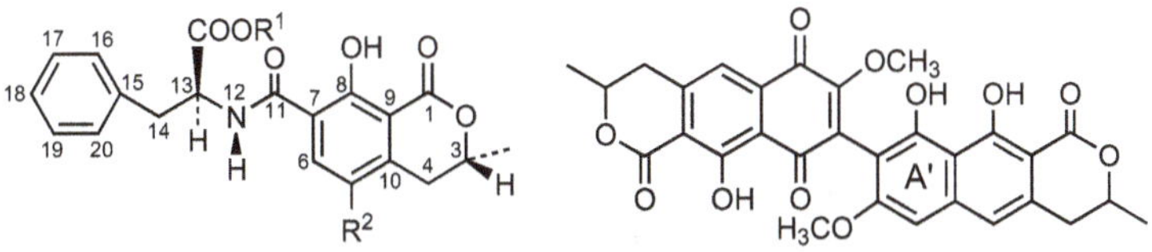

Ochratoxin A R^1=H, R^2=Cl
Ochratoxin B R^1=R^2=H
Ochratoxin C R^1=C_2H_5, R^2=Cl

Viomellein
Xanthomegnin (Ring A' p-chinoid)

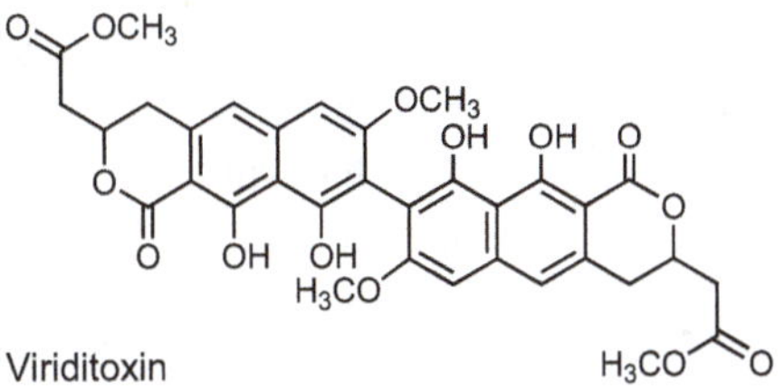

Viriditoxin

Fig. 3.11: Ochratoxins and related compounds.

The biogenesis of the base body of ochratoxins takes place from five acetate units (pentaketides). The formation of ochratoxin A includes polyketide synthases, non-ribosomal peptide synthetases, and, for the final step, a chloroperoxidase. Encoding genes have been identified in *A. carbonarius* (Bainier) Thom [8].

Producers of ochratoxins are Aspergillus species, especially *Aspergillus ochraceus* G. Wilh., and Penicillium species, especially *Penicillium verrucosum* Dierckx. The ochratoxin formation by *A. ochraceus* occurs at temperatures between 20 and 30 °C and at least 40% water content of the substrate. The ochratoxin-forming Penicillium species already thrive well at temperatures of 5–10 °C. They are therefore, in contrast to the heat-loving Aspergillus species, of particular importance for the contamination of foodstuffs in colder climates. Mention should also be made of *Aspergillus glaucus* as an ochratoxin producer. It already grows at very low water contents of the substrates.

Ochratoxins, predominantly ochratoxin A, and less frequently its less toxic dechloroanalogue ochratoxin B, are widespread in foods of plant and animal origin. They could be detected mainly in cereals and cereal products, but also in peanuts, other legumes, coffee beans, cacao beans, grapes, vine, and dark chocolate. They were also found in meat of pigs and chickens that had received feed containing ochratoxin A, thus also in sausage products (contamination rate 16–19%, content 0.1–920 µg/kg, [66]). Out of 180 maize samples in Ghana, 103 were positive for ochratoxins with levels from 4.00 to 97.51 µg/kg for ochratoxin A [1]. When barley containing ochratoxin A was used for the brewing process, 28–39% of the mycotoxin was detectable in the beer. Of 219 beer samples tested, 41% were contaminated with ochratoxin A [26]. The contamination rate of dried fruits, e.g., figs and raisins, ranged from 20% to 95%, with values up to 53.6 µg/kg ochratoxin A [27].

About 54% of the total ochratoxins ingested in Europe come from cereal products, 15% from red wine, 12% from coffee, 7.6% from beer, and the remainder, e.g., from pork or spices [10]. In Germany, ochratoxin A can be detected in almost all human blood samples (detection limit 60 ng/ml). The mean value is 0.27 µg/ml [98]. In the examination of breast milk, 11% positive samples were found with a content of 17–30 pg/g ([80, **26**], overview occurrence: [8, 85]).

Viomellein, xanthomegnin, viriditoxin, and related compounds are also dihydroisocoumarin derivatives (Fig. 3.11). Viomellein is formed by, e.g., *Aspergillus ochraceus, A. sulphureus* Desm., and *Penicillium aurantiogriseum* (*P. viridicatum* Westling), viriditoxin by *A. viridinutans* Ducker & Thrower and *A. brevipes* G. Sm., and xanthomegnin by Trichophyton, Aspergillus, and Penicillium species.

These fungi thrive on similar substrates as the ochratoxin formers. The compounds, along with citrinin, kojic acid, or penicillic acid, are often found accompanying ochratoxins. In wheat and barley, 1,100 µg/kg xanthomegnine and 1,800 µg/kg viomellein were found [99].

3.3.10.2 Pharmacology and Toxicology

Ochratoxin A is rapidly absorbed after peroral or inhalational uptake and is bound with high affinity to blood plasma albumin. This and the uptake into the enterohepatic circulation result in a very long residence time of ochratoxin in the organism (half-life in the human body 35 days) and an accumulation in the food chain. Various CYP450 isozymes are involved in the metabolization to 4-hydroxy-ochratoxin and other metabolites. CYP450 polymorphisms explain the different susceptibility of individual populations to ochratoxin A [33]. In monkeys, the elimination half-life was 19–21 days [109].

Mechanisms of action of ochratoxins include induction of oxidative stress, inhibition of protein biosynthesis, blockade of the respiratory chain, disruption of cellular calcium homeostasis, and formation of DNA adducts leading, for example, to single-strand breaks [**116**]. Epigenetic changes have been found [55, 68].

Ochratoxin A preferentially attacks the kidney and the liver. In the kidney, there is degeneration of tubules, interstitial fibrosis, and, at a late stage, hyalinization of glomeruli. Inhibition of glucose transport, of gluconeogenesis, and of glycogen synthesis and promotion of glycogen degradation occur in the liver. In addition, genotoxic, carcinogenic (Tab. 3.1), teratogenic, and immunosuppressive effects are of toxicological significance [111, **14, 26**].

The LD_{50} of ochratoxin A is 22 mg/kg BW, i.p., mice, 14 mg/kg BW, i.p., rat, and 9 mg/kg BW, p.o., guinea pig [134].

Viomellein, xanthomegnin, and viridotoxin are also renal and liver toxicants. Xanthomegnin has been shown to affect the respiratory chain. The LD_{50} of viridotoxin in mice is 2.8 mg/kg BW, i.p. [**14**].

Correlations between food contamination with ochratoxin A, blood ochratoxin A levels, and an increased incidence of kidney diseases including renal cell carcinoma have been demonstrated in humans [32, 135]. But the Balkan endemic nephropathy that was formerly attributed to ochratoxin A is considered today to be multifactorial and mainly caused by aristolochic acids (see [**161 = Vol 3**]). Advices on a possible promotion of Alzheimer disease by ochratoxin A exist [87].

Ochratoxin poisoning on a large scale occurred mainly in pigs and poultry [11, 57].

For ochratoxin A, the following maximum levels are currently set in the EU: 3 µg/kg in cereal products, 10 µg/kg in dried grapes, 2 µg/L in wine, 5 µg/kg in soluble coffee, 3 µg/kg in roasted coffee, 2 µg/kg in dried fruits (except grapes and figs), and 0.5 µg/kg in food for infants. Low limits are also set for different spices and herbal drugs, e.g., licorice [8, **39**].

3.3.11 Cytochalasans

Cytochalasans (Fig. 3.12) have a perhydro- or hexahydroisoindole ring system as an essential structural element. It contains in position 3 either a benzyl residue (cytochalasins and zygosporins) or a 3-methyl-indolyl residue (chaetoglobosins). In positions 8 and 9 are bound: a 11-membered (e.g., zygosporin E) or a 13-membered (e.g., chaetoglobosin A) alicyclic ring or a 14-membered lactone ring (e.g., cytochalasin B). More than 400 representatives of this group are known. Cytochalasans also include, among others, the cytotoxic and antibiotic alachalasins, pyrichalasins, chaetoconvosins, and trichalasins (overviews: [31, 59, 104, 141, **14, 138, 139**]). Because of their N-heterocyclic character, cytochalasans can also be classified as alkaloids.

The formation of cytochalasans has been demonstrated under in vitro conditions in different taxonomic orders, some of which are unrelated, including the genera Aspergillus, Chaetomium, Diaporthe, Helminthosporium, Penicillium, Phoma, Phomop-

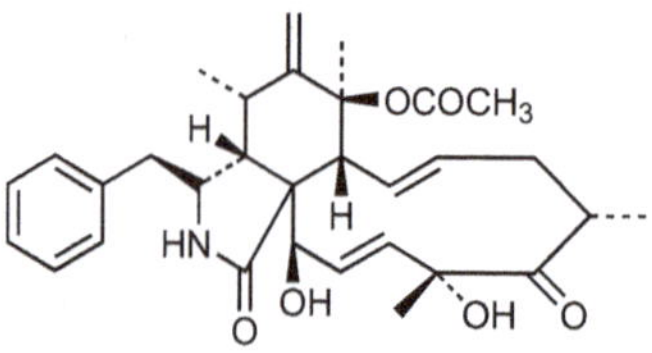

Zygosporin E

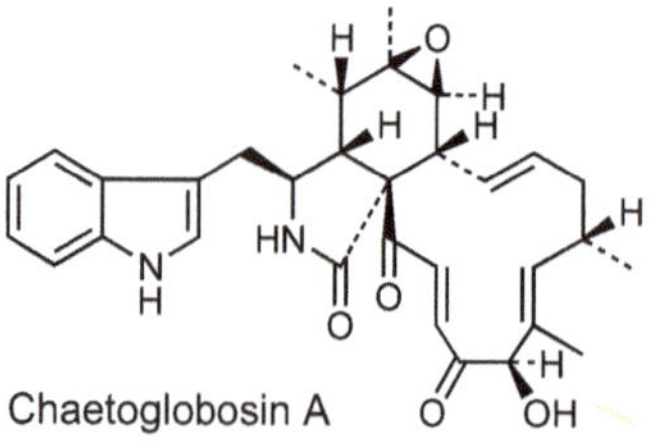

Chaetoglobosin A

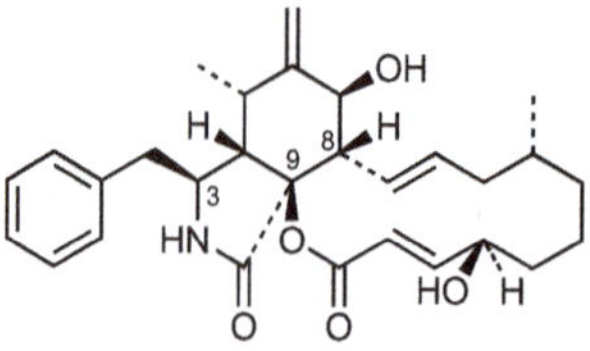

Cytochalasin B

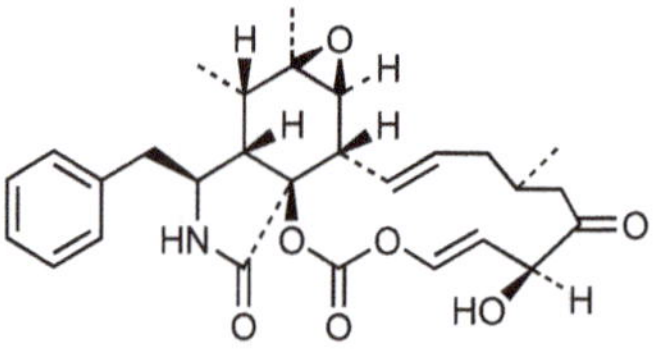

Cytochalasin E

Fig. 3.12: Cytochalasans.

sis, Rosellinia, and Zygosporium [59]. Largest proportion of compounds is described for the genera Chaetomium and Diaporthe [50].

Cytochalasans have rarely been isolated from food.

The biogenesis of the hybrid compounds occurs in a concerted action by a polyketide synthase and a non-ribosomal peptide synthetase (PKS-NRPS) through fusion of a polyketide chain with an amino acid-derived building block. The resulting acyclic precursor undergoes a Diels–Alder cyclization. The formed tricyclic core structure of the cytochalasans is further modified by oxidative rearrangements and cationic cyclizations [50, 107].

Cytochalasans exhibit a broad activity spectrum. Most prominent is the interaction with F-actin filaments resulting in inhibition of F-actin polymerization [59]. This leads to changes in all cellular processes where actin is involved, e.g., cellular motility including endocytosis, cytokinesis, and migration. Structure–activity-relationship investigations show the importance of the hydroxy group at C-7 and the NH function at

N-2 for the inhibition of actin polymerization [50]. Via prevention of cell division without disruption of nuclear division, multinucleated cells are formed.

Actin-independent activities are, e.g., inhibition of transporters such as the human glucose transporter 1 (hGLUT1), inhibition of development of biofilms of prokaryotes, antiviral, and anti-inflammatory activities [59].

The LD_{50} of cytochalasin E in rats is 9.1 mg/kg BW, p.o., and 2.6 mg/kg BW, i.p., and that of chaetoglobosin A is 6.5 mg/kg BW. s.c., in male mice and 17.8 mg/kg BW, s.c., in female mice [**14**].

The toxicological significance of cytochalasanes for humans is low. However, they can be used as tools for cytological investigations and are of interest because of their biofilm-inhibiting activities.

3.3.12 Fumonisins

3.3.12.1 Chemistry, Biogenesis, and Occurrence

The fumonisins have a C_{20} backbone formed from C-3 to C-20 on the acetate pathway, the C atoms 1 and 2 and the amino group at C-2 are derived from alanine. The C-chain also carries a variety of hydroxy, methyl, and tricarballyl ester groups. According to the differences in the nitrogen function and the length of the carbon backbone, the approx. 30 reported fumonisins are divided into four groups: A, B, C, and P, with B type fumonisins being the most common. The main representative fumonisin B1 (FB1, macrofusin, Fig. 3.13) accounts for 70–80% of all fumonisin contaminations. The less oxygenated representatives FB2, FB3, and FB4 are probably precursors of the more oxygenated FB1. Fumonisins are easily modified, e.g., during processing of food [63, 64].

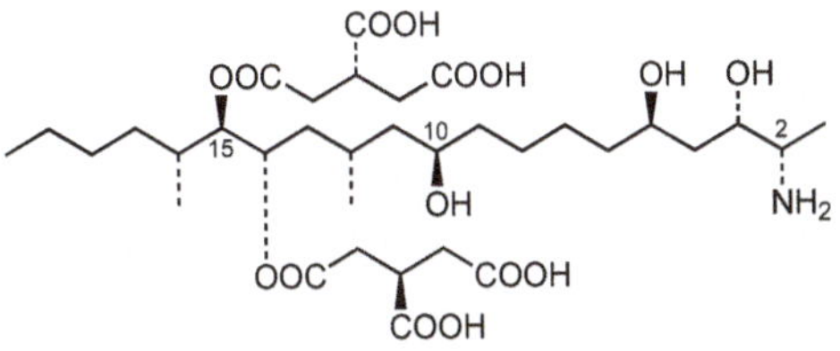

Fumonisin B1

Fig. 3.13: A representative of fumonisins.

The production of fumonisins is regulated by 17 coordinated genes [63, 93].

Since the discovery of the first fumonisin in South Africa in 1988 [75], intensive research began on these compounds, which are now among the most important mycotoxins.

Important producers are *Fusarium fujikuroi* Nirenberg (*F. moniliforme* J. Sheld.) and *F. proliferatum* (Matsush.) Nirenberg. Fumonisin production occurs primarily on corn cobs. Contamination has been detected worldwide. According to WHO about 50% of global maize and maize-based products are contaminated by FB1 to various extents. In particular, fumonisin contamination has become a critical issue in Sub-Saharan Africa, South and Latin America, and South/Southeast Asia. In South Africa, approx. 90% of maize contained FB1 at levels up to 118 mg/kg [18, 64, 136]. A mean level of 200 µg/kg was detected in maize grown in Central Europe [79]. On the basis of global occurrence data reported between 2007 and 2017, the incidence and maximum levels of fumonisins in raw cereal grains were 61% and 71,121 µg/kg [60]. The percentage of animal feeds contaminated by fumonisins was 47% in the Middle East and North Africa during the years 2012–2020 and 20–40% in animal feeds from Italy over the 5 years, 2018–2022 [67]. Heat treatment, discarding the cooking water, and other preparation measures can reduce fumonisin levels [79]. Fumonisins have also been detected in post-harvest fruits, e.g., in figs [64].

3.3.12.2 Pharmacology and Toxicology

Due to their structural similarity with sphinganine, the natural precursor of sphingomyelins, cerebrosides, and gangliosides, the fumonisins inhibit sphinganine *N*-acyltransferase (ceramide synthase, CerS) and thus the biosynthesis of sphingolipids, which are important as membrane building blocks and regulatory substances. This results in the accumulation of free sphingoid bases, especially sphinganine, disruption of membrane assembly and function of membrane proteins, and induction of oxidative stress and apoptosis [5, 18, 36]. The activation of the inflammasome NLRP3 [67] and epigenetic changes [5] contribute to the toxicity. The interference with the folate-binding protein (human folate receptor α) results in folic acid deficiency and consequent disturbances in embryonic development [71, 76].

The bioavailability of fumonisins after peroral intake is low. The compounds are degraded in the gastrointestinal tract into their partially hydrolyzed forms, most likely by the resident microbiota [105].

Fumonisin B1 has toxic effects on kidney, liver, nervous system, respiratory system, digestive system, immune system, and reproductive system [18, 21, 36] as well as mitogenic and tumor-promoting properties. The fumonisins are classified by IARC as 2 B carcinogen 'as possibly carcinogenic to humans' (Tab. 3.1). Chromosomal aberrations have been induced by high concentrations of fumonisin B1 (5–10 µg/g) in human lymphocytes [61]. Maternal FB1 can cause growth retardation and developmental abnormalities in the embryos of hamsters, rats, mice, and chickens [71].

The high incidence of neural tube defects and of other anomalies occurring in some countries (Guatemala, South Africa, China) may be due to the consumption of corn products containing fumonisins in the first months of pregnancy [4, 71, 76]. The increased incidence of esophageal cancer in regions of China, Iran, Mozambique, and South Africa may also be triggered by foods made from corn contaminated with fumonisins [20, 75, **116**].

In horses, feeding of moldy corn containing fumonisins induces a neurological disease, leukoencephalopathy (ELEM), which is often fatal. In experiments, this disease could be provoked by injection of fumonisin B1. In pigs, fumonisins have been blamed for pulmonary edema ('porcine pulmonary edema', PPE) [41, 64, 75, 95, 130].

The EFSA Panel on Contaminants in the Food Chain suggests a tolerable daily intake of 1.0 µg FB1/kg BW per day. The EU specifies a maximum level for total fumonisins (FB1 and FB2) in different human foods at between 200 µg/kg food for infants and young children to 4,000 µg/kg raw-unprocessed maize. The FDA specifies a maximum level of 2,000 µg/kg for maize and maize-based products intended for human consumption and higher values for animals. According to EU the maximum FB1 and FB2 concentration in swine and horse feed must not exceed 5 mg/kg feed and 60 mg/kg maize and maize products [64, 105, 112, **39**].

References

For numbers in bold, see cross-chapter literature p. 233.

[1] Ahiabor WK et al. (2024) Environ Health Insights 18: 1
[2] Aichinger G et al. (2021) Compr Rev Food Sci Food Saf 20(5): 4390
[3] Alameri MM et al. (2023) Toxins 15: 246
[4] Alvito P, Pereira-de-Silva L (2022) Toxins 14: 189
[5] Arumugam T et al. (2021) Crit Rev Toxicol 51(1): 76
[6] Asao T et al. (1965) J Am Chem Soc 87: 882
[7] Bacha SAS et al. (2023) Front Plant Sci 14: 1139757
[8] Ben Miri Y et al. (2024) Foods 13: 1184
[9] Benkerroum N (2019) Int J Envion Res Public Health 16(19): 3633
[10] Blank R. (2002) Z Umweltchem Ökotox 14: 104
[11] Bonerba E et al. (2024) Toxins 16: 487
[12] Brückner L et al. (2024) Mycotoxin Res 40: 709
[13] Butler WH (1974) in: Mycotoxins. Purchase JFH (ed), Elsevier, Amsterdam, p. 1
[14] Caceres I et al. (2020) Toxins 12: 150
[15] Cao W et al. (2022) Toxicol Mech Methods 32(6): 395
[16] Chávez R et al. (2023) J Fungi 9: 459
[17] Chen A et al. (2021) J Agric Food Chem 69(28): 7817
[18] Chen J et al. (2021) Molecules 26(17): 5238
[19] Chen T et al. (2022) Nutrients 14: 1027
[20] Come J et al. (2019) Pan African Med J 33: 187
[21] Creppy EE (2002) Toxicol Lett 127: 19
[22] Da Silva JVB et al. (2021) Molecules 26: 7312
[23] Dai Y et al. (2017) Food Chem Toxicol 109(Pt1): 683
[25] Degen GH et al. (2023) Toxins 15: 26; Correction: Toxins 15: 322
[26] Deutsche Gesellschaft für Ernährung (DGE e.V., 1996) Ernährungsbericht: 131
[27] Drusch S, Aumann J (2005) Adv Fruit Nutr Res 50: 33
[28] Dubey MK et al. (2018) Front Pharmacol 9: 288

[29] Eaton DL, Groopman JD (1994) The Toxicology of Aflatoxins: Hum Health, Veterinary, and Agricultural Significance, Acad Press, San Diego
[30] Elgioushy MM et al. (2020) Environ Sc Pollut Res Int 27(28): 35526
[31] Evidente A et al. (2003) J Nat Prod 66: 1540
[32] Fink-Gremmels J (1994) Ernährungsumsch 41: 226
[33] Fink-Gremmels J et al. (1995) Natural Toxins 3: 214
[34] Florea et al. (2017) Phytopathology 107(5): 504
[35] Fontaine KI et al. (2015) Food Control 47: 634
[36] Gao Z et al. (2023) Environ Pollut 320: 121065
[37] Gedek B (1983) Kompendium der Medizinischen Mykologie, Parey, Berlin, p. 283
[38] Gomes de Oliveira Filho JW et al. (2017) Food Chem Toxicol 110: 130
[39] Groopman JD et al. (1988) CRC Crit Rev Toxicol 19: 113
[40] Han X et al. (2022) Toxins 14: 386
[41] Haschek WM et al. (2001) Environm Health Perspect Suppl 109: 251
[42] Henry SH et al. (2002) Adv Exp Med Biol 504: 229
[43] Hou HF et al. (2015) PLoS ONE 10(5): e0125956
[44] Hsu P et al. (2024) Food Chem Toxicol 193: 115008
[45] Hu T et al. (1995) J Chem Soc Chem Commun: 2159
[46] Ismail A et al. (2019) Aflatoxins in Plant-based Foods: Phytochemistry and Molecular Aspects. In: Plant and Human Health, Vol. 2, Ozturk M, Hakeem KR (eds), Springer Nature Switzerland AG, p. 313
[47] Ivers RS et al. (1994) J Am Chem Soc 116: 1603
[48] Jiang Z et al. (2024) Front Microbiol 15: 1460690
[49] Jijima H et al. (1983) Chem Pharm Bull (Tokyo) 31: 362
[50] Kagho MD et al. (2024) J Nat Prod 87: 2421
[51] Kamdem LK et al. (2006) Chem ResToxicol 19: 577
[52] Kamle M et al. (2022) Toxins 14: 85
[53] Kang YW et al. (2022) Food Addit Contam Part A 39(2): 373
[54] Keiko I, Hitoshi N (2005) Mycotoxins 55: 35
[55] Khoi CS et al. (2021) Int J Mol Sci 22(20): 11237
[56] Krishnamachari KAVR et al. (1975) Lancet 1: 1061
[57] Krogh P (1978) Acta Pathol Microbiol Scand Sect A Suppl No 269
[58] Kyei NNA et al. (2020) Mycotoxin Res 36: 243
[59] Lambert C et al. (2023) Biomolecules 13: 1247
[60] Lee HJ, Ryu D (2017) J Agric Food Chem 65(33): 7034
[61] Lerda D et al. (2005) Food Chem Toxicol 43: 691
[62] Li L et al. (2021) Ecotoxicol Environ Saf 221: 112457
[63] Li M et al. (2020) Nat Prod Rep 37: 1568
[64] Li T et al. (2024) J Adv Res 60: 13
[65] Li W et al. (2024) J Fungi 10: 396
[66] Li X et al. (2021) Molecules 26(22): 6928
[67] Liao C et al. (2024) Vet Sci 11: 291
[68] Liu WC et al. (2022) Toxicon 213: 59
[69] Llewelly GC et al. (1998) Food Chem Toxicol 36: 1107
[70] Louro H et al. (2024) Arch Toxicol 98: 425
[71] Lumsangkul C et al. (2019) Toxins 11: 114
[72] Mafe AN, Büsselberg D (2024) Foods 13(21): 3502
[73] Mahato DK et al. (2021) Toxicon 198: 12
[74] Mahato DK et al. (2021) Toxins 13(2): 92
[75] Marasas WFO (2001) Environm Health Persp Suppl 109: 239

[76] Marasas WFO et al. (2004) S Afr J Nutr 134: 711
[77] Matthiaschk G (1990) Bundesgesundheitsbl 12: 581
[78] Mazibuko M et al. (2024) Arch Toxikol 98: 2143
[79] Meister U, Springer M (2004) J Appl Bot Food Quality 78: 168
[80] Miraglia M et al. (1995) Natural Toxins 3: 436
[81] Moloi TP et al. (2024) Toxicology 509: 153983
[82] Müller T (1987) Nahrung 31: 117
[83] Nagashima H (2001) Biochem Biophys Res Commun 287: 829
[84] Nahle S et al. (2020) Eur J Plant Pathol 159: 247
[85] Nazareth TDM et al. (2024) Foods 13: 1920
[86] Newberne PM, Butler WH (1969) Cancer Res 29: 236
[87] Niaz K et al. (2020) Environ Sci Pollut Res Int 27(36): 44673
[88] Nurshad A (2018) J Sci Food Agric 98(6): 2055
[89] Peach JT et al. (2024) Arch Toxicol 98: 3477
[90] Pohland AE (1993) Food Addit Contam 10: 17
[91] Probst C et al. (2007) Appl Environ Microbiol 73(8) 2762
[92] Puntscher H et al. (2018) Anal Bioanal Chem 410: 4481
[93] Qu Z (2024) mLife 3(2): 176
[94] Ráduly Z et al. (2020) Front Microbiol 10: 2908
[95] Rao ZX et al. (2020) Toxins 12(9): 604
[96] Rogowska A et al. (2019) Toxicon 162: 46
[97] Ropejko K, Twaruźek M (2021) Toxins 13: 35
[98] Rosner H et al. (2000) Arch Lebensmittelhyg 51: 104
[99] Saidomore KA et al. (1986) J Stored Prod Res 22: 81
[100] Saleh I, Goktepe I (2019) Food Chem Toxicol 129: 301
[101] Sargeant K et al. (1961) Nature 192: 1096
[102] Šarkanj ID et al. (2024) Toxins 16: 293
[103] Sava V (2004) Gene Expr 11: 211
[104] Scherlach K et al. (2010) Nat Prod Rep 27(6): 869
[105] Schertz H et al. (2018) Toxins 10: 150
[106] Silva LJG et al. (2021) Foods 10: 14
[107] Skellam E (2017) Nat Prod Rep 34(11): 1252
[108] Smith LE et al. (2017) Am J Trop Med Hyg 96(4): 770
[109] Stander MA et al. (2001) Arch Toxicol 75: 262
[110] Stob M et al. (1962), Nature 196: 1318
[111] Stoev SD (2022) Toxins 14: 380
[112] Stoev SD (2023) Toxins 15: 464
[113] Stopper H et al. (2005) Mutat Res 574: 139
[114] Straumfors A et al. (2015) World Mycotoxin J 8(3): 361
[115] Stroe MC et al. (2024) Mol Microbiol 121(1): 18
[116] Stupin Polančec D et al. (2024) Toxins 16: 321
[117] Tandon HD et al. (1978) Arch Pathol Lab Med 102: 372
[118] Thorpe CW (1982) in: Environmental Carcinogens: Selected Methods of Analysis. Egan H (ed), IARC Publ 44, Lyon, p. 311
[119] Trucksess MW et al. (1982) J Assoc Off Anal Chem 65: 884
[120] Turna NS et al. (2024) Crit Rev Food Sci Nutr 64(27): 9955
[121] Ueno I et al. (1995) Free Radical Res 23: 41
[122] Ueno Y (1982) in: IARC: Environmental Carcinogens: Selected Methods of Analysis. Egan H (ed), Lyon, p. 399

[123] Ueno Y (1985) CRC Crit Rev Toxicol 14: 99
[124] Urry WH et al. (1966) Tetrahedron Lett: 3109
[125] Van der Hoeven JCM et al. (1984) Mutat Res 136: 9
[126] Van der Merve KH et al. (1965) Nature 205: 1112
[127] Vanhee C et al. (2024) Foods 13: 1919
[128] Vidal A et al. (2019) Food Chem Toxicol 129: 249
[129] Visconti A (2003) Mycotoxicology Newsletter VII: (1). www.mycotoxicology.org
[130] Voss KA et al. (2001) Environm Health Persp Suppl 109: 259
[131] Wang F et al. (2024) Biosensors 14(7): 322
[132] Wang JS, Groopman JD (1999) Mutat Res 424: 167
[133] Wei C et al. (2020) Toxicon 184: 83
[134] WHO Genf (1979) Mycotoxins: Environ Health Criteria 11. www.who.int/publications/i/item/9241540710 (accessed 04/14/2024)
[135] Więckowska M et al. (2024) Cancers (Basel) 16(20): 3473
[136] Yli-Mattila T, Sungheim L (2022) Toxins 14: 419
[137] Zentai A et al. (2023) Toxins 15: 195
[138] Zhang H et al. (2021) Food Res Int 141: 110075
[139] Zhen H et al. (2024) Food Addit Contam Part A 41(10): 1344
[140] Zhong L et al. (2018) Toxins 10(11): 475
[141] Zhu H et al. (2021) Prog Chem Org Nat Prod 114: 1
[142] Zingales V et al. (2020) Food Chem Toxicol 146: 111802

3.4 Alkaloids as Mycotoxins

3.4.1 Ergoline Alkaloids as Toxins of Ergot and Ergot Relatives (Claviceps Species)

The representatives of the family Clavicipitaceae, order Hypocreales, are fungi of the phylum Ascomycota, class Sordariomycetes. They are preferentially parasites of sweet grass species and cereals. About 400 of the about 600 genera of sweet grasses (Poaceae) can be infected by the ergot fungus, additionally some rushes (Juncaceae), sedges (Cyperaceae), representatives of greenbriers (Smilacaceae), and pine plants (Pinaceae) [8, 32].

Claviceps species are distinguished by the plants that they infect, the shape of the fruit bodies (perithecium), the wintering forms (sclerotia), and the forms of their spores (conidia). Because these characteristics do not always allow a clear differentiation of the clans from each other, the data of the number of the assumed species varies widely (up to 50 species).

Claviceps purpurea (Fr.) Tul., ergot fungus (Ph. 3.4), *C. paspali* F. Stevens & J.G. Hall., and *C. fusiformis* Loveless are of high toxicological interest.

Claviceps purpurea grows frequently on Poaceae of the tribes Hordeeae (important cereal and forage grass-producing clades), Aveneae (oats are rarely infested), Festuceae, and Agrosteae. Very important is its frequent occurrence on rye, triticale, wheat, and on barley. Infected grasses on roadsides and on field margins are probably starting points for infestation of cereal plants on fields. The fungus accompanies rye in all its growing regions and is also found on wild grasses worldwide. *C. paspali* lives

Ph. 3.4: *Claviceps purpurea*, ergot (source: Michael Hoffmann).

almost exclusively on the tropical genus Paspalum, Poaceae. *C. fusiformis* parasitizes on tropical Pennisetum species [8].

Other Clavicipitaceae species also live as endophytes or epiphytes of plants, e.g., in the bindweed family, Convolvulaceae (see **[161 = Vol. 3**, 84]).

The cycle of infection with *Claviceps purpurea* begins with the germination of the ascospores, spread by the wind, on unfertilized ovaries of flowering grasses. The developing hyphae of the fungi overgrow and penetrate the ovaries of the floret within 6–8 days. From about the tenth day after infection, the sugary sieve tubes of the infected ovaries begin to seep out their sap at the upper end of the ovaries mixed with conidia which are released by constriction of the hyphae. This colorless sticky fluid is called honeydew. By running down, splashed by the wind or by spreading by insects, the honeydew infects further ovaries of the host. After completion of the honeydew formation, the cotton like mycelium is transformed into a spindle-shaped, brown to blackish-purple overwintering form (sclerotium). It consists of tightly intertwined hyphae formed by the glumes of the florets of the host plant and reaches different lengths depending on the type of host plant, a few millimeters in grasses, up to approx. 3 cm in rye, and up to approx. 8 cm in corn. The sclerotia fall down during the harvest of the cereals or spontaneously in the autumn and overwinter on the ground. After a cold period in the winter, 6–15, rarely up to 60, pink little mushroom-shaped stromata grow out of the sclerotium in the next spring. The flask-shaped perithecium (closed fruit bodies) contains, embedded in a sterile palisade-like layer (hymenium), eight meiotically derived ascospores in elongated, conical tubes (asci). The spores may be injected into the air to initiate new infections. In addition, asexual reproduction is possible by constriction of exospores (conidia) from the hyphae [8, 24, 32].

As a result of intensive seed cleaning and of the reduction of the areas of field margins and roadsides as reservoirs of dropped down sclerotia and of honeydew producing grasses, the incidence of ergot infections declined.

It is noteworthy that ergot alkaloids can be transferred from the infected ovary to other plant parts so that even non-infected healthy grains of the infected rye plant can contain small, toxicologically harmless quantities of the alkaloids [92].

A large part of the ergot can be removed from the harvested grains by processing and by the milling process. The remaining alkaloid content is further reduced (more than 20–50%) during baking, frying, or by ammonization [82].

The alkaloid content of the ergot sclerotia is up to 1%. Which of the about 65 known alkaloids quantitatively dominates depends mainly on the species and race of the fungus and on the kind of the infected plant.

Sclerotia of *Claviceps purpurea* contain almost exclusively ergopeptines, especially ergotamine. Clavine alkaloids (Fig. 3.14) occur in very small amounts or are absent. However, clavine alkaloids are often the main alkaloids in saprophytic ergot cultures.

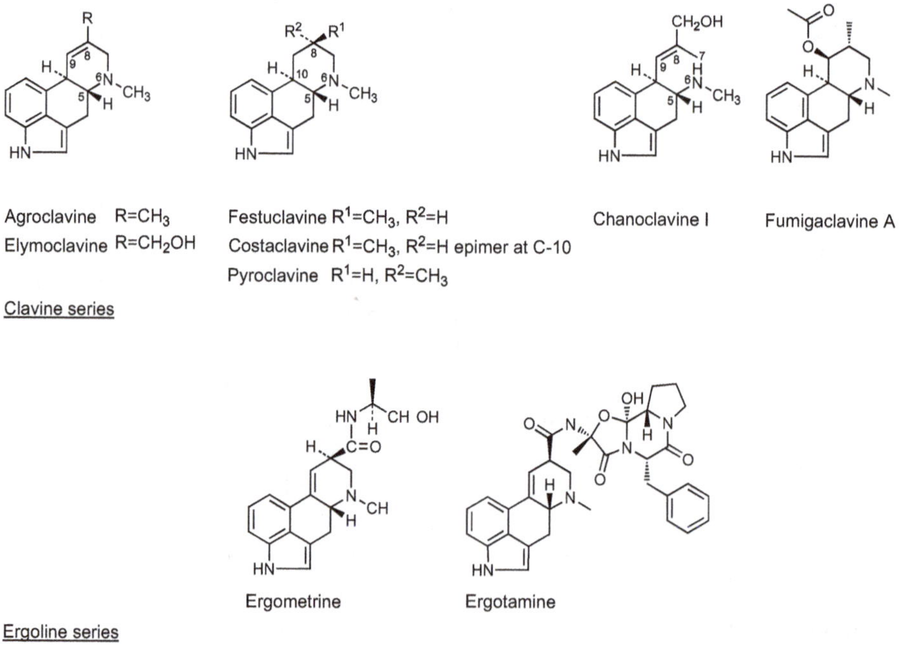

Agroclavine $R=CH_3$
Elymoclavine $R=CH_2OH$

Festuclavine $R^1=CH_3$, $R^2=H$
Costaclavine $R^1=CH_3$, $R^2=H$ epimer at C-10
Pyroclavine $R^1=H$, $R^2=CH_3$

Chanoclavine I

Fumigaclavine A

Clavine series

Ergometrine

Ergotamine

Ergoline series

Fig. 3.14: Ergoline alkaloids of *Claviceps purpurea*.

C. paspali preferentially forms simple amides of lysergic acid, paspalic acid (Fig. 3.15), and 10-hydroxypaspalic acid, e.g., lysergic acid-α-hydroxyethylamide, ergine (lysergic acid amide), and 10-hydroxy-paspalic acid amide. In addition, diterpene indole alkaloids occur, some of which with tremorgenic effects. *C. fusiformis* produces mainly clavine alkaloids ([24], for details of chemistry and pharmacology of ergot alkaloids, see [**161 = Vol. 3**]).

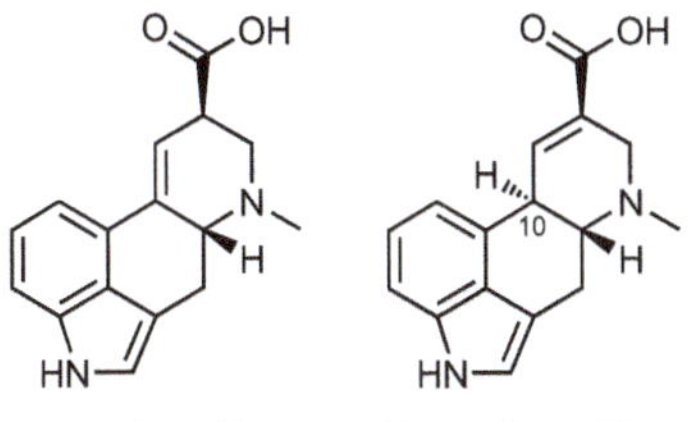

Lysergic acid Paspalic acid

Fig. 3.15: Base bodies of ergoline amides of *Claviceps paspali.*

The alkaloids accompanying bioactive substances in the sclerotia of ergot are numerous amines, the toxic, yellow-colored ergochromes, and some degradation products of native ergot alkaloids that are formed by other lower fungi. The ergochromones, e.g., secalonic acid A (Fig. 3.16), are C–C-linked dimers of xanthone derivatives [42, 96].

In the Middle Ages, flour from cereals containing ergot sclerotia led to large-scale poisoning, named ergotism, in all European countries. Since the end of the sixth century, about 300 mass illnesses of ergotism can be traced. Often thousands of people, mainly from the poorer population, were affected. The first well-documented epidemic was in 944–945, when about 20,000 people, living in Paris and the Aquitane region of France, were concerned. About one-half of them died. An epidemic in Limoges, a French city, in 994 caused 40,000 deaths, and an epidemic in Cambrai, a French town, in 1129 caused 12,000 deaths [32, 75].

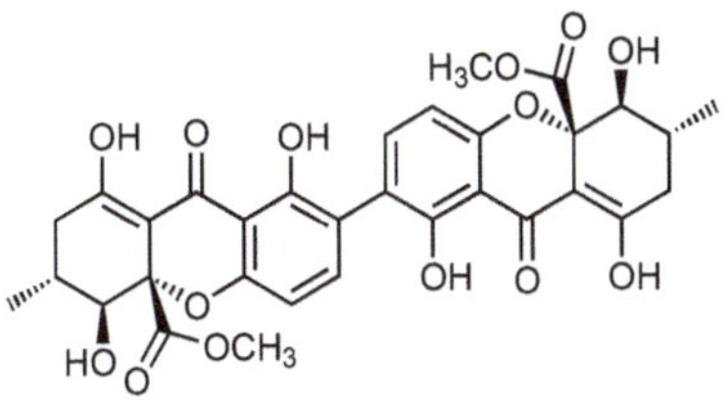

Secalonic acid A

Fig. 3.16: An ergochromon.

In 1692 in North America, women with hallucinations and strange behavior, who were possibly poisoned with ergot, were executed as witches (Witches of Salem [8, 14,

50, 75]). In ancient Greece, ergot was probably deliberately used as a hallucinogen (Eleusinian Mysteries [**147**]).

The cause of the ergotism was mostly rye flour made from grains contaminated with ergot sclerotia that was used for baking bread. Mass poisonings occurred especially in years when drought during the honeydew period favored ergot infestation (for history, see [51, 75]).

Although the first considerations about the connection of ergotism with the contamination of rye by ergot were already made in antiquity, certainty was only gained through the work of the French mycologist Louis René Tulasne (1815–1885) in the year 1853. He revealed the correlation between the infected rye and ergotism [41].

Despite this realization epidemics still occurred in the twentieth century, e.g., in 1951 in Pont Saint Esprit in France (more than 200 diseased persons) and in 1977/78 in the Wollo district in Ethiopia (93 diseased persons, including 47 deaths). Occasionally, acute poisoning occurred during abortion attempts with ergot [32].

With the increasing popularity of 'alternative' diets with consumption of cereals from plants cultivated without use of pesticides, often not sufficiently cleaned grains, and without using a baking process ('muesli'), the risk of ergot poisoning increases. Thus, deaths have occurred in younger time, e.g., in southern France [19, 38, 51, 87, 92].

A young girl who consumed 'muesli' made from cereals containing ergot sclerotia (one sample studied contained 12% ergot) every morning showed symptoms of intoxication such as headache, eye pain, and visual disturbances [67].

A case of ergot poisoning due to inhalation of grain dust is also known. The affected patient experienced pain in both feet after walking only a short distance. The elevated ergotamine level in the blood plasma did not decrease until several months after interruption of exposure [70].

The lethal dose of ergot sclerotia to humans is estimated to be 5–10 g, p.o. Chronic poisoning may occur with ingestion of as little as 10 sclerotia daily ([**43, 94**], for pharmacology and toxicology, see [**161 = Vol. 3**]).

The symptoms of acute poisoning by ergot are highly variable depending on the alkaloid quantity and the alkaloid spectrum of the ingested or inhaled material. In most cases, vomiting, diarrhea, and severe abdominal pain occur. Central manifestations include headache, dizziness, anxiety, delirium, hallucinations, and temperature elevation. Blood pressure drop may lead to collapse and cardiac arrest. In pregnant women abortion may be induced.

In acute cases of poisoning by ergot, vomiting must be induced, or gastric lavage should be performed as soon as possible. Solutions of sodium sulfate are suitable as laxative, activated charcoal as adsorbent. In severe cases, combination of hemodialysis or peritoneal dialysis with forced dialysis may be successful. In cases of vasospasms, vasospasmolytic agents must be given under blood pressure control, and hemostaseologic measures must be performed. Convulsions can be treated with diazepam under controlled intubation. Adrenalin should not be administered if blood pressure drops (adrenalin reversal!). If respiratory paralysis is imminent, intubation and ventilation should be performed. In case of micturition disorders, bladder emptying is required. To avoid chronic poisoning, medication with ergot alkaloids must be interrupted at the first appearance of symptoms of poisoning [**94**].

☹ Chronic mass poisonings occurred in the past as ergotismus gangraenosus or as ergotismus convulsivus. Both forms began with general malaises and paraesthesias such as numbness and feeling of 'pins and needles' ('tingling sickness') first in the hands and feet, later over the whole body.

Ergotismus gangraenous was characterized by painful spasms, circulatory disturbances, damage of the blood vessel endothelium, thromboses, and finally gangrenes of the affected parts of the body, first of the fingers and toes. It often led to loss of entire extremities and blindness. Cause was the overreaction of α-adrenergic receptors of blood vessels induced by ergot alkaloids. This ergotism form was called 'Saint Anthony's fire' or 'Ignis sacer' (holy fire), referring to the severe burning sensations of the limbs (fire!) and to Saint Anthony (patron saint against fire, epilepsy, and infectious diseases [30, 75]).

Ergotismus convulsivus was characterized by severe cramps, which could turn into permanent cramps leading to crippling. Mental disorders, stupefaction, and rage addiction occurred. This kind of ergotism was called 'St. Vitus Dance' (Saint Vitus, one of the 14 holy helpers, patron saint of dancers) because of the unusual movements of the sick due to the very painful cramps.

Both forms of ergotism usually ended fatally.

In the fifteenth century, the Order of Saint Anthony (founded 1089 in France, home monastery in Saint Antoine l' Abbaye) cared for thousands of people suffering from ergotism in about 400 hospitals. In these hospitals, the sick people were fed bread made from wheat or spelt, which was much less affected by ergot fungus than rye. In the best case, a cure could be achieved [30, 75].

In pregnant women, ergot poisoning almost always caused abortion. It is remarkable that breast feed-infants of poisoned mothers did not show any poisoning symptoms.

In Europe, ergotismus convulsivus occurred East of the Rhine and ergotism gangraenosus West of the Rhine. Presumably, the Rhine was a boundary for two races of ergot fungus with different alkaloid spectra [16, 29, 36, 38, 40, 49, 83].

An analysis of ergot alkaloids in food and feed samples collected between 2011 and 2016 in 15 different European countries found the highest alkaloid level in rye and rye-containing commodities. In humans, the mean chronic dietary exposure was highest in toddlers and other children with maximum daily intake of 0.47 µg/kg BW. In animals the exposure varied between 0.31 µg/kg BW daily in cattle and 8.07 µg/kg BW in piglets [3].

Poisonings of animals by grasses or cereals containing sclerotia of Claviceps species or grasses with endophytic parasitizing Claviceps species have been often described. Experimental investigations with weanling pigs showed that already the intake of 1.2 g sclerotia/kg feed for 28 days led to signs of poisoning, e.g., alteration of the liver tissue, inflammatory infiltrates, vacuolization, necrosis of hepatocytes as well as reduced villi height and number of mucus producing cells in intestinal epithelium [48]. In feeding trials with pigs, a slight decrease in feed intake and a slightly reduced body weight gain were observed for a content over about 10 g sclerotia/kg in the feed (about 5 mg/kg total alkaloids corresponding). At higher doses, given to pregnant animals, reduced litter size occurred when ingested in the first third of gestation. Lack of development of the udder and shortened gestation were observed in the second third. In cattle, necrosis of the ears, claws, teats, udder, and tail tip, and abortion may occur. Abortions or stillbirths are also common in horses. In poultry, necrosis after feeding contaminated cereals was first mentioned around 1100 B.C. [9, 15, 17, 32, 43, 49, 58, 61, 79, 83].

The permitted maximum content of sclerotia in unprocessed cereal grains (except maize and rice) in the EU is 0.2 g/kg. The permitted alkaloid content in ground rye products is presently 500 µg/kg (from 1 July 2028: 250 µg/kg), in ground products from other cereals 50 or 150 µg/kg, depending on ash content, in cereal-based baby food for infants 20 µg/kg [**39**].

The use of ergot for medicinal purposes is mentioned for the first time in Europe in the 'Kreuterbuch' of Adam Lonitzer from 1582. Today, ergot itself is no longer used in modern medicine but pure alkaloids and some of their hemisynthetic derivatives are applied, for example, salts of ergometrine against postnatal bleedings and salts of ergotamine against acute migraine. The compounds are obtained mainly from saprophytic cultures of the fungus. The semisynthetic lysergic acid diethylamide (LSD) was and is used as hallucinogen (further details, see [**161 = Vol. 3**]).

3.4.2 Alkaloids as Toxins of Endophytic Fungi Living in Sweet Grasses (Poaceae)

Many sweet grasses can be toxic because of their infestation with mycotoxin producing fungi.

Whereas the colonization of grasses by the parasite *Claviceps purpurea* (see Section 3.4.1) is partly harmful to the infested plants (reproduction of the host is restricted) and the advantage for the infected plants (rejection of robbers) is small, endophytes are useful for their hosts due to their toxins that protect against predators, insects, and plant parasitic nematodes. The endophyte benefits from the connection by greater access to nutrients and by enhanced tolerance to biotic and abiotic stress, such as salts, drought, waterlogging, cold, heavy metals, and low nitrogen stress. Thus, this relationship has a mutualistic character.

Examples of such 'joint ventures' are fungi living endophytic in, e.g., *Lolium perenne* L., perennial ryegrass (Ph. 3.5), *L. rigidum* Gaud., annual ryegras, *L. temulentum* L., darnel (Ph. 3.6), *L. arundinaceum* (Schreb.) Darbysh (*Festuca arundinacea* Schreb.), tall fescue, *Cynodon dactylon* (L.) Pers., Bermuda grass, *Paspalum dilatatum* Poir., dallisgrass, and *Achnatherum robustum* (Vasey) Barkworth (*Stipa robusta* Vault), sleepy grass [9, 10, 77].

Ph. 3.5: *Lolium perenne*, perennial ryegrass, host of endophytic fungi (source: 49pauly/iStock/Getty Images Plus).

Ph. 3.6: *Lolium temulentum*, darnel, host of endophytic fungi (source: H. Zell, wikimedia, CC-BY-SA-3.0).

The endophytes are transmitted vertically via seeds and horizontally via spores. Unlike Claviceps species, which remain confined to the hosts flower and form the bulk of their mycelium outside the plant, the endophytes grow through the entire tissue of the grasses.

Ph. 3.7: *Epichloe typhina* on grass (source: Gerhard Koller, Mushroom Observer, wikimedia, CC-BY-SA-3.0).

Endophytic fungi do not significantly affect the growth of the grasses. On the contrary, some infected grasses have been observed to have better assertiveness toward uninfected competitors. However, endophytes suppress seed formation in some grasses [71, 76].

Epichloe typhina (Pers.) Brockm. (Ph. 3.7), Clavicipitaceae) is an endophyte in *Lolium arundinaceum* and in *Achnatherum robustum*. The fungus produces the clavine alkaloid ergine (lysergic acid amide) and several ergopeptines, e.g., ergovaline, ergosine, ergonine, and more rarely α-ergoptine and ergocornine (0.4–22 mg total alkaloids per kg green mass, 0.1–4.0 mg ergopeptines in *L. arundinacea*, about 20 mg lysergic acid amide in *A. robustum*), and the corresponding isolysergic acid derivatives (see **[161 = Vol. 3]**). Additionally, it produces the indole diterpene lolitrem B that causes the so-called 'rye grass staggers' in animals [9, 68, 71].

Species of the Poaceae genera Agrostis, Calamagrostis, Cynodon, Danthonia, Eragrostis, Festuca, Lolium, Panicum, and Poa are frequently attacked by Balansia species (17 species are known, Clavicipitaceae, e.g., *Balansia epichloe* (Kunze) Diehl). These fungi have been found in over 100 Poaceae species, many of them in forage grasses or pasture weeds. Often up to 60% of grasses in pastures are infected [5]. The main spread of the fungi is through the infected grass fruits. Balansia species produce clavine alkaloids, e.g., agroclavine, chanoclavine I, and elymoclavine, and the simple lysergic acid amide ergometrine (Fig. 3.14), besides a proline-free ergopeptine, ergobalanisine [4]. Pyrrolizidine alkaloids, mainly *N*-formyl-loline and *N*-acetyl-loline, Fig. 3.17), have also been detected. Their occurrence in the plant is linked to the presence of parasitic fungi [34]. It remains unclear whether the pyrrolizidine alkaloids are formed as phytoalexins by the infested plant or, more likely, whether they are metabolic products of the endophytic fungi. The toxicological significance of these pyrrolizidine alkaloids is unclear.

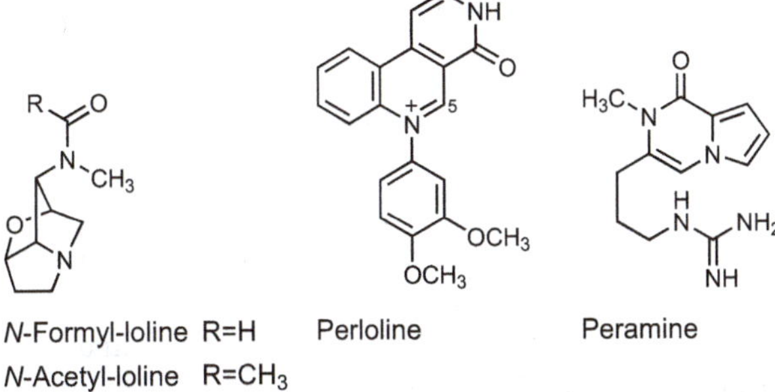

Fig. 3.17: Alkaloids of fungi-infected *Lolium temulentum* and other plants.

Lolium temulentum (Ph.3.6), a grass native to the Mediterranean region, was in earlier times widespread throughout Europe. Now it is only rarely found in cereal fields. The presence of the endophyte *Gloeotinia temulenta* (Prill. & Delacr.) M. Wilson, Noble & E.G. Gray (*Endoconidium temulentum* Prill. & Delacr.) and the occurrence of pyrrolizidine alkaloids, presumably depending on this, were observed in the infected plants. Lolines, peramine (Fig. 3.17), and two unnamed concomitant alkaloids were detected in the fruits. A 2,9-diazaphenanthren alkaloid, perloline (Fig. 3.17), was found in the herb. However, perloline is probably not responsible for the poisoning symptoms caused by infected *L. temulentum*.

In the past, bread baked with flour, made from grains of infected *L. temulentum*, often caused mass poisoning. Linseed oil pressed from linseed contaminated with these fruits and beer that was produced from contaminated cereals were possibly also the cause of intoxication. Due to the rarity of *L. temulentum*, such cases are today unlikely to occur.

In humans, poisoning by infected *L. temulentum* is characterized by pupil dilation, vomiting, dizziness, headache, disturbance of equilibrium, mental confusion, speech disorders, drowsiness, bradycardia, convulsions, and raving fits. Death occurs by respiratory paralysis [**86**]. Fescue-associated edema has been associated with tall fescue varieties infected with a specific strain of *Epichloe coenophiala* (Morgan-Jones & W. Gams) C.W. Bacon & Schardl (*Neotyphodium coenophialum* [9, 10]).

Some of the compounds produced by the endophytes are toxic to higher animals. For example, cattle that graze on sweet grass infected by *Epichloe festucae* Leuchtm., Schardl & M.R. Siegel (*Neotyphodium lolii* (Latch. N.J. Chr. & Samuels) Glenn, C.W. Bacon & Hanlin) may develop uncontrollable spasms. Infected *L. arundinaceum*, which is widespread in Europe, can cause poisoning of grazing animals, known as 'summer syndrome' (occurring in summer) or 'fescue foot' or 'fescue toxicosis' (occurring in colder season of the year). Presumably, the fungi produce different alkaloids during different periods of the year; therefore, the symptoms vary depending on the season. They include reduced food intake, weight loss, decreases in milk production, temperature rise, and increased respiratory rate; in the colder months, paralysis and gangrene also occur.

'Sleepy grass' and 'drunken horse grass' are other poisoning syndromes in livestock. They are caused by Eriocoma (Achnatherum) species, infected with Neotyphodium species, which produce ergot alkaloids [31, 35]. Grazing on infected sleepy grass, which occurs primarily in the southwestern United States, causes drowsiness to deep somnolence.

Lolium perenne, native to Europe and spread worldwide, is the most common utilized pasture grass on dairy farms in Australia. The grass can be attacked by *Pseudopithomyces chartarum*, a filamentous fungus, that produces sporidesmin A, a 3,6-epidithio-2,5-dioxopiperazine antibiotic (Fig. 3.18). Sporidesmin containing grass causes the so-called 'facial eczema' in sheep, characterized by liver damage and following secondary photosensitization. It occurs particularly in New Zealand [9, 71].

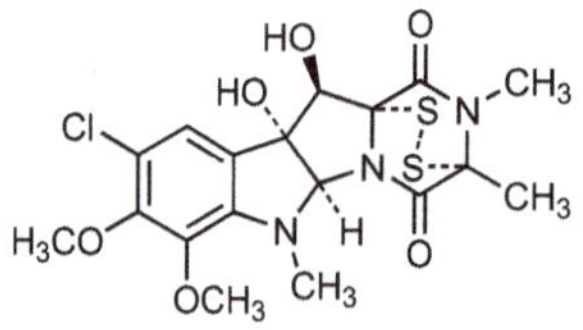

Sporidesmin A

Fig. 3.18: Sporidesmin A.

3.4.3 Toxins of *Penicillium roqueforti*, Starter for Production of Blue Veined Cheese

The production of alkaloids and other toxins by strains of *Penicillium roqueforti* is remarkable because selected strains of this mold are used as starter strains for production of blue veined cheese (Ph. 3.8), e.g., French Roquefort, French Bleu, Italian Gorgonzola, English Stilton, Spanish Cabrales, Danish Danablue, and Valdeón.

P. roqueforti strains (more than 50 strains were investigated in artificial culture) may produce about 20 secondary metabolites of at least seven different families. Sixteen of these metabolites are mycotoxins.

The formation of mycotoxins occurs especially by free living *P. roqueforti*. Under microaerobic conditions, only a few metabolites are formed and only in small quantities. They hardly accumulate in the cheese. The produced compounds include roquefortine A (isofumigaclavine A, up to 5 mg/kg in Roquefort cheese), roquefortine B (isofumigaclavine B), roquefortine C (up to 7 mg/kg in Roquefort cheese [12, 25, 46, 52, 72, 78, 80]). Roquefortine A and roquefortine B are ergoline alkaloids of the clavine series, roquefortine C has a pyrazino [1′,2′:1,5] pyrrolo[2,3-b] indole base body, roquefortine D is a pyrroloindole (Fig. 3.19).

Roquefortine A shows in animal assays weak muscle-relaxing, antidepressant, and local anesthetic activities. Roquefortine C (50–100 mg/kg BW, i.p.) causes convul-

Ph. 3.8: Blue veined cheese (source: etienne voss/iStock/Getty Images Plus).

Roquefortine A R=AcO
Roquefortine B R=HO
Roquefortine C
Roquefortine D

Fig. 3.19: Mycotoxins of *Penicillium roqueforti.*

sions. The acute toxicity of roquefortines is very low (LD_{50}, mouse, i.p., roquefortine A: 340 mg/kg BW, roquefortine B: 1,000 mg/kg BW, roquefortine C: 169–184 mg/kg BW, [31]).

Some strains of *P. roqueforti* form, additionally to roquefortines, in culture further alkaloids, e.g., festuclavine (Fig. 3.14), and the marcfortines A–C (2-oxo-indole alkaloids of unusual structure [52, 69]).

Besides, some nitrogen-free toxins, e.g., botryodiplodine [73], PR toxin [54, 90], mycophenolic acid [25, 39], and patulin (Fig. 3.20, see Section 3.3.1), are produced in saprophytic culture.

Botryodiplodine possesses genotoxic and mutagenic activity. It induces cross-links of DNA and inhibits the biosynthesis of nucleic acids. The LD_{50} ranges from 40 to 50 mg/kg BW in mice. It seems not to be present in the cheese ready to eat [25, 73].

The contamination of the blue veined cheese with PR toxin (see Section 3.2.2) is very unlikely due to its high instability in the complex media of the cheese, where it is rapidly degraded to PR amide and PR imine [55].

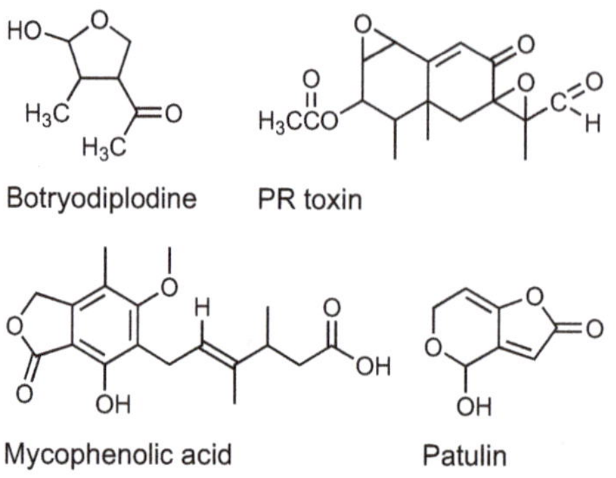

Botryodiplodine PR toxin

Mycophenolic acid Patulin

Fig. 3.20: N-free mycotoxins of *Penicillium roqueforti.*

Mycophenolic acid has low cytotoxic effects ([10, 20, 24, 25], see Section 3.3.1). The levels of patulin (see Section 3.3.1) in cheese found in Europe are mostly below the limits set by the European Commission [89].

Because people normally consume only small amounts of cheese, the risk of intoxication from fungal toxins of the cheese is generally regarded as low.

The concentration of the mycotoxins may become dangerous when the cheese is left for too long time unrefrigerated, or when it is despite refrigerator storage not consumed within several weeks. In those cases the cheese should not be consumed.

3.4.4 Toxins of *Penicillium camemberti*, Starter for Production of Soft-Ripened Cheese

All strains of *Penicillium camemberti* Thom, which are used to produce soft-ripened cheese (Camembert-style cheese), form α-cyclopiazonic acid (Fig. 3.21) under laboratory conditions. This indole derivative is also produced by other Penicillium species, e.g., by *Penicillium aurantiogriseum* Dierckx. (*P. cyclopium* Westling, *P. viridicatum* Westling, *P. puberulum* Bainier), *P. griseofulvum* Dierckx, *P. roqueforti*, and by Aspergillus species, e.g., by *Aspergillus flavus* [13, **26**]. The fungi produce α-cyclopiazonic acid also in contaminated corn, in peanuts (till 6.5 mg/kg, [88]), and in mold-ripened salami, raw ham, and soft cheese [13, 78].

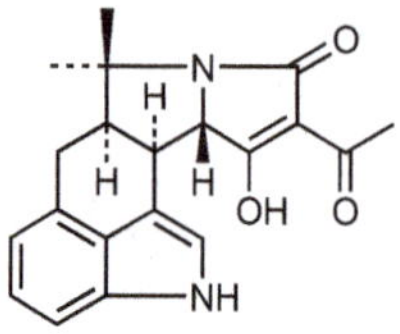

α-Cyclopiazonic acid

Fig. 3.21: An indole alkaloid found in cheese.

The medium level of the mycotoxin found in soft-ripened cheese was 48 µg/kg; it was detected also in blue veined cheese [47, 64].

High doses of α-cyclopiazonic acid cause signs of hepatotoxicity (increased activity of glutamate pyruvate transaminase, glutamate oxaloacetate transaminase, γ-glutamyl transpeptidase) and neurotoxicity and lead in animal assays to degenerative changes of several organs. About 11–14 mg/kg BW led to death in mice [60].

The occurrence of α-cyclopiazonic acid in foods, e.g., in cheese, is too low to be of health concern. However, in cheese stored for 5 days at 25 °C concentrations as high as 4 mg/kg have been found [64]. Besides, α-cyclopiazonic acid is the cause of Kodo poisoning.

3.4.5 α-Cyclopiazonic acid as Cause of Kodo Poisoning

Kodo millet, *Paspalum scrobiculatum* L., is grown in Nepal, India, and other Asian countries and in West Africa. It is a staple food by the people on the Deccan Plateau in North India. The grains of the plant are frequently infested with *Aspergillus tamarii* Kita, more rarely with other Aspergillus, Penicillium, or Claviceps species. The fungi produce α-cyclopiazonic acid (Fig. 3.21) in high concentrations. The compound is suggested to be responsible, at least partly, for the so-called Kodo poisoning that is caused by consumption of infected kodo millet in daily nutrition. Kodo poisoning is a serious health threat [2, 18, 44].

Symptoms of acute kodo poisoning are nausea, vomiting, delirium, depression, and unconsciousness. Liver dysfunction, lesions of the myocardium, and impaired heart function follow [44].

If humans and animals stop eating the infected grains early, the prognosis is good. To avoid poisoning, suitable pre- and post harvest management of the kodo millet is necessary [18].

3.4.6 Indole Alkaloids as Tremorgens

Some toxins produced by filamentous fungi are known as tremorgens because of their ability to induce tremor (involuntary trembling) in experimental animals at sublethal doses. These compounds, of which more than 50 are known [12, 81, 85, **26, 138, 139**], are in most cases diterpene indole alkaloids. They possess a cyclic diterpene base body derived from geranylgeranyl diphosphate and an indole moiety derived from tryptophan. Paspaline occurs as intermediate in biosynthesis. The tremorgens are produced by Aspergillus, Acremonium, Claviceps, Emericella, and Penicillium species. The tremorgenic indole derivatives include:

- fumitremorgins A and B (also known as fumitremorgens) and verruculogens (Fig. 3.22), formed by *Aspergillus fumigatus* Fresen. and other Aspergillus and Penicillium species; they are diprenylated diketopiperazine derivatives, composed of tryptophan and proline [12, 94, **26**];
- tryptoquivaline (Fig. 3.22), formed by *Aspergillus fumigatus* Fresen. and *Neosartorya clavata* (Desm.) Pitt & A.D. Hocking (*Aspergillus clavatus* Desm.), derived from a tetrapeptide consisting of tryptophan, anthranilic acid, valine, and alanine [94, 95];
- paxilline and related compounds (Fig. 3.22), formed by *Penicillium paxilli* Bainier, *Epichloe festucae var. lolii* (Latch, M.J. Chr. & Samuels) C.W. Bacon & Schardl, and Emericella species [7, 28, 59, 63, 74, 86, 94, 96];
- aflatrem (α,α-dimethylallylpaspalinine) and β-aflatrem (Fig. 3.22), produced by *Aspergillus flavus* [21];
- penitrems A–F (Fig. 3.22), formed by *Penicillium solitum* Westling; penitrem A is the most abundant penitrem; it inhibits BK channels (voltage-activated K^+ channel receptors, BK = big potassium, important in neuronal excitability) and due to this proliferation, migration, and invasion of cancer cells; it interferes with the release of GABA and glutamate and causes neuronal death [1, 71, 93];
- lolitrems A–N (Fig. 3.22, main representative lolitrem B) and related compounds from *Epichloe festucae var. lolii*, biogenetically related to the penitrems; BK channel inhibitors, inhibit the release of proinflammatory cytokines [27, 33, 68, 94];
- janthitrems A–M (Fig. 3.23), formed by *Penicillium simplicissimum* (Oudem.) Thom (*P. janthinellum* Biourge), resemble the penitrems [65, 91];
- paspalitrems, paspaline, paspalinine (Fig. 3.23) and paspalicine, formed by *Claviceps paspali* [45] and *Penicillium paxilli*, block BK channels but possess only weak tremorgenic activity [45, 71];
- shearinines A–C (Fig. 3.23), isolated from *Penicillium myczynskii* K.W. Zaleski (*Eupenicillium shearii* Stolk & D.B. Scott), shearinines D–K from a Penicillium ssp., endophytic in the mangrove plant *Aegiceras corniculatum* (L.) Blanco, and shearines L–M, formed by the fungus *Escovopsis weberi* J.J. Muchovej & Della Lucia; shearinines are very similar to janthitrems; they exhibit insecticidal and antifungal activity [71];
- terpendoles A–L (Fig. 3.23) and related compounds, isolated from cultures of *Albophoma yamanashiensis* Tak. Kobay., Masuma, Omura & Kyoto Watan.; inhibit acyl-CoA:cholesterol acyl transferase (ACAT); possibly protective against atherosclerosis; 6,7-dehydroterpendole A elicits the most intense tremors among the terpendoles; terpendole E is an inhibitor of human kinesin, a target for antitumor compounds [71];
- sulpinines A–C (Fig. 3.23) and related compounds, isolated from the sclerotia of *Aspergillus sulphureus*; exhibit insecticidal activities; sulpinine A is cytotoxic toward human cancer cells [71];

– emindoles (Fig. 3.23), first isolated from a Emericella spp., later from many other fungal species together with related compounds, e.g., the asporycins (naphthofuran derivatives), some with insecticidal activity (overview: [37, 71]).

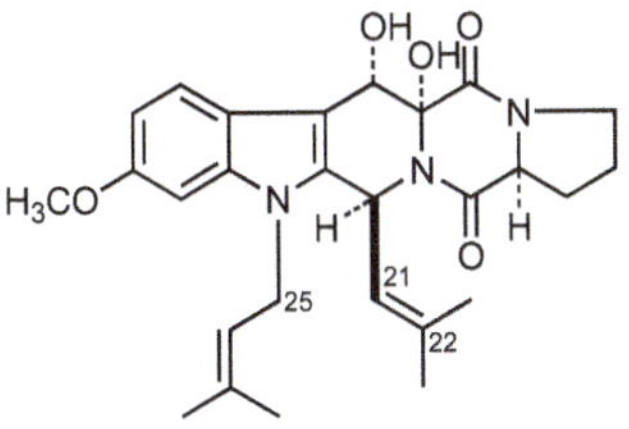

Fumitremorgin B
Verruculogen (21,22 Dihydro-22,25 dioxy-derivative)

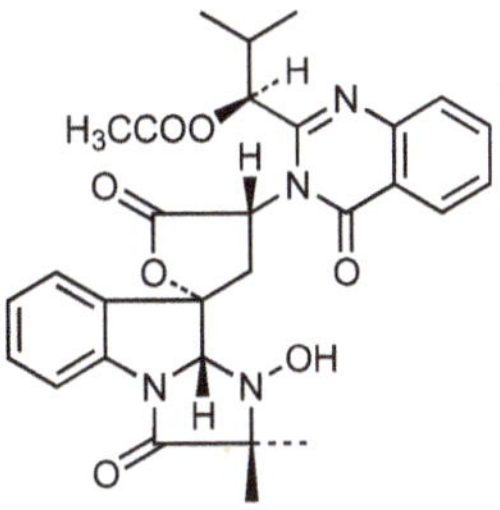

Tryptoquivaline

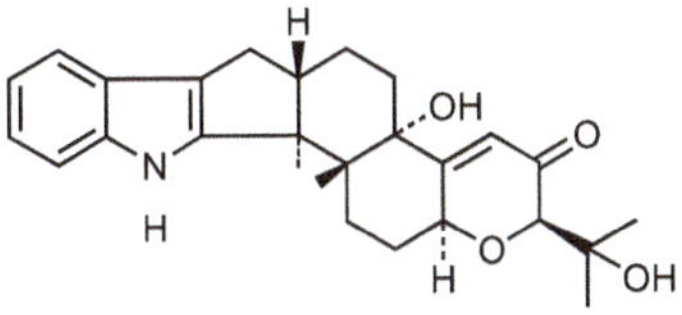

Paxilline

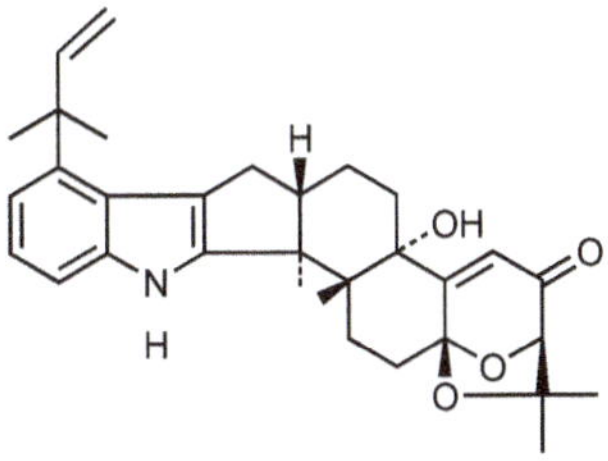

Aflatrem

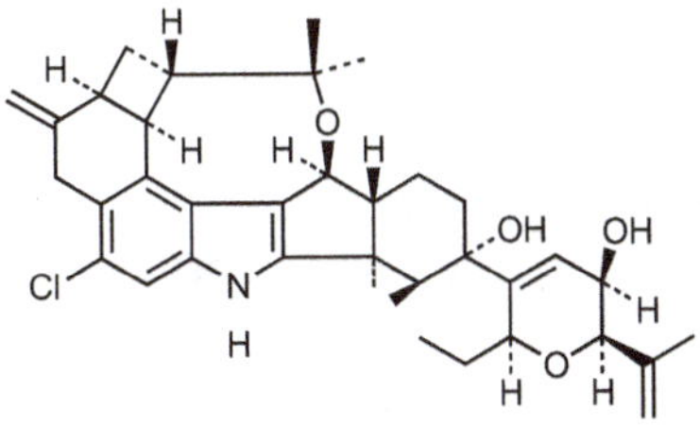

Penitrem C

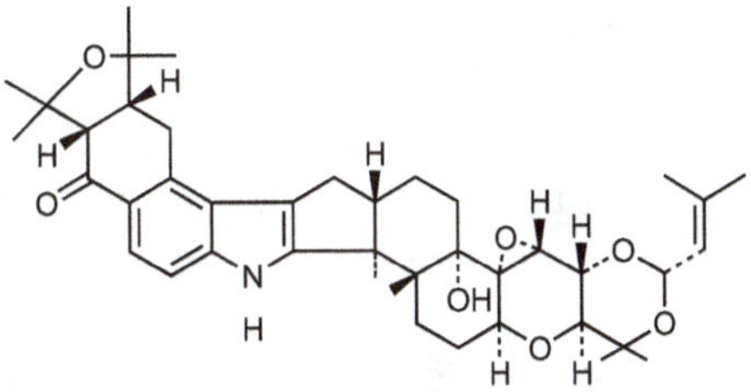

Lolitrem B

Fig. 3.22: Tremorgens, Part 1.

Except *Epichloe festucae* var. *lolii* (see Section 3.4.2) and *Claviceps paspali* (see Section 3.4.1), the substrates of tremorgen-producing filamentous fungi are very diverse. Silage is frequently infested. The fungi have also been detected in food, e.g., in peanuts, maize, nuts, meat products, and cheese. Many of these filamentous fungi are also found in the soil. The tremorgens formed in the soil are presumably partly taken up by the plant roots [74].

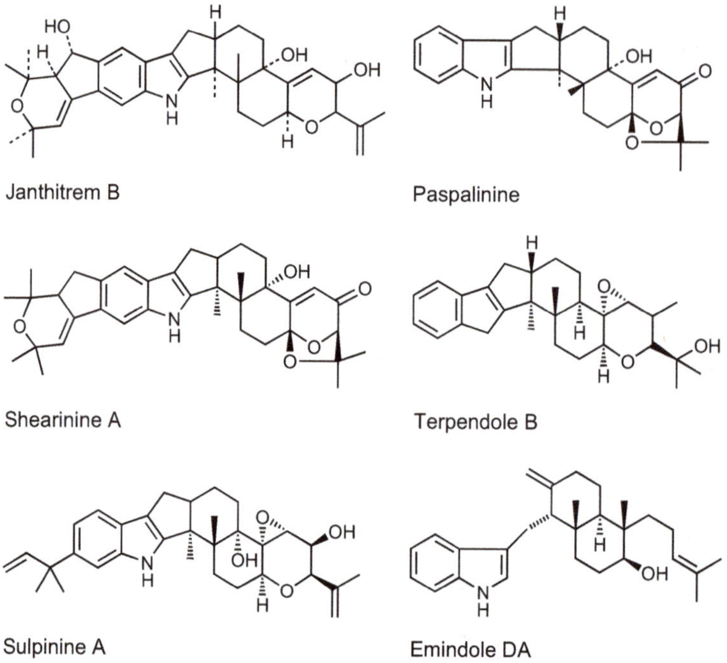

Janthitrem B

Paspalinine

Shearinine A

Terpendole B

Sulpinine A

Emindole DA

Fig. 3.23: Tremorgens, Part 2.

The tremor-inducing effect is one of the neurotoxic effects of the compounds and is caused by specific effects on the CNS. The mechanisms of action discussed are the reduction of the glycine level in the brain by the penitrems [74], the agonistic attack on receptors for GABA by tremorgens with a nitrogen atom (paspalitrems, penitrems, lolitrems [81]), the inhibition of inhibitory interneurons of the spinal cord, and the blocking of BK channels by, e.g., lolitrem E, penitrem A, and paspalicine [71].

Data on the tremor-inducing dose and LD_{50} of tremorgens are summarized in Tab. 3.4.

Acute tremorgenic mycotoxicoses in humans have not yet been observed. However, involvement of tremorgens in various human diseases, e.g. extrinsic allergic alveolitis, is suspected [81].

Poisoning by tremorgens occurs particularly in grazing animals through the ingestion of tremorgen-containing grasses. It causes considerable animal losses in some countries, e.g., in Australia, New Zealand, and South Africa, since 1992 also in the south of the United States (California). Poisoning by tremorgens has also been reported in European countries, e.g., in Great Britain and Italy. The intoxications are generally referred to as 'staggers' [7]. Known clinical pictures are 'perennial ryegrass staggers' ('marsh staggers'), particularly common in New Zealand and in the USA, caused by fungus-infected *Lolium perenne*, and 'paspalum staggers', caused by ingestion

Tab. 3.4: Efficacy and toxicity of tremorgens.

Tremorgen	ED_{50} Tremor-inducing activity (mg/kg BW)	LD_{50} (mg/kg BW)	References
Fumitremorgin A	0.18 mice, i.p.	0.19 mice, i.v.	[94]
Fumitremorgin B	3.5 mice, i.p.		[94
Verruculogen	0.39 mice, i.p.	2.4 mice, i.p.	[94
Paxilline	25 mice, p.o.		[94
Aflatrem	<14 chicken, p.o.		[94
Penitrem A	0.19 mice, i.p. 0.5 mice, p.o.	1.1 mice, i.p. 10 mice, p.o.	[74, 53]
Penitrem B		5.8 mice, i.p.	[74]

of fungus-infected Paspalum species [6, 11, 12, 22, 26, 27, 57, 81]. The tremorgenic mycotoxin poisoning of a dog in the Czech Republic was possibly caused by ingestion of compost in the garden. The vomit contained the tremorgens paxilline, penitrems, and roquefortine C [62].

In addition to tremors, tremorgenic poisoning is characterized by overexcitability, coordination disorders, stumbling movements, head shaking, and the inability of the animals to rise after falling. At higher doses, death occurs after convulsions. The animals usually recover within a short time without permanent damage [11].

Tremorgenic mycotoxins or their derivatives are of interest as possible pesticides and for treatment of glaucoma and cancer (BK channel inhibition!).

References

For numbers in bold, see cross-chapter literature p. 233.

[1] Abraham W-R (2018) Curr Med Chem 25(2): 123
[2] Antony M et al. (2003) J Ethnopharmacol 87(2–3). 211
[3] Arcella D et al. (2017) EFSA J 15(7) e04902
[4] Bacon CW et al. (1985) Mycologia 77: 418
[5] Bacon CW et al. (1986) Agron J 78: 106
[6] Botha CJ et al. (1996) J S Afr Vet Assoc 67(1): 36
[7] Bourke CA (1995) Aust Vet J 72(6): 228
[8] Bové F (1970) The Story of Ergot. S. KARGER Verlag, Basel
[9] Canty MJ et al. (2014) Ir Vet J 67(1): 21
[10] Chávez R et al. (2023) J Fungi 9(4): 459
[11] Cheeke PR (1995) J Anim Sci 73: 909
[12] Cole RJ (1981) J Food Protect 44: 715

[13] Cole RJ (1986) in: Diagnosis of Mycotoxicoses. Richards JL, Thursten JR (eds) Martinus Nijhoff Publ, Dordrecht, p. 91
[14] Coporael LR (1976) Science 192: 2
[15] Coufal-Majewski S et al. (2016): Front Vet Sci 3:15
[16] Creutzig A, Alexander K (1985): Dtsch Med Wochenschr 110: 1420
[17] Dänike S (2015) Toxins 7(6): 2006
[18] Deepika C et al. (2022) J Food Sci Technol 59(7): 2517
[19] Dorner WG et al. (1985) Chem Rdsch 34: 38
[20] Dubey MK et al. (2018) Front Pharmacol 9: 288
[21] Duran RM et al. (2007) Appl Microbiol Biotechnol 73: 1158
[22] Eich E et al. (1984) Biochem Pharmacol 33: 523
[23] EU Verordnung 2024/1808. https://eur-lex.europa.eu/legal-content/DE/ALL/?uri=CELEX:32024R1808 (accessed 12/13/2024)
[24] Florea et al. (2017) Phytopathology 107(5): 504
[25] Fontaine KI et al. (2015) Food Control 47: 634
[26] Gallagher RT (1980) Tetrahedron Lett 21: 235
[27] Gallagher RT (!981) N Z Vet J 29. 189
[28] Gallagher RT, Clardy J (1980) Tetrahedron Lett 21: 239
[29] Garcia GD et al. (2000) J Vascul Surgery 31(6): 1245
[30] Grzybowski A et al. (2021) Clin Dermatol 39(6): 1088
[31] Guerre Ph (2015) Toxins 7(3): 773
[32] Haarmann Th et al. (2009) Mol Plant Pathol 10(4): 563
[33] Hattori M et al. (1990) Phytother Res 4: 66
[34] Jenett-Siems K (1994) J Nat Prod 57(9): 1304
[35] Jones TA et al. (1983) Crop Sci 23: 1135
[36] Kolta KS (1987) Beitr Gesch Pharm 39(38): 97
[37] Kozák L et al. (2019) Appl Microbiol Biotechnol 103(4): 1599
[38] Kruse H et al. (1989) Pharm Ztg 134: 321
[39] Lafont P et al. (1979) Appl Environm Microbiol 37: 365
[40] Langguth W (1980) Dtsch Apoth Ztg 120: 319
[41] Liu M et al. (2022) Can J Plant Physiol 44(6): 783
[42] Lünne F et al. (2021) J Nat Prod 84(10): 2630
[43] Mainka S et al. (2005) Arch Anim Nutr 59(6): 377
[44] Mamatha Bhanu LS et al. (2023) Biomed J Sci Tech Res 50(5): 008014
[45] Mantle PG (1975) in: Filamentous Fungi 1. Smith JE, Barry DR (eds), E, Arnold, New York., p. 281
[46] Maragos CM (2022) Food Addit Contam Part B 15(1): 1
[47] Maragos CM et al. (2023) Food Addit Contam Part B 16(1): 14
[48] Maruo V et al. (2018) Toxins 10(5): 183
[49] Mattossian MK (1981) Med History 25: 73
[50] Mattossian MK (1982) Am Sci 70: 355
[51] Militz M (1996) Pharm Ztg 141(19): 720
[52] Mioso R et al. (2015) J Appl Microbiol 118(4): 781
[53] Moldes-Anaya A et al. (2012) Toxicon 60(8): 1428
[54] Moreau S et al. (1976) Tetrahedron Lett: 833
[55] Moreau S et al. (1980) Appl Environ Microbiol 39: 770
[56] Mueller-Schweinitzer E et al. (1978) Handb Exp Pharmacol 49: 87
[57] Munday BL et al. (1985) Aust Vet J 62: 207
[58] Munoz MC et al. (1986): Rev Salud Anim 8(3): 245
[59] Musuku A et al. (1994) J Nat Prod 57(7): 983

[60] Nishie K et al. (1985) Food Chem Toxicol 23(9): 831
[61] Noble JW (1985) Aust Vet J 62(12): 432
[62] Novotna T et al. (2023) Veterinarny Medicina 68(12): 483
[63] Nozawa K et al. (1889) Chem Pharm Bull Tokyo 37(5): 1387
[64] Ostry V et al. (2018) World Mycotoxin J 11(1): 135
[65] Penn J et al. (1993) Phytochemistry 32: 1431
[66] Petrovski J et al. (1992) Nat Toxins 1(2): 84
[67] Pfänder HJ et al. (1986) Dtsch Ärztebl 8: 2013
[68] Porter JK (1995) J Anim Sci 73(3): 871
[69] Prange T et al. (1982) Tetrahedron Lett 22: 1977
[70] Proksa B et al. (1986) Tetrahedron Lett 27: 5413
[71] Reddy P et al. (2019) Toxins 11(5): 302
[72] Reinholds I et al. (2020) Foods 9(1): 93
[73] Renauld F et al. (1985) Tetrahedron 41: 955
[74] Richards JL et al. (1986) in: Diagnosis of Mycotoxicoses. Richards JL, Thursten JR (eds) Martinus Nijhoff Publ, Dordrecht, p. 5
[75] Schäfer SG (2021) Die Dosis macht das Gift. Verl. Quelle und Meyer, Wiebelsheim
[76] Schardl CL (2001) Fungal Genet Biol 33(2): 69
[77] Scharl Ch L et al. (2007) Phytochemistry 68(7): 980
[78] Schiefer HB (1990) Can J Physiol Pharmacol 68: 987
[79] Schrenk D et al. (2024) EFSA J 22(1): e8496
[80] Scott PM, Kennedy BPC (1976) J Agric Food Chem 24: 865
[81] Selala MI et al. (1989) Drug Chem Toxicol 12: 237
[82] Silva A et al. (2023) Molecules 28(20): 7233
[83] Smakoszn et al. (2021) Toxins 13(7): 492
[84] Steiner U, Leistner E (2018) Planta Med 84(11): 751
[85] Steyn PS, Vleggaar R (1985) Prog Chem Org Nat Prod 48: 1
[86] TePaske MR Gloer JB (1992) J Nat Prod 55: 1080
[87] Thun M (1989) Z Phytother 10(3): 10
[88] Van Rensburg SJ (1984) Food Chem Toxicol 22:293
[89] Vidal A et al. (2019) Food Chem Toxicol 129: 249
[90] Wei RD et al. (1975) Tetrahedron Lett 31: 109
[91] Wilkins AL et al. (1992) J Agric Chem 40: 1307
[92] Wolff J, Richter W (1989) Bundesforschungsanstalt für Getreide- und Kartoffelverarbeitung. Detmold 4, p. 103
[93] Yamaguchi T et al. (1993) Phytochemistry 32(5): 1177
[94] Yamamoto Y, Arai K (1986): in **99**, XXIX: p:185
[95] Yamazaki M et al. (1978) ref. C. A. 90, 11758 u
[96] Yamazaki M et al. (1971) Chem Pharm Bull 19: 199

3.5 Peptides as Mycotoxins

Some mycotoxins are cyclic peptides. Well studied are islanditoxin and cyclochlorotine (Fig. 3.24). Both compounds are formed by *Talaromyces islandicus* (Sopp) Samson (*Penicillium islandicum* Sopp). They have rarely been detected in food. Cyclochlorotine is a pentapeptid containing the unusual amino acids β-phenylalanine, 2-aminobutyrate, and 3,4-chloroproline. The non-ribosomal peptide synthetase CctN and DUF3328 proteins, which mediate chlorination, hydroxylation, and O/N transacylation, are involved in its biosynthesis [13, 18].

The liver is the main target organ of both mycotoxins. Islanditoxin causes centrilobular necrosis; cyclochlorotine mainly damages peripheral regions. The LD_{50} values for mice are 3.6 mg/kg BW, s.c., for islanditoxin and 0.47 mg/kg BW, s.c., and 6.55 mg/kg BW, p.o., for cyclochlorotine [1, 7].

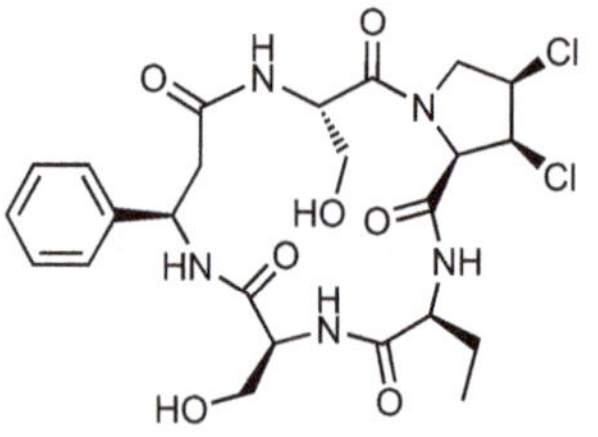

Cyclochlorotine

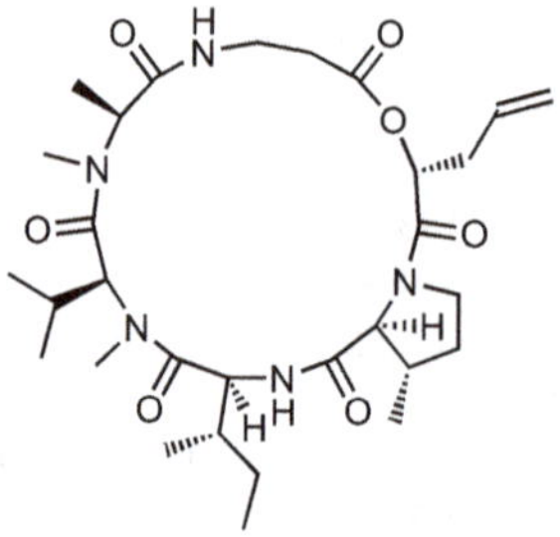

Roseotoxin B

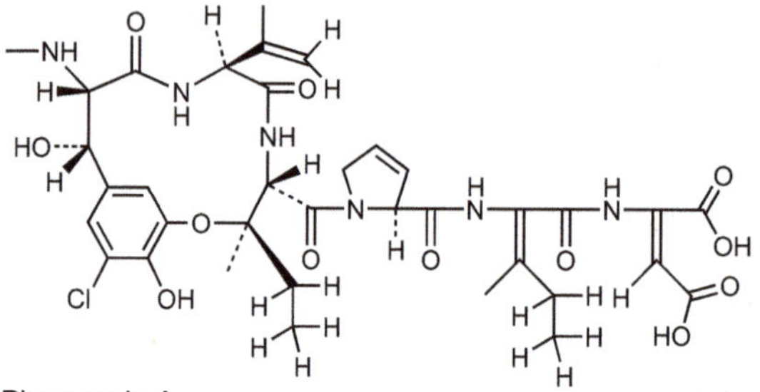

Phomopsin A

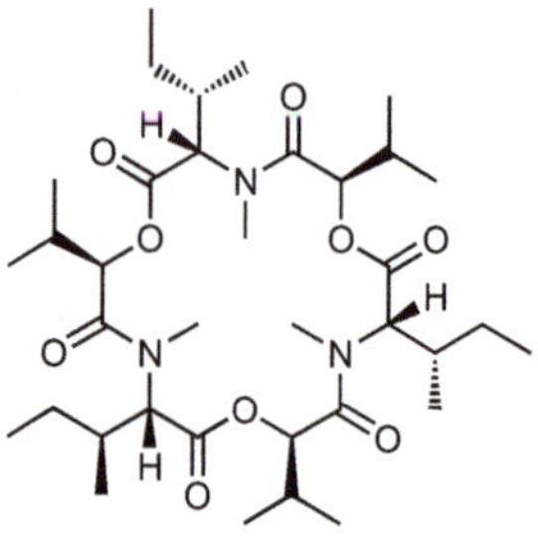

Enniatin A

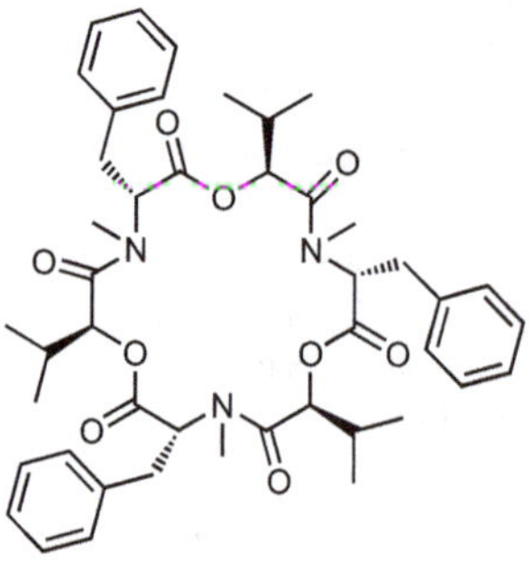

Beauvericin

Fig. 3.24: Peptides as mycotoxins.

The cyclic depsipeptides roseotoxin B and roseotoxin S (Fig. 3.24) are formed by *Fusarium sambucinum* (*F. roseum* Link) in addition to trichothecenes (see Section 3.2.1). Roseotoxin B is composed of five amino acids and one hydroxy acid, roseotoxin S of three amino acids and two hydroxy acids [**138, 139**]. Roseotoxin B inhibits the PDGF-B/PDGFR-β pathway in liver cells, which is a driver for liver fibrosis and a target for the treatment of this disease [21].

Of economic interest is the lupinosis in grazing animals, caused by the infestation of seeds of living lupines and during storage with *Diaporthe toxica.* The fungus may also grow on other seeds such as pea seeds [15].

This fungus produces several toxic peptides, above all phomopsin A (Fig. 3.24). Phomopsin A is a hexapeptide that has a 13-membered ring with an ether bridge. It belongs to the ribosomally synthesized and post-translationally modified peptides (RiPPs) whereby UstYa proteins, a group of oxidases, contribute to the modification [20]. Possibly the fungi can also produce quinolizidine alkaloids that are normally present in lupins [5, **161=Vol. 3**].

The phomopsins, phomopsin A is 2–5 times more toxic than phomopsin B, bind at rhizoxin binding site of microtubuli and inhibit tubulin polymerization. They cause severe liver damage. For sheep, the toxic dose starts at 2.5 µg/kg BW, the LD_{50} is about 40 µg/kg BW (in rats 5 mg/kg BW [4, 16]).

Lupinosis occurs particularly in sheep, cattle, and horses and causes considerable animal losses in Australia and some areas of Africa and New Zealand. In Germany, it was frequently observed in sheep in the late nineteenth century. Symptoms of acute poisoning include reluctance to eat, separation from the herd, constipation, jaundice, apathy or aggressiveness, and possibly also kidney damage. Death occurs within 2–14 days, at lower doses within 2 months. Surviving animals show hepatogenic photosensitization [2, 3, 4, 9].

Due to the increasing importance of lupins as a protein-rich food for humans and animals, not only the content of quinolizidine alkaloids in lupins (see [**161 = Vol. 3**]), but also that of mycotoxins should be kept as low as possible [5, 17]. Phomopsin A was found in 7 of 40 lupin samples from Australia and Europe but not in tissue samples from animals that were fed lupin-containing feed [19]. It was not found in 42 samples of lupin and lupin-containing food products collected in the Netherlands from 2011 to 2012 [10]. In Australia, the limit for phomopsins in lupine seeds and products is 5 µg/kg [17].

Enniatins (ENNs, Fig. 3.24) are cyclic hexadepsipeptides. They are produced by *Fusarium avenaceum* (Fr.) Sacc. (*Gibberella avenacea* R.J. Cook), *F. poae,* and other Fusarium species that grow mainly on cereals. They were detected in 76% of unprocessed grain samples collected in Europe between 2000 and 2013. ENNs A, A1, B, and B1 are the most frequently detected among the 29 enniatins known presently. Maximum reported levels in cereal-based food were 42, 125, 832, and 980 µg/kg for ENNA, ENNA1, ENNB, and ENNB1, resp.

Due to their ionophoric properties enniatins facilitate the transport of mono- or divalent cations such as K^+ or Ca^{2+} across membranes and change their physiological concentrations. They interact with proteins such as some enzymes and transporter proteins. Whereas enniatins exhibit cytotoxic activities in vitro, their in vivo toxicity seems to be low (1 mg/kg BW/day, p.o., mice, for 6 days without toxic effect [11]).

Beauvericin (Fig. 3.24) is a cyclic hexadepsipeptide produced by *Fusarium proliferatum*, *F. oxysporum*, and other Fusarium species. It occurs worldwide on feed and food products. It was detected in 54% of unprocessed grain samples collected in Europe between 2000 and 2013 with a maximum concentration in grain of 6,400 µg/kg. Like the ennatiatins, it is an ionophore and acts in vitro cytotoxic and as an enzyme and transporter inhibitor. The LD_{50} values in mice are high (>100 mg/kg BW, p.o., >10 mg/kg BW, i.p. [11]).

Acute exposure to enniatins and to beauvericin seems to be not a concern to animal and human health [11]. However, the ingestion of these mycotoxins in feed for animals or in animal-derived food for humans over long periods of time may possibly pose a health risk [6, 12, 14]. The compounds have negative effects on reproductive functions in animals [8].

References

For numbers in bold, see cross-chapter literature p. 233.

[1] Ainshie RD et al. (1985) J Org Chem 50: 2859
[2] Allen JG, Hancock GR (1989) J Appli Toxicol 9: 83
[3] Allen JG, Randall AG (1993) Aust Vet J 70: 283
[4] Aust SD (1986) in: Diagnosis of Mycotoxicoses. Richard JL, Thurton JR (eds) Martinus Nijhoff Publ., Dordrecht: p. 81
[5] Buccioni F et al. (2024) Toxins 16: 481
[6] Caloni F et al. (2020) Environ Toxicol Pharmacol 75: 103349
[7] Chang Z et al. (2004) J Nat Prod 67: 1356
[8] Chiminelli I et al. (2022) J Appl Toxicol 42(12): 1901
[9] Culvenor CCJ et al. (1989) Tetrahedron 45: 2351
[10] De Nijs M et al. (2013) Food Addit Contam Part A 30(10): 1819
[11] Gruber-Dorninger C et al. (2017) J Agricult Food Chem 65: 7052
[12] Hasuda AL, Bracarense APFRL (2024) Toxicon 239: 107534
[13] Jiang Y et al. (2021) Org Lett 23(7): 2616
[14] Křižova L et al. (2021) Toxins 13: 32
[15] Kunz BM et al. (2021) Mycotoxin Res 37: 63
[16] Li Y et al. (1992) Biochem Pharmacol 43(2): 219
[17] Pereira A et al. (2022) Molecules 27: 8557
[18] Schafhauser T et al. (2016) Environ Microbiol 18(11): 3728
[19] Schryvers S et al. (2024) Food Addit Contam Part A 41(8): 885
[20] Sogahata K et al. (2021) Angew Chem Int Ed Engl 60(49): 25729
[21] Wang X et al. (2020) Cell Death Dis 11(6): 458

3.6 Further Mycotoxins

Butenolide (Fig. 3.25) is produced by various Fusarium species. The fungi thrive on grasses, especially on Festuca species. The mycotoxin triggers oxidative stress. High doses damage in animal studies myocard and liver. It is unclear whether levels that can be reached by dietary exposure would cause adverse effects [4]. The LD_{50} is 43.6 mg/kg BW, i.p., and 275 mg/kg BW, p.o., mice.

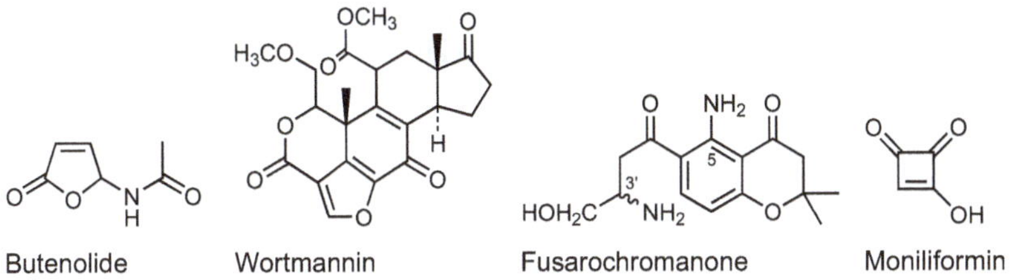

Butenolide Wortmannin Fusarochromanone Moniliformin

Fig. 3.25: Further mycotoxins.

Wortmannin (Fig. 3.25) is a steroid derivative that is probably biogenetically derived from lanosterol. The compound belongs to the viridin group of antibiotics along with wortmannolone, viridin, viridiol, and other compounds. It was first isolated from the culture broth of *Talaromyces wortmannii* (Klöcker) C.R. Benj. by Brian's group in 1957 [1]. The structure was elucidated in the late 1960s by the competing research groups of MacMillan [10] and Petcher [12]. The main producer is *Talaromyces funiculosus* (Thom) Samson, N. Yilmaz, Frisvad & Seifert (*Penicillium funiculosum* Thom).

Wortmannin reacts with various biological targets. Best studied is the covalent and hydrogen bonding to a lysine residue (Lys-802) of the ATP recognition region of phosphatidylinositol 3-kinase (PI3K). Inhibition of phosphorylation at the 3-position of the inositol ring by wortmannin results in the interruption of signaling necessary for cell division and differentiation and in the inhibition of autophagy. In addition, wortmannin inhibits the serine-specific auto-kinase activity of mTOR (mammalian target of rapamycin). This effect contributes to its anticancer properties. Viridiol also prevents autophagosome formation [6].

Wortmannin and related compounds exhibit cardio-, hepato-, and nephrotoxic effects. Hemorrhages and necrosis of various organs have been found in animal studies [3, 14].

Wortmannin, together with fusarochromanone (Fig. 3.25), are thought to interfere with chick maturation in the egg ('avian tibial dyschondroplasia'). The compounds are produced by *Fusarium equiseti* (Corda) Sacc. occurring on corn [3]).

Wortmannin serves as research tool and as lead compound for the synthesis of potential new cytostatic drugs [6]. Fusarochromanone is of interest as inhibitor of angiogenesis [9].

The cyclobutene derivative moniliformin (Fig. 3.25), presumably a diketide, is produced especially by *Fusarium fujikuroi* Nirenberg *(F. moniliforme* J. Sheld.) and *F. avenaceum*, frequently found on grain maize or on wheat. It is a strong organic acid and occurs in nature typically as sodium or potassium salt. Finnish cereal samples collected in 2005 were positive for moniliformin with a mean level of 190 µg/kg and a maximum level of 850 µg/kg [4]. Norwegian grain samples, collected from 2000 to 2002, contained maximal concentrations of 380 µg/kg in barley, 210 µg/kg in oats, and 950 µg/kg in wheat [13].

Moniliformin irreversibly inhibits thiamin pyrophosphatase–dependent enzymes, among them pyruvate dehydrogenase, thereby blocking the citric acid cycle. The impaired energy supply of the cells first affects the cardiac muscle cells. Immunosuppressive effects have also been detected. Genotoxicity is controversial. Symptoms of poisoning in animals include decreased feed intake, weight loss, myocardial damage, ascites, and visceral hemorrhage. Lethal doses (LD_{50} mouse: 21–29 mg/kg BW, p.o.; chicken 4.5 mg/kg BW, p.o.; female mink 2.2–2.8 mg/kg BW, p.o.) cause heart damage and respiratory paralysis. Poultry are the most affected animals [2, 4, 5, 7, 8, 11, **116**].

References

For numbers in bold, see cross-chapter literature p. 233.

[1] Brian PW et al. (1957) Trans Brit Mycol Soc 40: 365

[2] Broomhead JN et al. (2002) Avian Dis 46: 901

[3] Bryden WL et al. (2004) in: Poisonous Plants and Related Toxins, Acamovic T et al. (eds.) CABI Publishing, Wallingford, UK, p. 63

[4] Gruber-Dorninger C et al. (2017) J Agricult Food Chem 65: 7052

[5] Harvey RB et al. (2001) J Food Protect 64: 1780

[6] Hassan AMIA et al. (2024) Int J Mol Sci 25: 7459

[7] Knasmueller S et al. (1997) Mutat Res 391: 39

[8] Li YC et al. (2000) Poultry Science 79: 871

[9] Luces CA, Wuthier RE (2003) Abstracts, 225th ACS Nat Meeting, New Orleans 2003, CHED-869, Publisher: Am Chem Soc, Washington DC

[10] MacMillan J et al. (1968) J Chem Soc Chem Commun 1968: 613

[11] Morgan MK et al. (1999) Vet Hum Toxicol 41: 1

[12] Petcher TJ et al. (1972) J Chem Soc Chem Commun 1972: 1061

[13] Uhlig S et al. (2004) Food Addit Contam 21: 598

[14] Wipf P, Halter RJ (2005) Org Biomol Chem 3: 2053

Part II: **Bacterial Toxins**

4 Toxins of Human Pathogenic Bacteria

4.1 General

Bacteria are ubiquitous prokaryotic microorganisms that mostly consist of one biological cell and can reproduce independently.

Although some bacterial toxins, such as, e.g., diphtheria toxin, elucidated by Friedrich Loeffler, Émil Roux, and Alexandre Yersin in 1888, tetanus toxins, elucidated by Knud Helge Faber in 1889, and botulinus toxin, elucidated in 1896 by Emile Pierre van Ermegem, are known for more than 100 years, intensive research into the structure and mechanisms of action only began in the second half of the twentieth century (for history, see [5, 90]).

The toxins of microorganisms are either components of the cell wall and released at larger scale only in the case of lysis of the cell wall (endotoxins) or they are intracellular toxins, that are produced to be secreted into the environment (exotoxins).

Endotoxins are mostly complex lipopolysaccharides that occur as components of the cell wall of gram-negative bacteria.

Exotoxins are polypeptides and proteins that are produced by gram-positive and by gram-negative bacteria. They are mostly not heat resistant. They include membrane damaging toxins, superantigen-toxins that hyperactivate the immune system, and AB-toxins with enzymatic and other functions.

Some of the bacterial toxins are used as experimental tools in cell biology and pharmacology [2, 34, 115], in the form of toxoids for vaccine production and, in few cases (botulinum toxins and diphtheria), as drugs.

Poisoning caused by bacterial toxins must not be confused with infections caused by pathogenic bacteria.

Cyanobacteriota is an important phylum of bacteria whose representatives are capable to form both high-molecular metabolites and low-molecular poisonous compounds. The cyanobacterial toxins are described separately from those of other bacteria in a separate chapter (see Section 5).

4.2 Bacterial Endotoxins

Endotoxins are components of the outer regions of the bacterial cell wall. As amphiphilic molecules, they play a key role in the cell wall barrier function. In higher animals and in human beings soluble antibodies are formed against them during the whole span of bacterial infection. They are released to a small extent from the intact cell into the medium, but the largest part is released during lysis of the bacteria (overview: [64, 140]).

https://doi.org/10.1515/9783110728576-004

The endotoxins of gram-negative bacteria have been particularly well studied. They are complex lipopolysaccharides (LPS). Their molecular mass is between 10 and 90 kDa. They can be heated up to 120 °C for several hours without losing their toxicity. They are composed of a hydrophilic and a lipophilic part. The hydrophilic part is a so-called O-polysaccharide. It consists of the core oligosaccharide and an O-specific chain. The lipophilic part is the component lipoid A [20, 21].

The O-specific chain consists of several repeating oligosaccharide units, which are built of 5-galactose and *N*-acetyl-D-glucosamine. The core region is characterized by the unusual ketose 3-deoxy-D-manno-2-octulosonic acid (KDO), which can serve as an antigen determinant. In most cases, however, antibodies are only formed against the O-chain; the other parts of the molecule are masked in the intact lipopolysaccharide.

Lipoid A, the endotoxic principle of lipopolysaccharides, contains a bi-phosphorylated disaccharide composed of 2β(1->6)-linked D-glucosamine residues. This disaccharide is linked to 5 or 6 molecules of ester- or amide-like linked 3-hydroxy fatty acid residues, which in turn may be esterified at their hydroxy groups with fatty acids [23, 47, 48, 95]. Structure-activity studies made it clear that a maximum endotoxic effect is exerted by a molecule containing a 2β(1->6)-linked D-glucosamine disaccharide to which two phosphate residues and at least 5 and at most 6 fatty acid groups are bound in a defined arrangement. The presence of a 3-acyloxyacyl structure is essential for the effect. Even small modifications reduce the effect [96].

The lipopolysaccharides trigger the release of endogenous mediators, e.g., interleukins, leukotrienes, and tumor necrosis factor in macrophages and monoclonal cells of the host organism, mediated via toll-like receptors [96]. The toxic effects of lipopolysaccharides result from the effects of the released factors. Even the injection of very small amounts of lipopolysaccharides (1–2 ng/kg BW, i.v.) leads to fever (pyrogens), chills, nausea, vomiting, changes in the white blood count, consumption coagulopathy, shock, and possibly death in humans. Hemorrhagic reactions occur at the injection site, with repeated injections leading to necrosis (Sanarelli–Shwartzman phenomenon). Perorally ingested lipopolysaccharides are not absorbed and therefore do not cause symptoms of intoxication even in high concentrations [51].

Beneficial effects of lipopolysaccharides for the host organism include their immunostimulatory activities and the resulting improved defense against bacterial infections [96].

In gram-positive bacteria teichoic acids act as activators of the immune system. They consist of chains of polyols, e.g., ribitol, mannitol, and glycerol, linked through phosphodiester bridges and substituted by sugars and D-alanine. They can cause similar symptoms like lipopolysaccharides, but only in concentrations 100–1,000 times higher [95].

For humans, endotoxins are initially important as markers for the existence of an infection and for triggering the formation of antibodies. However, if the immune defense is inadequate, e.g., after serious injuries or operations, they are released during bacteriolytic processes and contribute significantly to the course of the disease.

Poisoning by bacterial endotoxins results, e.g., from a local bacterial infection, especially with gram-negative bacteria, or from parenteral administration of drugs containing lipopolysaccharides.

Symptoms include fever, chills, nausea, vomiting, confusion, tachycardia, tachypnea, thrombocytopenia, leukocytopenia, and, in the case of local penetration, hemorrhagic disease and necrosis at the site of infection. In severe cases, penetration of the bacterial pathogens from the infection site into the bloodstream leads to sepsis ('blood poisoning'). It is one of the main causes of death in hospital intensive care units, especially in elderly or weakened patients. The toxins can persist in the tissue and cause chronic inflammation, e.g., carditis and synovitis [54].

Sepsis is treated causally by sanitizing the source of infection, applying high doses of broad-spectrum antibiotics, and later, after identification of the pathogen, by administration of specific antibiotics and symptomatic treatment. Intensive care measures to support organ function and to prevent organ failure and complications are particularly important [54, 133].

All injection and infusion solutions must be tested for the absence of endotoxins prior to administration in medicine [140].

Due to the lack of or low absorption from the gastrointestinal tract, the presence of endotoxins in food is not of toxicological significance [51].

Bacterial endotoxins are used for diagnostic and experimental purposes.

4.3 Bacterial Exotoxins

4.3.1 General

Exotoxins are formed in bacterial cells and are either secreted from the intact bacteria or released when the bacteria are lysed. Each toxic protein provides the bacterium with a selective advantage in host-pathogen interaction and can be a target for a monoclonal antibody to treat the respective infection [58].

Exotoxins are often single-chain polypeptides or proteins. They are usually formed in the form of inactive precursor proteins (protoxins) and activated by limited proteolysis. The molecular mass of the proteins is between 2 kDa and 3 kDa for the thermostable enterotoxins of *Escherichia coli* and 300 kDa for the *Clostridium difficile* toxins A and B.

Many toxins occur as oligomeric complexes consisting of two or more non-covalently linked subunits. Cholera toxin and the heat-labile enterotoxins I and II of *E. coli*, for example, form heterohexamers (1A + 5B complex). The A subunit, relative molecular mass 28 kDa, transfers ADP-ribosyl residues from NAD to target proteins. The five identical 11.8 kDa B subunits mediate the specific binding of the toxin to the ganglioside GM1 on the surface of the intestinal target cells. The most complex bacterial toxin known to date is the pertussis toxin produced by Bordetella species. It is also an A1 + B5 hexamer.

The spatial structures of several toxins have been determined, e.g., of diphtheria toxin, cholerae enterotoxins, botulinus toxin A, and some staphylococcal superantigens [5, 73, 90].

The toxins can be divided into those that act on intracellular targets after uptake into the cell and those that damage the bacterial cell membrane.

The genes encoding the structures of the toxins are localized on the chromosomes of the bacteria or on mobile genetic elements, e.g., on plasmids, on transposons, or on bacteriophages. Many of these genes are already sequenced.

4.3.2 Toxins with Intracellular Targets (A/B Type Toxins)

Toxins with intracellular targets, e.g., cholera toxin, botulinus toxin, tetanus toxin, pertussis toxin, bordetella adenylate cyclase toxin, and shiga toxins, consist of two different functional units (A and B), which can be localized in two different domains on one polypeptide chain (e.g., diphtheria toxin) or on different chains. The B unit of the toxin (the haptomer) binds to a surface structure of the attacked cell of the infected person and mediates the uptake of the A part into the cell. The A-unit (the toxomer) causes damage to the cell after penetration. The process of uptake is subject to very complex mechanisms and includes endocytosis, translocations, and release processes into the cytosol. In the case of diphtheria toxin and probably also other exotoxins, a cytosolic translocation factor complex is involved in the uptake into the cell [63, 128].

The uptake mechanism of the toxins of *Bacillus anthracis*, the causative agent of anthrax, has been relatively well studied. The pathogen produces three factors, the edema factor (EF), the lethal factor (LF), and the protective antigen (PA). PA binds to anthrax toxin receptors of the host cells (ANTXR 1 and ANTXR 2) where it is cleaved into two fragments. After release of a small fragment, the bound large fragment forms a ring-shaped heptameric aggregate called a 'prepore'. EF and/or LF bind to the 'prepore' and change its conformation. The 'prepore' is thus transformed into a membrane-penetrating pore and the destructive proteins EF and LF enter the cytosol [26, 116].

The cell-damaging effect of the A-units is caused by different enzymatic activities depending on the toxin. Diphtheria toxin, *Pseudomonas aeruginosa* exotoxin A, cholera toxin, heat-labile enterotoxins of *Escherichia coli* and pertussis toxin are ADP-ribosyltransferases that catalyze the transfer of adenosine ribosyl phosphate residues from NAD to target proteins. In the case of diphtheria toxin and pseudomonas toxin, this transfer takes place to elongation factor 2 (EF-2), which is necessary for protein biosynthesis. In the case of cholera toxin, the heat-labile enterotoxins of *E. coli*, and pertussis toxin, the transfer is made to a cGMP-dependent regulatory protein (G protein) of adenylate cyclase [50]. Inactivation of these target proteins follows. The Bordetella adenylate cyclase toxin and the EF toxin from *Bacillus anthracis* are adenylate cyclases. They trigger an intracellular flooding with cAMP and thus a loss of regulation of the affected cells. The A1 chain of shiga toxins (verotoxins, from *Shigella dysenteriae*) and of shiga-like toxins (shigella toxin like, STL) from strains of *E. coli, Salmonella typhimurium*, and *Campylobacter jejuni* enzymatically inactivates the 60 S subunit of ribosomes (RNA-*N*-glycosidase: [5, 49, 73, 90]).

4.3.3 Toxins Acting on the Cell Membrane

Cytolysins ('membrane damaging toxins', MDT) do not penetrate the cell but damage the cell membrane from outside and thus cancel its barrier function. Other toxins interfere with signal transduction (overviews: [73, 90, 129]).

About 35% of the bacterial toxins known to date are cytolytic active. They are either enzymatic active and destroy the phospholipid bilayer of the cell membrane, solubilize membrane components due to their surfactant properties or form pores through the cytoplasmic membrane. Cytolysins include perfringolysin O from *Clostridium perfringens*, leucocidins from *Staphylococcus aureus*, and aerolysin from *Aeromonas hydrophila*. The α-toxin (α-hemolysin) of *S. aureus* is a major virulence factor in human infections; it causes, e.g., boils, tonsillitis, conjunctivitis, mastitis, and colpitis, due to its membrane-destroying effect. It is secreted as a monomer and initially forms a heptameric structure, the prepore, on the surface of the host membrane. A conformational change causes the prepore to form a membrane-penetrating pore [14, 88, 126].

The toxins that influence signal transduction bind to membrane receptors. For example, the heat-stable enterotoxins of *E. coli* activate guanylate cyclase. The increase in the cGMP level in the cell stimulates chloride secretion and/or inhibits NaCl reabsorption and thus triggers diarrhea. Another example of this group of toxins are the superantigens, which include those of *S. aureus* and *Streptococcus pyogenes*. They bind to molecules of the MHC II (major histocompatibility complex II) of antigen-presenting cells outside the classic antigen-binding sites and at the same time to T cells (T cell receptor, TcR). The resulting excessive activation of T cells leads to the increased release of cytokines and other factors that trigger inflammation. More than 20 superantigens are known from *S. aureus*, among them the enterotoxins A to E and the toxic shock syndrome toxin 1 (TSST-1).

Some well-studied bacterial exotoxins will be described in more detail in the following.

4.3.4 Enterotoxins of *Staphylococcus aureus* and Other Bacteria

Enterotoxins are usually exotoxins. They are released by their producers into the intestinal lumen, or they are secreted by bacteria living on food and may be ingested by humans with food. They act preferentially in the intestine.

Pathogens that produce enterotoxins and can cause bacterial food poisoning include, e.g., *Bacillus cereus* (produces the heat-resistant dodecadepsipeptide cereulide with strong emetic activity), *Clostridium perfringens* (see Section 4.3.5), *Escherichia coli* (enterohemorrhagic strains, EHEC, produce shiga-like toxins, verotoxins), *Salmonella enteritidis, S. typhimurium, S. cholerae-sius, Shigella dysenteriae* (produce shiga toxins), *Sh. sonnei, Staphylococcus aureus* (see this Section), *Streptococcus faecalis, Vibrio cholerae* (see Section 4.3.6), *V. parahaemolyticus,* and *Yersinia enterocolitica* [6, 41, 97, **97, 117**].

Staphylococcus aureus is a gram-positive, spherically shaped, facultatively anaerobic, catalase positive, not movable bacterium (Ph. 4.1). It is a member of the microbiota of the body and is found on the skin and in the mucous membranes of the upper respiratory tract of humans and animals. Between 15% and 40% of the healthy human population carry *St. aureus*. *St. aureus* can become an opportunistic pathogen.

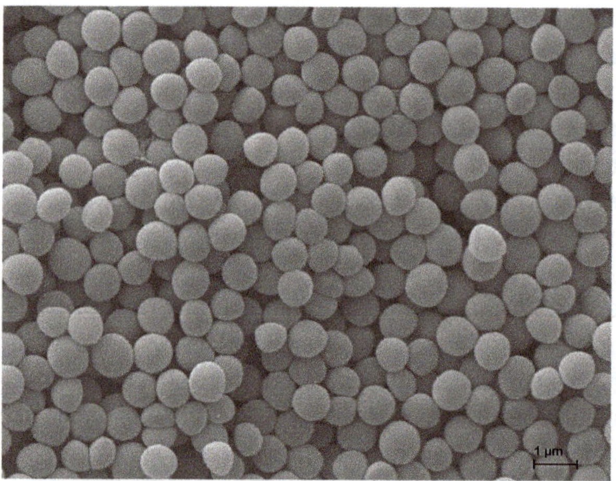

Ph. 4.1: *Staphylococcus aureus*, SEM image, magnification 8,000 times (source: provided by Florian Bonn, Dep. of Microbial Proteomics, Inst. of Microbiology, and imaged by Rabea Schlüter, Imaging Center of the Dep. of Biology, Univ. of Greifswald; acknowledgement to Stefan Bock for technical assistance).

The multiplication of *S. aureus* and its toxin formation are possible at temperatures between 7 and 45 °C, optimal between 30 and 37 °C. The bacteria are resistant to drying out and largely tolerant to pH changes. They are capable to form biofilms, e.g., on medical implants, and can cause biofilm-associated infections [1, 76, 88, 97].

Toxins that are involved in the pathogenicity of *St. aureus* are

- Staphylococcal pore-forming toxins (PFTs, create pores in the host cell membrane), among them
  - hemolysins, lyse erythrocytes, α-hemolysin is an important virulence factor and contributes to severe infections such as pneumonia and osteomyelitis;

  - Panton-Valentine leukocidin (PVL), targets white blood cells, linked primarily to skin and soft tissue infections;
  - phenol-soluble modulins (PSMs), aggressive virulence factors and cytolysines,
  - epidermal cell differentiation inhibitor (EDIN) exotoxins, disturb endothelial barrier, promote bacterial colonization and host tissue invasion.
- Exfoliative toxins (ETs), possess protease activity and cause epidermal dissociation of the human epidermis, responsible for staphylococcal scalded skin syndrome,
- Superantigens (Sags), T cell mitogens, induce the secretion of proinflammatory cytokines and possibly multiorgan failure, among them
  - staphylococcal enterotoxins (SEs);
  - toxic shock syndrome toxin 1 (TSST-1) 1).

Only the enterotoxins and TSST-1 will be discussed in more detail here.

Staphylococcal enterotoxins are 20–30 kDa proteins that disturb intestinal activity and induce staphylococcal food poisoning (SFP). At present, more than 24 enterotoxins of *St. aureus* are differentiated serologically and named alphabetically, e.g., SEA–SEV. They are water-soluble, stable even when heated to 100 °C for 30 min, and resist hydrolysis by gastric and jejunal enzymes [79]. Clinical signs of SFP have been linked to inflammatory mediators such as leukotriene B4 and prostaglandin E2 that are produced in response to SEs [1, 25, 88, 90, 124].

SEs, especially SEB, are one of the most common sources of food poisoning. It occurs when improperly handled foodstuffs are contaminated with *St. aureus*, which in turn produces and releases SEs into the food that is subsequently ingested. The outbreaks often affect whole groups of people who have eaten the same meal. In humans, ingestion of less than 1 µg SEA is sufficient to trigger symptoms of poisoning [57, 122, 123, **117**].

Food contamination is usually caused by workers in the food industry who harbor the bacteria on their skin, in small pustules, in the nasopharynx or in the intestine. Flies, cockroaches, and ants can also be the carrier. The main substrates are raw milk [111] and dairy products, ice cream, meat, sausages, fish preserves, potato salad, and vegetable products [43, 54, 122, **117**].

Poisoning may be also the result of infections with *St. aureus* (enterotoxicosis). The incubation period after ingestion of toxins is 2–6 h, by infections 4–10 days [97, 124].

☠ One to six hours after eating food containing *S. aureus* enterotoxins occur nausea, vomiting, extremely frequent discharge of watery stools, abdominal pain, and circulatory problems. Fever only occurs in very severe cases. As a rule, the symptoms subside within a few hours and disappear after a day at the latest. Severe intoxication and deaths are rare and only occur in individuals with weakened immune system [1, 57, 122, **117**].

Apart from plenty of fluids and restriction of food intake, treatment is usually not necessary. In severe cases, parenteral fluid and electrolyte replacement and shock control may be necessary [54].

TSST-1 (Mr about 23 kDA) stimulates the release of proinflammatory cytokines from the host T cells and macrophages. These mediators cause the toxic shock syndrome (TSS), which is characterized by high fever, rash, desquamation, hypotension, and hypovolemic shock. It can progress to multiorgan failure. TSST-1 was first identified in 1981 as the cause of menstrual toxic shock syndrome triggered by contaminated tampons. It is rare but maybe life-threatening [1, 7, 90].

Poisoning by bacterial enterotoxins can best be avoided by taking hygienic measures to prevent the transmission of bacteria from humans into food, e.g., food should be touched as little as possible with the hands ([12], overview on new antitoxin treatments, e.g., with antibodies or nanoparticles: [1]).

Staphylococci are increasingly developing multiple resistance to antibiotics (MRSA = methicillin resistant *Staphylococcus aureus*). Resistance results from the formation of enzymes (β-lactamases) that degrade the β-lactam antibiotics or from the formation of a changed penicillin-binding protein PBP2a, genetically based on the mecA gene. MRSA pose a great hazard as a cause of nosocomial infections. Community acquired MRSA (cMRSA) occur outside of hospitals and can be responsible for infections in previously healthy people [24, 61, 131].

4.3.5 Enterotoxins of *Clostridium perfringens*

Clostridium perfringens is a gram-positive, rod-shaped, non-flagellated, mostly pore-forming, anaerobic bacterium (Ph. 4.2). It forms spores that are resistant to stressful environments, such as high temperatures. It is an important causative agent of gas gangrene and a potential trigger of food poisoning. It can multiply rapidly in many foods, especially in meat products including poultry, less frequently in fish, dairy products, and vegetables, at temperatures between 20 and 50 °C and at low oxygen partial pressures. It is found in the intestinal flora of humans and animals. Its spores can survive in soil or sediments and are widespread in the wild [29]. Reproductive bacteria die when heated to 70 °C or more for at least 2 min. The heat resistance of the spores varies widely and can be up to 60 min at 100 °C.

C. perfringens is classified according to its pathogenicity factors into the serotypes A to F. Sources of contamination with *C. perfringens* type A and F that are most relevant for food poisoning are dust, contaminated water, soil particles, and human and animal feces.

Clostridium-perfringens-enterotoxin (CPE) is the primary toxin responsible for food poisoning of humans and animals. It is a single-chain protein consisting of 319 amino acids and three domains: I, II, and III. Domain I binds to the claudin receptor

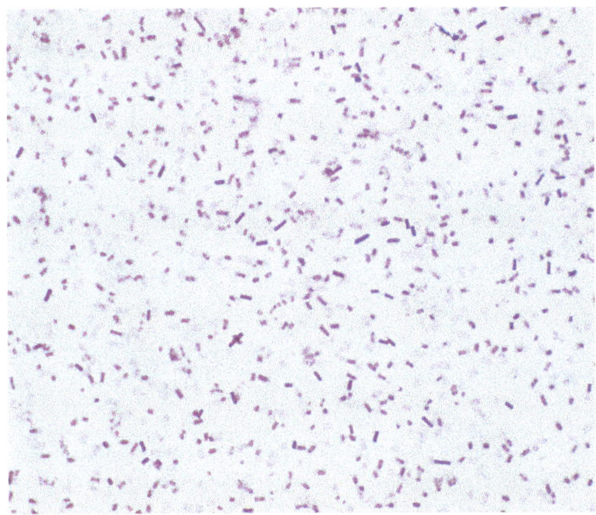

Ph. 4.2: *Clostridium perfringens* (source: J. Douan & K. Becker, Universitätsmedizin Greifswald).

(components of the tight junctions between epithelial or endothelial cells at the cell surface), while domains II and III are responsible for the formation of pores in the cell membrane. Consequently, tight junctions between cells are disrupted. The increased permeability of cell membranes leads to loss of fluids and ions and osmotic imbalance. CPE is resistant to trypsin [18, **117**].

Further major toxins of *C. perfringens* are

- alpha-toxin (CPA, *cpa/plc* gene, 370 amino acids), lecithinase, necrotizing;
- beta-toxin (CPB, *cpb* gene, 336 amino acids), necrotizing, responsible for intestinal fire;
- epsilon-toxin (ETX, *etx* gene), necrotizing;
- iota-toxin (ITX, *iap, ibp* binary genes), necrotizing.

Additionally, several accessory toxins, neuramidases, and hyaluronidases occur [18].

Food poisoning by *C. perfringens* that occurs after ingestion of contaminated food or feed is the second leading cause of foodborne bacterial diseases in the USA (one million illnesses yearly) and Europe's fourth leading cause [18].

☠ Symptoms following the consumption of food contaminated with *C. perfringens* occur after an incubation period of 8–24 h; they include diarrhea, abdominal pain, possibly nausea, and vomiting, rarely fever, and headache. The symptoms usually pass within 24 h.

Other serotypes of *C. perfringens* spread rapidly in the tissue in the event of major soft tissue damage, open fractures, gunshot wounds, or stab wounds. After an incubation period of 1–4 days, severe myonecrosis develops. The transfer of the toxins into the systemic circulation can lead to damage of vital organs. Each toxin type is associated with a specific disease: For example, toxinotype A, characterized by the presence of *cpa*+ toxin type toxin genes, is linked to gas gangrene in humans and animals. Toxinotype F with *cpa*+ and *spe*+, is associated with food poisoning and antibiotic-associated diarrhea [18, 29, 117].

Treatment of poisoning by enterotoxins of *Clostridium perfringens* must be symptomatic. In most cases, the symptoms subside after 1 to 2 days even without treatment.

4.3.6 Enterotoxins of *Vibrio cholerae*

The toxigenic strains O1 and O139 of *Vibrio cholerae* are the causative agents of the highly contagious diarrheal cholera. *V. cholerae* is a gram-negative, curved rod-shaped bacterium with a terminal flagellum (Ph. 4.3). It can switch between motile and biofilm lifestyles.

It occurs preferably in brackish water of estuaries or in coastal seawater, where it colonizes, e.g., crustaceans and shellfish. If these are consumed, the pathogens colonize the human intestine. They are spread further via the fecal–oral route. The risk of spreading is particularly high under poor hygienic conditions. The TCP adhesin (receptor for CTX phage) is of decisive importance for the interaction of the pathogen with the epithelial cells of the intestinal mucosa. It consists of hydrophobic bundled filaments that belong to the group of type IV pili [35, 54, 58, 59, 103, 107, 114, 135].

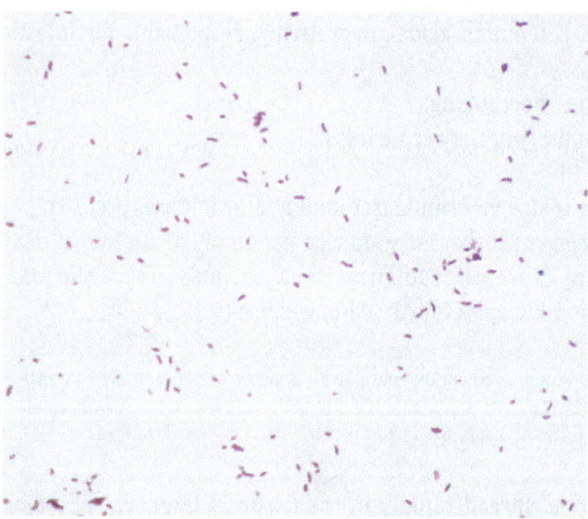

Ph. 4.3: *Vibrio cholerae* (source: J. Douan & K. Becker, Universitätsmedizin Greifswald).

The cholera enterotoxin (CTX), secreted by *V. cholerae*, is composed of an A subunit (Mr 27 kDa, 240 amino acid residues) and 5 B subunits (Mr 11 kDa, 103 amino acid residues) (AB5 type of ADP-ribosyltransferases). At least three other CT-related enterotoxins are known. The heat-labile enterotoxins (LT) of *Escherichia coli* and Salmonella species are very similar in structure. The heat-stable toxins (ST) of *E. coli*, on the other hand, have much smaller molecular masses than the heat-labile cholera toxins.

Cholera toxin is expressed by a DNA segment of *V. cholerae* that originally comes from the filamentous, lysogenic bacteriophage *Vibrio virus* CTXϕ [107].

The B subunits of the cholera toxin (cholera genoid) act as haptomers by binding to the ganglioside GM 1 of the mucosal epithelium (toxin receptor) and mediate the uptake of the A chain into the cells of the intestinal mucosa. Upon entry into the cells, the toxin is proteolytically cleaved by reduction of the S–S bridges into the enzymatically active cleavage piece of the A chain A1 (Mr 21.8 kDa) and the inactive subunit A2 (Mr 5.4 kDa). A1 transfers the ADP-ribose residue of NAD to the GTP-containing α-subunit of a G-protein, inhibits its GTPase activity and leads to a permanent activation of adenylate cyclase [38, 130]. In addition, CTX is also thought to increase the production of the prostaglandins PGE1 and PGE2 and the secretion of the vasoactive intestinal peptide (VIP) [54].

As a result of the unregulated activation of adenylate cyclase, there is a very strong increase in the cAMP level of the cells of the intestinal wall. Mediated by protein kinases, e.g., protein kinase A, this leads to excessive secretion of water and electrolytes into the intestinal lumen while at the same time reabsorption is inhibited. As a result, death from dehydration can occur. Through their effect on the enteric nervous system, cholera toxins also stimulate intestinal peristalsis [38, 130]. While no effects were detectable after the application of 2.5 µg of purified cholera toxin, 5 µg resulted in severe diarrhea in 4 out of 5 people examined with an excreted amount of 1–6 L of stool [60]. The LD_{50} in mice is 250 µg/kg BW [4].

Cholera occurs especially in regions of Africa and Asia in countries with poor sanitary condition. In 2023, more than 535,321 cases and 4,007 deaths from 45 countries have been recorded by the WHO. Researchers estimate the number of cases between 1.3 and 4 million cases and 21,000–143,000 deaths per year worldwide [139]. In Germany, nine travelling-associated infections (Pakistan, Kamerun, Irak) have been reported in 2023. Eight out of the nine people were not immunized. Three cases in February 2025 were caused by contaminated water brought from Ethiopia [102].

The incubation time is 12 h to 6 days after infection. Most people show no or mild symptoms. Nevertheless, they can spread the bacteria by their feces for 1–10 days. In severe cases, cholera manifests itself as febrile enteritis with 'rice water stools', muscle cramps, oliguria, exsiccosis, and circulatory failure [35, 58]. The clinical course of an infection depends heavily on the causative strain (classical type, type 01, type 0139, El Tor biotype) and the general condition of the patient.

Patients with mild symptoms can be treated with oral fluid substitution. Severe cases need i.v. fluid substitution, oral rehydration, and antibiotics [58, 135]. If treatment is adequate, the fatality rate is below 1%. Active immunization is possible and lowers the number of resistant infections [54, 65, 75]. Population's access to safe water, basic sanitation, and hygiene is essential to prevent infections [135].

4.3.7 Neurotoxins of *Clostridium tetani*

The causative agent of tetanus, *Clostridium tetani,* is a gram-positive, spore-forming, strictly anaerobic, rod-shaped bacterium. Its terminals are characteristic formed like 'drumsticks' (Ph. 4.4). Its spores are found in soil, ash, in the intestinal tracts and feces of animals and humans, and on the surfaces of skin and rusty tools such as nails and barbed wire. The spores are very resistant to heat and most antiseptics. They can survive for years. Tetanus cannot be transmitted from person to person. Infections result from wound contamination with *C. tetani* spores [87, 127, 138].

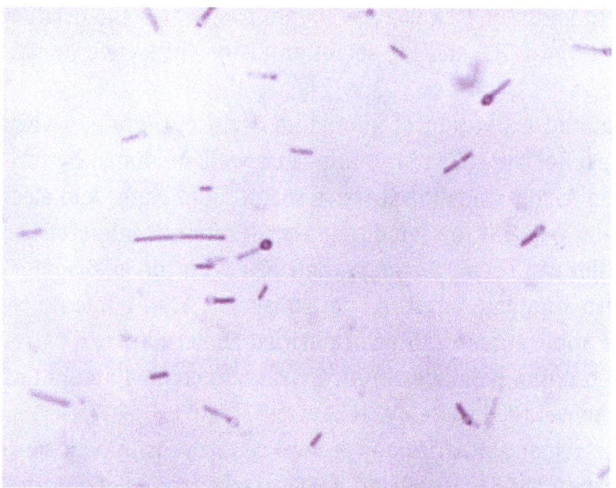

Ph. 4.4: *Clostridium tetani* (source: Institut für Mikrobiologie und Infektionsimmunologie, Charité – Universitätsmedizin Berlin).

The contaminating tetanus spores germinate in wounds under anaerobic conditions into vegetative bacteria that multiply and produce intracellularly the tetanus neurotoxin (TeNT). At the time of manifestation of the disease, the wound may already have healed [70, 87].

TeNT (tetanospasmin, Mr 150 kDa, 1,315 amino acid residues) resembles in primary structure the botulinus toxin. It is actively secreted from the bacterial cell or released upon lysis of the bacterial cells and cleaved into a light fragment (Mr 50 kDa, zinc-containing) and a heavy fragment (Mr 95 kDa). Both fragments remain connected by a S–S bridge (toxin BE, not homogeneous, isotoxin mixture). The light fragment probably functions as a toxomer, the heavy fragment as a haptomer. The genetic information for the toxin is located on a large plasmid of 75 kb. Also worth mentioning is the hemolysin tetanolysin [54, 69, 78, 87, 106, 132].

Tetanospasmin initially binds to the presynaptic membrane of nerve cells. After cleavage of the S–S bridge, the light fragment enters the axon by endocytosis. After axonal transport and transsynaptic migration from the motor neurons into the interneurons, it cleaves VAMP (vesicle associated membrane protein, synaptobrevin), a membrane protein of the small synaptic vesicles that regulates exocytosis and thus neurotransmitter release. This prevents the release of inhibitory transmitters, above all, glycine and γ-aminobutyric acid (GABA). The toxin thus leads to death by spastic convulsions, in high doses also by paralysis [71, 78, 92, 106, 132].

With an LD_{50} of 2 ng/kg BW, s.c., mouse, the two-chain toxin BE is about twice as toxic as the single-chain toxin S (LD_{50} 4 ng/kg BW, s.c., mice, [132]). In humans, less than 1 µg tetanospasmin is probably lethal [54].

The main risk of tetanus for adults is from injuries sustained during gardening and agricultural work, open traffic accident injuries, stab and gunshot wounds, and non-sterile syringes in drug addicts. In Germany, there were less than 15 cases per year during the last years. In tropical countries where tungiasis, sand flea disease, occurs, the fleas can bring in bacteria from the soil when they enter the skin [36]. Tetanus neonatorum originates from an umbilical cord infection or from unhygienic conditions during delivery. In 2018, approx. 25,000 newborns died from neonatal tetanus (1988: 787,000), especially in low-income countries [138].

The incubation period of tetanus is between 3 and 21 days, rarely several months. A striking symptom is the progressive stiffness of the masticatory muscles. Subsequently, the facial muscles become paralyzed (facies tetanica, risus sardonicus) and other areas are affected. Even minor stimuli such as light and touch reinforce the seizures, which the affected person experiences with unimpaired consciousness. If tetanus is generalized, spasticity of the muscles of the trunk and upper and lower limbs follows. Despite therapy, around 13% of those treated die by cardiac arrhythmia, malignant hypertrophy, or respiratory failure. Tetanus neonatorum always ends fatally [36, 54, 87, 104].

Treatment involves high doses of tetanus antiserum (human tetanus hyperimmunoglobulin), excision of the entry site to prevent further toxin formation, administration of drugs to control muscle spasms and to maintain vital functions, accommodation in a dark room, and administration of analgesics and antibiotics (metronidazol) to prevent complications. Active immunization can prevent infection [36, 53, 54, 104, 138].

Animals can also be infected. The most susceptible are horses, guinea pigs, monkeys, and birds. Carnivores, e.g., cats and dogs, are less vulnerable. Sheep and goats are quite resistant. Poikilothermic animals are resistant when kept at low temperatures (<18 °C, [87]).

4.3.8 Neurotoxins of *Clostridium botulinum*

The causative agent of botulism, *Clostridium botulinum*, is a gram-positive, strictly anaerobic, flagellated, spore-forming rod-shaped bacterium (Ph. 4.5). It is like *C. tetani* a so-called soil bacterium. The pathogen occurs in a variety of species and strains. The spores are extremely resistant and can survive in food, which is insufficiently heated. Reactivation of the spores is possible even after decades [30]. It will not grow in acidic conditions.

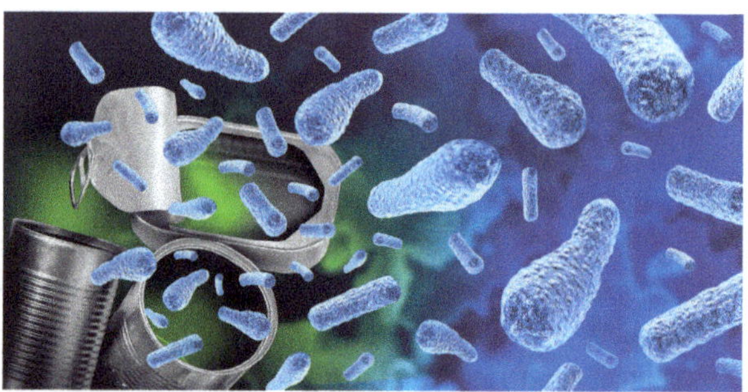

Ph. 4.5: *Clostridium botulini* (source: wildpixel/iStock/Getty Images Plus).

C. botulinum and some other species of the genus Clostridium, e.g., *C. butyricum, C. barati*, and *C. sporogenes*, produce botulinum neurotoxins (BoNTs). The toxins are thermolabile and are inactivated by exposure to 80 °C for 10 min. The nomenclature is not consistent. Presently, eight serotypes (A to H) are described. They occur in numerous subtypes or mosaic forms, e.g., A1 to A8. Types A, B, E, and rarely F cause human botulism. Types C, D, and E cause illness in other mammals, birds, and fish [31, 83, 134].

The toxin-producing bacteria can be divided into four groups according to their physiological/biochemical performance. Group I, proteolytic effective, includes representatives that produce toxin types A, B, and F. Representatives with toxin types B, E, and F are classified in group II, non-proteolytic effective. Group III includes strains that produce toxin types C and D. Strains with toxin type G are also referred to as *Clostridium argentinense*. Only groups I and II with toxins A, B, E, and, to a lesser extent F, are of pathogenetic relevance for humans [30, 72, 82, 85, 86, 92, 94].

The habitat for most types is the soil. The pathogen multiplies at temperatures above 10 °C and pH values between 4.8 and 8.0 under anaerobic conditions, e.g., in vacuum packaging, in canned products, and in tins, which are often characterized by a bombage after contamination. Substrates are foods of animal origin such as smoked sausage products (lat.: botulus = sausage), raw ham, smoked or cured meat, preserved fish such as salt or smoked fish, and marinades as well as vegetables preserved in the household, e.g., green beans, spinach, asparagus, and tomatoes. Insufficiently acidified foods stored in jars, such as sour beans, in beet and dairy products, can also contain toxins [30, 53, 89, 134]. The spores are not killed at 100 °C but can be inactivated by combined thermal treatment and irradiation [77].

The pure botulinum toxins have molecular masses of 150 kDa. Like tetanus toxin, botulinum toxins are activated by fragmentation into 2 fragments (Mr approx. 100 kDa and 50 kDa resp.), which remain connected by a S–S bridge. Here, too, one fragment probably has the task of acting as a haptomer and introducing the toxomer into the cell. Both fragments are non-toxic on their own. The toxins also resemble tetanus toxin in their primary structure [53, 82, 92, 108, 113].

BoNTs are released together with several auxiliary proteins as progenitor toxin complexes (PTCs) to become highly potent oral poisons. The genes for the neurotoxins and the associated proteins are located on one of two bacterial gene clusters, the HA cluster or the OrfX cluster. Both clusters encode not only for the toxins, but also for the nontoxic non- hemagglutinin (NTNH). NTNH protects the neurotoxins in a pH-sensitive complex against the acidic conditions in the stomach. The HA cluster encodes for a 760 kDa multiprotein complex that includes the neurotoxin and the HA proteins (hemagglutinins HA70, HA17, HA33, [55, 56]). The OrfX cluster encodes additionally to the neurotoxin the OrfX/P47 proteins OrfX1, OrfX2, OrfX3, and P47. These associated proteins alone or in combination with each other are nontoxic but in combination with the neurotoxin and with NTNH they act as 'trojanic horse' to increase the oral toxicity of the neurotoxin. The formation of the complex OrfX2-NTNH-BoNT/E and thus the activation takes place only after proteolytic cleavage of OrfX2 by digestive enzymes in the gastrointestinal tract of the host [40, 101].

The neurotoxic botulinus toxins A, B, C1, E, and F are zinc-containing metalloproteinases. After interacting with gangliosides and the vesicle proteins synaptogamine I and II, which act as receptors, they penetrate the peripheral cholinergic nerve endings by endocytosis (double receptor concept). There they enzymatically cleave one or two of the three core proteins of the neuroexocytotic apparatus (SNARE = soluble *N*-ethylmaleimide-sensitive fusion protein attachment protein receptors). Types A and E cleave the protein SNAP-25 (synaptosomal associated protein), C cleaves preferentially syntaxin, and the toxins B, D, F, and G cleave synaptobrevin (VAMP = vesicle associated membrane protein). Thus, the toxins prevent the formation of the SNARE complex that is an essential step in the fusion of neurotransmitter-loaded vesicle with the synaptic membrane. Consequently, the neurotransmitter release mainly at peripheral cholinergic nerve terminals of the skeletal and autonomic nervous system is persistently inhibited. Impulse transmission is only possible again when the nerve endings are regenerated [31, 74, 92, 121].

Botulinum toxins are among the most potent toxins known. The lethal dose for humans is about 1 µg/kg BW, p.o. The lethal dose of botulinum toxin A for a person of 70 kg BW is estimated at 90–150 ng, p.o., 700 ng, inhaled, and 70 µg, p.o. [30, 121].

Between 2012 and 2021, an average of 94 cases of human botulism were annually notified in EU countries [31]. There are different forms of botulism related to the forms of entry of the toxin into the body.

Foodborne botulism is a serious, potentially fatal disease. It is caused by consumption of improperly processed toxin-containing food. Person to person transmission does not occur. Homemade canned, preserved or fermented foodstuffs are a common source. Poisoning has been caused, for example, by eating mascarpone cream (a product that resembles whipped cream made from a heavy cream cheese [115]) and smoked salmon [32] or drinking carrot juice [19]. Twenty-two-month-old twins died after eating contaminated vegetable porridge [39].

In September 2023, health authorities in France reported 15 cases of suspected botulism including one death. Of these cases, 10 have been hospitalized, with eight patients being admitted to an Intensive Care Unit. All suspected cases consumed the same products, sardines in jar, at the same restaurant. The presence of botulinum toxin in serum samples and in the sardines was confirmed [139].

The symptoms of foodborne botulism appear after a latency period of 12–36 h (3 h up to 8 days are possible). The disease is often characterized by the 4 Ds: Diplopie, Dysphagie, Dysphonie, Disarthrie. The symptoms include nausea, dry mouth, vomiting, diarrhea (only in about every 2nd patient), increasing constipation, dizziness, hoarse speech, visual disturbances such as blurred vision, accommodation paralysis, double images, and mydriasis, as well as swallowing and breathing disorders. Increasing muscle weakness leads untreated to death from respiratory paralysis. The lethality rate in untreated cases is greater than 50%. The activity of the neurotoxins can last up to 12 weeks so that the pareses can persist for months. Treated recovery is complete, although it may take a few months [9, 85, 101, 134].

In the case of wound botulism and infant botulism, the toxins are formed within the human organism. They are not destroyed by gastric passage.

In wound botulism the bacteria find suitable anaerobic conditions for toxin production in deep wounds. The toxins cause neurotoxic poisoning symptoms, the gastrointestinal symptoms are absent. The incubation time is 4–14 days. Wound botulism is associated with substance abuse, particularly when injecting black tar heroin [101, 134]. In Great Britain, 6 cases of wound botulism were registered for the first time in 2000, and by March 2004 there had already been 51 further cases [16, 27].

Infant botulism (infantile botulism, neonatal botulism) occurs in children under 6 months of age with non-fully developed microbiota after ingestion of the spores, which may be contained in honey, for example. The bacteria colonize in the gut and release toxins. The toxins are absorbed and can very quickly lead to death. It has also been observed in breastfed infants. Infant botulism may be one cause of sudden infant death syndrome [9, 39, 82, 101, 113, 134].

Symptoms of infant botulism (incubation time about 10 days) include increasing constipation, loss of appetite, an altered cry, and a striking loss of head control. Do not feed honey to infants before the age of 1 year [101, 134].

In extremely rare cases, colonization of the intestine with *C. botulinum* has also been described in adults and older children [101].

Inhalation botulism is associated with accidental (e.g., for laboratory staff who carry out animal experiments [101]) or intentional (bioterrorism) events which result in release of the toxins in aerosols [134].

Iatrogenic botulism is caused by improper cosmetic or medicinal use of botulinum toxin preparations [120]. In the United States, four patients were injected with more than 2,000 times the usual dose of a botulinum toxin A preparation due to a dilution error [22]. The patients were hospitalized with increasing weakness, respiratory disorders, and facial paralysis. Intensive treatment, especially with an antiserum, saved all patients.

In 2023, 34 cases of iatrogenic botulism have been reported from Germany, Austria, Switzerland, and France in persons that had travelled to Türkiye to get intragastric injection of botulinum toxin for weight reduction [31].

Hospitalization is essential if botulism is suspected. Treatment must be initiated as soon as possible. It consists of intensive gastric and intestinal emptying, i.v. and i.m., administration of polyvalent botulinus antitoxin (only in the first 48 h and only for food and wound botulism, not for infant botulism), of cholinesterase inhibitors, e.g., neostigmin i.v., and intensive care measures, especially artificial respiration. Wound botulism requires immediate wound debridement and antibiotic therapy [54, 84, **117**]. As a precaution, the contents of bombed cans should always be discarded; in case of doubt, food must be heated to 100 °C for 15 min before consumption. Suspicious food should be discarded!

Among farm animals, cattle are primarily affected by acute botulism, but also poultry. Visceral or chronic botulism is an etiologically unexplained disease of cattle with a variety of symptoms that is reported since the late 1990s. A direct link between the occurrence of *Clostridum botulinum* and chronic disease incidence in dairy herds has been assumed but not yet confirmed [11, 13].

Botulinus toxins type A (BOTOX) and type B are used in very low concentrations, i.m. or s.c., for disorders caused by hyperactivity of cholinergic nerve endings. These include neuropathic pain [28], headache [8], chronic migraine [44, 52], depression [44], dysphagia [81], hyperhidrosis [68], blepharospasmus, and dystonia [33]. They are also used to treat strabismus [15], bruxism, and facial spasms [37, 52, 66]. Spastic movement disorders, also in children, and bladder dysfunction are further indications.

In dermatology and cosmetics botulinus toxins are used to smooth skin wrinkles by paralyzing the mimic muscles [37]. Daxibotulinumtoxin A (Daxxify™) is a long-lasting preparation. It consists of a 150-kDa botulinum toxin A and a proprietary stabilizing excipient peptide [109, 112, 119].

The risk of the formation of neutralizing antibodies during long-term botulinum toxin therapy may be reduced by long intervals between treatments, by application of low doses [3], and by use of highly purified toxins.

Botulinum toxins are used experimentally to study the processes at motor end plates.

4.3.9 Cytotoxins of *Corynebacterium diphtheriae*

Corynebacterium diphtheriae, C. ulcerans, and *C. pseudotuberculosis* are the producers of diphtheria toxin [91].

C. diphtheriae is a gram-positive, short, clumsy rod-shaped bacterium (Ph. 4.6) that preferentially colonizes the mucous membranes, especially the pharyngeal organs, of humans during infections and secretes its toxin into the bloodstream. *C. ulcerans* occurs in animals, e.g., hedgehogs, and *C. pseudotuberculosis* occurs also in animals, rarely in humans. The bacteria can spread from person to person when the infected person coughs or sneezes [136] or by contact with contaminated material.

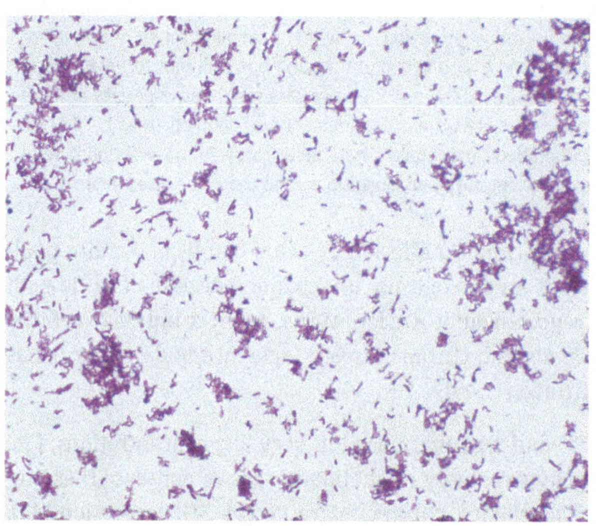

Ph. 4.6: *Corynebacterium diphtheriae* (source: J. Douan & K. Becker, Universitätsmedizin Greifswald).

The pathogens form a single-chain protoxin, the diphtheria toxin (DT, Mr 58kDa, 535 amino acid residues), which is cleaved by limited proteolysis into a 2-chain toxin (fragments A, Mr 21 kDa and fragment B, Mr 37 kDa) linked by a S–S bridge. One fragment is the haptomer (B), the other the toxomer (A). After endocytosis, the toxomer is released by reductive cleavage of the S–S bond in the target cell. The primary structure of the diphtheria toxin is partly similar to those of the exotoxin A of *Pseudomonas aeruginosa*. The genetic information for the toxin is located on a bacteriophage DNA that is integrated into the genome of the toxigenic strains of *C. diphtheriae*. Non-toxigenic strains can acquire the ability to produce the toxin by infection with the tox^+ phages [80, 99, 141, 142].

Receptor for diphtheria toxin is a heparin-binding growth factor, which partly resembles the endothelial cell growth factor (EGF) or its precursor. The toxomer catalyzes the transfer of an adenosine ribosyl phosphate residue from NAD to a posttranslational modified histidine residue (diphthamide residue) of the elongation factor 2 (EF-2), inactivates the elongation factor, leads to the stop of protein biosynthesis and thus to the death of the cell. Even one (!) toxin molecule can kill a cell. Necrosis of the affected tissues occurs, especially in the liver and heart [80].

The lethal dose for humans and sensitive mammals, e.g., monkeys, rabbits, and guinea pigs, is about 70 µg/kg BW [141].

Infections by *C. diphtheriae* occur worldwide, mostly in subtropical and tropical countries and in East Europe. The most common manifestation is pharyngeal (respiratory) diphtheria. As a result of under-vaccination and bad living conditions (damaged health infrastructure, overcrowding in residental camps, homelessness, drug consumers), outbreaks of respiratory and of wound (skin) diphtheria occur with increasing frequency also in Central Europe. In Germany, 29 cases have been reported in 2024 [105], 133 in 2023 [98]. Human infections by *C. ulcerans* were reported mainly from Western States and caused by contact with animals or consumption of raw milk products [99, 136].

After an incubation period of 1–7 days, respiratory diphtheria is usually characterized by an acute febrile course, headaches, plaque on the severely swollen tonsils, combined with difficulty swallowing, risk of suffocation, cough, circulatory failure, myocarditis, and paralysis. Complications can include inflammation of the heart and nerves. For unvaccinated people and without proper treatment, diphtheria can be fatal in around 30% of cases. Young children are at higher mortality risk. Wound diphtheria is mainly caused by nontoxigenic strains [54, 99, 136].

Diphtheria antitoxin and antibiotics (penicillin, erythromycin) must be administered early for treatment. Active immunization in childhood and regular booster vaccinations offer effective protection [54, 136].

Parts of diphtheria toxin are used to produce an antitumor preparation (Tagraxofusp), consisting of a truncated diphtheria toxin and human interleukin-3. IL-3 binds to its receptor CD123 that is overexpressed in some tumor cells, the molecule is endocytosed, the toxin is released in the cell and leads to apoptosis. The drug is used to treat BPDCN (basic plasmacytoid dendritic cell neoplasm [10]).

4.3.10 Toxins of *Bordetella pertussis*

The most common pathogen of whooping cough (pertussis) is *Bordetella pertussis*, a small ovoid, gram-negative rod-shaped bacterium (Ph. 4.7) that settles on the mucous membrane of the upper respiratory tract of humans and destroys the ciliated epithelium. Transmission occurs by droplet infection within a distance of up to about 1 m by coughing, sneezing, or speaking. *Bordetella parapertussis* and *B. holmesii* can also cause whooping cough but mostly with milder symptoms.

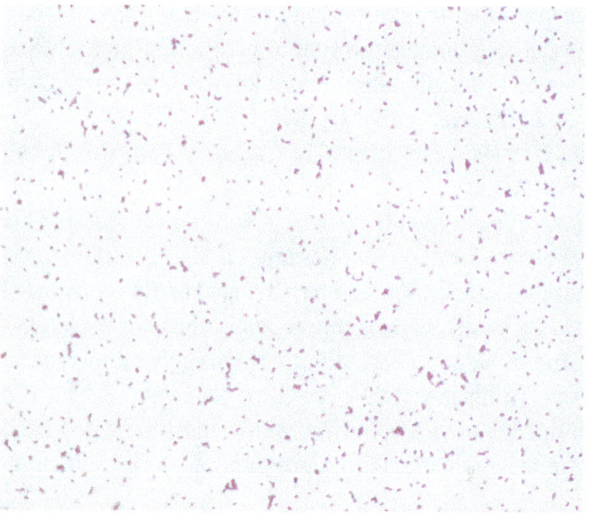

Ph. 4.7: *Bordetella pertussis* (source: J. Douan & K. Becker, Universitätsmedizin Greifswald).

The bacterium secretes some toxins, the most important of which are bordetella adenylate cyclase toxin (BACT, CyaA) and pertussis toxin (PT). Also worth mentioning are the tracheal cytotoxin (TCT) and a dermonecrotic toxin. Adhesins, e.g., the rod-shaped filament hemagglutinin FHA, mediate that the pathogens remain on the cilia of the ciliated epithelium of the upper respiratory tract [54, 62, 67, 118].

Bordetella adenylate cyclase toxin is a one-chain toxin with a molecular mass of 216 kDa. After limited proteolysis of the toxin the A-chain (molecular mass 43 kDa) is formed. It is a calmodulin-dependent adenylate cyclase. There is not yet complete clarity about the invasion mode. It is probable that this is also the A/B model. After the toxin is uptaken into the cells of humans or animals, there is an excessive production of cAMP, several hundred times more than normal, and thus multiple changes in cellular functions arise. Among other things, a decrease in phagocytosis activity of cells, of oxygen radical formation, and of cytotoxic activity of NK cells occur [45, 125].

Pertussis toxin (PT, Mr 94 kDa) is a hexameric protein consisting of five molecules that form the haptomer and an A chain (Mr 26 kDa). The A-chain is an ADP-ribosyl transferase that transfers ADP-ribose from NAD to proteins of the host cell, resulting, e.g., in altered signal transduction [42, 46, 93]. Via ADP ribosylation of G proteins, the pertussis toxin also leads to increased intracellular cAMP production [17, 125]. The effects triggered by pertussis toxin include leukocytosis with absolute lymphocytosis, spleen swelling, development of anaphylactic shock, and increased sensitivity to histamine.

Tracheal cytotoxin is a small-molecular-weight glycopeptide composed of glucosamine, muramic acid, alanine, glutamic acid, and diaminopimelic acid (1:1:2:1:1), which is formed as a 'by product' of bacterial cell wall synthesis. It inhibits the movement of cilia and destroys tight junctions in the epithelium.

The contribution of the dermonecrotic toxin to the clinical appearance of whooping cough is not yet fully understood. The single-chain polypeptide is only released when the cells are lysed and leads to vasoconstriction and skin necrosis when injected into mice [54].

In 2018, there were more than 151,000 cases of pertussis globally. The RKI counts an average of 13,000 cases per year in Germany and assumes a high number of unreported cases [110]. In the year 2024, more than 24,000 cases were reported. Infection occurs in all age groups, with the highest proportion of cases in infants under 1 year of age.

The incubation period for whooping cough is 1–2 weeks, mostly 9–10 days. The disease occurs in the Stadium catarrhale (1–2 weeks), Stadium convulsivum (4–6 weeks), and Stadium decrementi (6–10 weeks). Main symptoms are intermittent coughing fits, cyanosis, shortness of breath, and the production of large amounts of viscous mucus. There is no or only low fever. Many patients only have persistent cough [100].

Whooping cough is treated with antibiotics (macrolides) in the early stages; otherwise treatment is mainly symptomatic. Active immunization is recommended [100].

References

For numbers in bold, see cross chapter literature p. 233.

[1] Ahmad-Mansour N et al. (2021) Toxins 13(10): 677
[2] Aktories K (ed.) (1997) in: Cell Biology and Pharmacology. Chapman & Hall, London, Glasgow, Weinheim, New York, Tokyo, Melbourne, Madras
[3] Albrecht P et al. (2019) Neurology 92: 1 and 93(17): 767
[4] Alouf JE (1982) Toxicon 20: R 211
[5] Alouf JE (2005) in: **132**, p. 1
[6] Ashton AC et al. (1990) Toxicon 28: 963
[7] Atchade E et al. (2024) Antibiotics 13(1): 96
[8] Becker WJ (2020) Toxins 12(12): 803

[9] Bell C, Kyriakides A (2000) Clostridium botulinum: A Practical Approach to the Organism and its Control in Foods. Blackwell Science, London

[10] Beziat G, Ysebaert L (2020) Onco TargetsTher: 13:5199

[11] BfR (2020) www.bfr.bund.de/cm/350/hinweise_fuer_verbraucher_zum-botulismus_durch_lebensmittel.pdf (accessed 02/06/2025)

[12] BfR (2021) www.bfr.bund.de/cm/350/verbrauchertipps-schutz-vor-lebensmittelbedingten-erkrankungen-durch-bakterielle-toxine.pdf (accessed 02/05/2025)

[13] BfR (2023) www.bfr.bund.de/de/selten_aber_vermeidbar_fragen_und_antworten_zum_botulismus-70355.html (accessed 02/05/2025)

[14] Bhakdi S et al. (1984) Verh Dtsch Ges Inn Med 90: 346

[15] Bort-Marti AR et al. (2023) Cochrane Database Syst Rev 3(3): CD006499

[16] Brett MM et al. (2005) Epidemiol Infect 133: 575

[17] Burns DL (1988) Microbiol Sci 5: 285

[18] Camargo A et al. (2024) Emerg Microbes Infect 13(1): 2341968

[19] Centers for Disease Control and Prevention (CDC, 2006) MMWR (Morb Mortal Weekly Rep) 55: 1098

[20] Chatterjee SN, Chaudhuri K (2003) Biochem Biophys Acta 1639(2): 65

[21] Chatterjee SN, Chaudhuri K (2006) Biochem Biophys Acta 1762(1): 1

[22] Chertow DS et al. (2006) JAMA 296: 2476

[23] Chetty C, Schwab JH (1984) in: Handbook of Endotoxin 1: Chemistry of Endotoxin (ed. Rietschel ET), Elsevier, Amsterdam, p. 376

[24] Cheung GYC et al. (2021) Virulence 12(1): 547

[25] Cieza MYR et al. (2024) Pathogens 13(8): 676

[26] Collier RJ, Young JA (2003) Ann Rev Cell Dev Biol 19: 45

[27] Cooper JG et al. (2005) Eur J Emerg Med 12: 251

[28] Dekhne A et al. (2023) Cureus 15(10): e46848

[29] DocCheck Flexicon. https://flexicon.doccheck.com>de> Clostridium_ perfringens (accessed 02/08/2025)

[30] DocCheck Flexicon. https://flexicon.doccheck.com>de> Clostridium_ botulinum (accessed 02/08/2025)

[31] Dorner MB et al. (2023) Euro Surveill 28(23): pii = 2300203

[32] Dressler D (2005) Nervenarzt 76: 763

[33 Dressler D et al. (2021) J Neural Transm (Vienna) 128(4): 531

[34] Fabbri A et al. (2008) Current Medicinal Chemistry 15(11): 1116

[35] Feldmeier H (2024) Dtsch Apoth Ztg 164 (25): 2162 Cholera

[36] Feldmeier H (2024) Dtsch Apoth Ztg 164(32): 2798 Tetanus

[37] Fernández-Nunezv T et al. (2019) Med Oral Pato Cir Bucal 24(4): e416-e-e4214

[38] Finkelstein RA (1986) in: Immunochemical and Molecular Genetic Analysis of Bacterial Pathogens 40 (Eds. Owen P, Foster TJ) Elsevier, Amsterdam, p: 85

[39] Fischer D et al. (2004) Klin Pädiatr 216: 31

[40] Gao L et al. (2025) Nature Struct Mol Biol. doi.org/10.1038/s41594-024-01479-0

[41] Granum PE, Brynestad S (1999) in: The Comprehensive Sourcebook of Bacterial Protein Toxins (eds. Alouf JE, Freer JH) Academic Press, London, ego, Boston, New York, Sydney, Tokyo, Toronto, p. 669

[42] Gross R et al. (1989) Mol Microbiol 3: 119

[43] Halpin-Dohnalek MI, Marth EH (1989) J Food Prot 52: 267

[44] Herd CP et al. (2018) Cochrane Database Syst Rev 6(6): CD011616

[45] Hewlett EL et al. (1985) in: Pertussis Toxin (eds. Sekura RD et al.), Academic Press, New York, p. 241

[46] Hewlett EL et al. (1989) J Biol Chem 264: 19379

[47] Hinshaw LB (1985) Handbook of Endotoxin 2: Pathophysiology of Endotoxin. Elsevier, Amsterdam

[48] Homma JY et al. (1984) Bacterial Endotoxin: Chemical, Biological and Clinical Aspects, Verlag Chemie, Weinheim
[49] Jackson MP (1990) Microbial Pathogenesis 8: 235
[50] Jobling MG, Holmes RK (2005) in: **132**, p. 9
[51] Jülicher B et al. (1989) Arch Lebensmittelhyg 40: 79
[52] Kepcczýnska K, Domitrz I (2022) Toxins 14(9): 619
[53] Khiav LA, Zahmatkesh A (2022) Iran J Basic Med Sci 25(9): 1059
[54] Köhler W et al. (eds.) (2001) Medizinische Mikrobiologie, 8th ed., Urban und Fischer, München
[55] Lee K et al (2013) PloS Pathog 9(10): e1003690
[56] Lee K et al. (2014) Science 344 (6190): 1405
[57] Lee Wong AC, Bergdoll MS (2002) in: Foodborne diseases (eds. Cliver DO, Riemann HP), Academic Press, Amsterdam, p. 230
[58] Lemichez E, Barbieri JT (2013) Cold Spring Harb Perspec Med 3(2): a013573
[59] Leung T, Matrajt, L (2021) PloS Negl Trop Dis 15(5): e0009383
[60] Levine MM et al. (1983) Microbiol Rev 47: 510
[61] Liu K et al. (2024) Front Cell Infect Microbiol 14: 133682
[62] Locht C, Antoine R (1999) in: **2**, p. 130
[63] Lord JM et al. (1999) Cell Microbiol 1(2): 85
[64] Mamat U et al. (1999) in: **9**, p. 179
[65] Mandal Sh et al. (2011) Asian Pac J Trop Med 4(7): 573
[66] Martina E et al. (2021) Toxins 13(2): 120
[67] Mattoo S; Cherry JD (2005) Clin Microbiol Rev 18: 326
[68] McConaghy, Fosseman D (2018): Am Fam Physician 97(11): 729
[69] McGuinness WA et al. (2017) Yale J Biol Med 90(2): 269
[70] Megighian A et al. (2021) J Neurochem 158(6): 1244
[71] Mellanby J (1984) Bacterial Protein Toxins. Academic Press, London, p. 403
[72] Meurens F et al. (2023) Front Microbiol 13: 1099184
[73] Middlebrook JL, Dorland RB (1984) Microbiol Rev 48(3): 199
[74] Montecucco C, Molgó J (2005) Current Opinion in Pharmacol 5: 274
[75] Montero DA et al. (2023) Front Med (Lausanne) 10: 1155751
[76] Moormeyer DE, Bayle KW (2017) Mol Biol Microbiol 104(3): 365
[77] Munir MT et al. (2023) Foods 12(8): 1580
[78] Niemann H et al. (1988) Zentralbl Bakteriol Mikrobiol Hyg Abt 1, Suppl 17 (Bact Protein Toxins): 29
[79] Olivera D et al. (2018) Toxins 10(6): 252
[80] Oram DM, Holmes RK (2005) in: **132**, p. 99
[81] Panebianco M et al. (2020) Neurol Sci 41(1): 367
[82] Parkinson NG, Ito K (2002) in: Foodborne diseases (eds. Cliver DO, Riemann HP), Academic Press, Amsterdam, p. 249
[83] Peck MW et al. (2017) Toxins 9(1): 38
[84] Pfausler B et al. (2024) Botulismus, S1-Leitlinie. www.dgn./org/leitlinien (accessed 02/08/ 2025)
[85] Pirazzini M et al. (2017) Pharmacol Rev 69(2): 200
[86] Pirazzini M et al. (2022) Arch Toxicol 96(6):11521
[87] Popoff MR (2020) J Vet Diagn Invest 32(2): 184
[88] Prévost G (2005) in: **132**, p. 243
[89] Proctor RA (2015) Eur Cell Mater 30: 315
[90] Proft T et al. (2005) in: **132**, p. 179
[91] Prygiet M et al. (2022) Pathogens 11(11): 1264
[92] Quinn CP, Minton NP (2001) in: Clostridia: Biotechnology and Medical Applications (eds. Bahl H, Dürre P) Wiley-VCH, Weinheim, New York, Chichester, Brisbane, Singapore, Toronto, p. 211

[93] Raptis A et al. (1989) Infec Immun 57: 1725
[94] Rawson AM et al. (2023) Virulence 14(1): 2205251
[95] Rietschel ET (1984) Handbook of Endotoxin 1: Chemistry of Endotoxin, Elsevier, Amsterdam, p. 138, 376
[96] Rietschel ET, Brade H (1987) Infection 15: 133
[97] RKI. www.rki.de/DE/Aktuelles/Publikationen/RKI-Ratgeber/Ratgeber_Staphylokokken_MRSA.html (accessed 02/10/2025)
[98] RKI. www.rki.de/DE/Aktuelles/Publikationen/Epidemiologisches-Bulletin/2024/45_24.pdf (accessed 02/08/2025
[99] RKI. www.rki.de/DE/Aktuelles/Publikationen/RKI-Ratgeber_diphtherie.html (accessed 02/08/2025
[100] RKI. www.rki.de/DE/Aktuelles/Publikationen/RKI-Ratgeber_pertussis.html (accessed 02/08/2025
[101] RKI. www.rki.de/DE/Aktuelles/Publikationen/RKI-Ratgeber_botulismus.html (accessed 02/08/2025
[102] RKI.6www.rki.de/DE/Aktuelles/Publikationen/Epidemiologisches-Bulletin/2025/09_25.pdf (accessed 06/08/2025)
[103] RKI. www.rki.de/DE/Themen/Infektionskrankheiten/infektionskrankheiten-A-Z/C/Cholera/cholera-node.de (accessed 05/04/2025)
[104] RKI. www.rki.de/DE/Aktuelles/Publikationen/RKI-Ratgeber/Ratgeber/Ratgeber_Tetanus.html (accessed 05/04/2025)
[105] RKI. www.rki.de/DE/Aktuelles/Publikationen/Epidemiologisches-Bulletin/2025/18_25.pdf (accessed 05/21/2025)
[106] Rossetto O, Montecucco C (2005) in: **132**, p. 149
[107] Safa A et al. (2020) AIMS Microbiol 6(2): 144
[108] Sakaguchi G et al. (1984) Bacterial Protein Toxins. Academic Press, London, p. 435
[109] Salame N et al. (2023) Skin Therapy Lett 28(4): 1
[110] Schlenger R (2020) Dtsch Apoth Ztg 160(39): 3752
[111] Shalaby M et al. (2024) Food Microbiol 118: 104405
[112] Shaterian N et al. (2022) Pain Res Manag Mar 31: 2022: 3284445
[113] Shone CC (1987) in: Natural Toxicants in Food Progress and Prospects. (ed. Watson DH), Ellis Howard, Chichester England, p. 11
[114] Silva AJ, Benitez JA (2016) PloS Negl Trop Dis 10(2): e 0004330
[115] Simini B (1996) Lancet 348: 813
[116] Singh Y, Liang X, Duesbery NS (2005) in: Ü 108 = Proft, p. 285
[117] Smedley JG et al. (2004) Rev Physiol Biochem Pharmacol 152: 183
[118] Sock I (2006) Med Mschr Pharm 29: 206
[119] Solish N et al. (2021) Drugs 81(18): 2091
[120] Souayah N et al. (2006) Neurology 67: 1855
[121] Stahlmann R (2012) Dtsch Apoth Ztg 152(34): 4104
122] Stengel G (1989) Fleischwirtschaft 69: 1258 and 70: 307
[123] Sur D et al. (2007) Indian J Med Res 125: 772
[124] Tam K, Torres VJ (2019) Microbiol Spectr (2): 10
[125] TenBroek EM, Confor DL (1987) J Toxicol Toxin Rev 6: 99
[126] Thelestam M, Blomqvist L (1988) Toxicon 26: 51
[127] Thwaites CL et al. (2015) Lancet 385(9965): 362
[128] Trujillo C et al. (2006) Neurotox Res 9: 63
[129] van der Goot GF (2001) Pore-forming Toxins. Springer, Berlin
[130] van Heyningen S (1984) Bacterial Protein Toxins. Academic Press, London, p. 347
[131] Vestergaard et al. (2019) Microbiol Spectr 7(2): 10
[132] Weller U et al. (1988) Arch Pharmacol 338: 99

[133] WHO (2024) www.who.int/news/item/30-01-2024-guidelines-on-the-clinical-management-of-sepsis (accessed 05/03/2025)

[134] WHO www.int/news-room/fact-sheets/detail/botulism (accessed 02/07/2025)

[135] WHO www.who.int/news-room/fact-sheets/detail/cholera (accessed 02/02/2025)

[136] WHO www.who.int/news-room/Fact-sheets/detail/diphtheria (accessed 02/02/ 2025)

[137] WHO www.who.int/health-topics/pertussis#tab=tab_1 (accessed 02/02/2025)

[138 WHO www.who.int/news-room/fact-sheets/detail/tetanus (accessed 02/02/2025)

[139] WHO www.who.int/emergencies/disease-outbreak-news/item/2023-DON489 (accessed February 7, 2025)

[140] Williams KL (2007) Endotoxins: Pyrogens, LAL Testing and Depyrogenation, 3rd ed., Informa Healthcare, New York, London

[141] Word WHJ (1987) Trends Biochem Sci 12: 28

[142] Zhao JM, London E (1988) Biochemistry 27: 3398

5 Toxins of Cyanobacteria

5.1 General

Cyanobacteria (Bacteria, Cyanobacteriota) constitute a large and diverse phylum of autotrophic gram-negative bacteria capable of oxygenic photosynthesis [25]. The name 'cyanobacteria' derives from the Ancient Greek κύανος (kyanos), 'blue', referring to the characteristic blue-green coloration of many cyanobacterial genera. This pigmentation results from the combination of green chlorophylls and the blue accessory phycobiliprotein pigment phycocyanin, sometimes complemented with the red phycoerythrin. Although often informally referred to as 'blue-green algae', cyanobacteria are prokaryotic and thus not classified among the eukaryotic algae. To date, more than 6,000 cyanobacteria species have been described [11].

Cyanobacteria are thought to have originated approx. 3.5 billion years ago during the Archean Eon [22, 24], making them among the earliest known organisms to perform oxygenic photosynthesis [25]. Their activity played a pivotal role in shaping the Earth's atmosphere, culminating in the Great Oxidation Event, which enabled the emergence of aerobic life. Furthermore, cyanobacteria are considered ancestral to plant plastids, which arose through primary endosymbiosis, highlighting their importance in the evolution of photosynthetic eukaryotes [10, 15].

Cyanobacteria are ubiquitous and thrive in nearly all ecosystems, including marine, freshwater, and terrestrial environments, and are highly diverse in terms of their morphology, physiology, and metabolism [25]. Beyond their role as primary producers, they fulfill key ecological functions. For example, certain genera possess the ability to fix atmospheric nitrogen, contributing to soil fertility and supporting plant growth through leaf and root colonization [8]. From a (eco)toxicological perspective, however, freshwater cyanobacteria are of particular concern due to their capacity to form blooms and produce toxins: Planktonic species can regulate their depth in the water column to maintain optimal environmental conditions by using gas vesicles. Sudden changes in water convection, increased nutrient availability, or other environmental factors can trigger mass proliferation and lead to surface accumulations, which are often concentrated near the shoreline by wind action [2, 12]. Cyanobacterial genera commonly associated with bloom formation include Aphanizomenon, Dolichospermum, Microcoleus, Microcystis, Oscillatoria, and Planktothrix. However, even when the species composition of a cyanobacterial bloom is identified, its toxicity cannot be reliably predicted, as both toxin-producing and non-toxic strains can exist within the same species.

The first documented case of lethal cyanobacterial intoxication dates back to 1878, when George Francis reported a Nodularia bloom accumulating along a shoreline and forming 'a thick scum like green oil paint, [. . .] as thick and pasty as porridge', which was implicated in the deaths of livestock including sheep, horses, dogs,

https://doi.org/10.1515/9783110728576-005

and pigs [9]. Since then, cyanobacteria have frequently been associated with harmful effects on wildlife, either through the direct ingestion of cyanobacterial biomass or the consumption of water contaminated with cyanobacterial toxins. Due to their prevalence in drinking water sources and recreational water bodies, and their potential health risks to humans and animals, freshwater cyanobacteria and their toxins, collectively referred to as cyanotoxins, have been the focus of intensive research since the 1960s [1, 4, 21], and excellent review articles and books have been published on this topic [3, 5–7, 20].

Environmental drivers such as anthropogenic nutrient enrichment (eutrophication), climate change-induced warming, and rising atmospheric CO_2 levels contribute to the increasing dominance of cyanobacteria in aquatic ecosystems, enhancing the frequency, severity, and duration of cyanobacterial blooms, commonly referred to as harmful algal blooms (HABs), and the associated risks of cyanotoxin production [12, 18, 19]. Care must be taken to avoid contact with water blooms and the ingestion of drinking water or food contaminated with toxins [6, 13].

Some cyanotoxins rank among the most potent natural toxins known [3]. Their toxicity has drawn not only ecological and public health attention but also interest in their potential use as biological warfare agents. Indeed, two cyanotoxins, saxitoxins and microcystins, are listed among the ten biological toxins included in the German War Weapons Control Act.

Pharmacologically active specialized metabolites from cyanobacteria are attracting increasing interest as potential drugs, e.g., for the treatment of cancer, while others are used as experimental tools [14, 16, 17, 23]. Some cyanobacteria, e.g., Spirulina species and *Aphanizomenon flos-aquae*, are components of food supplements. It is urgently necessary to ensure that these are toxin-free.

Cyanotoxins are often categorized based on their primary toxicological target (e.g., neurotoxins, hepatotoxins, and dermatotoxins). In this book, however, they will be classified according to their biosynthetic origin: polyketide, amino acid, alkaloid, and peptide toxins.

References

[1] Botes DP et al. (1985) J Chem Soc Perkin Trans I 1: 2747
[2] Brandenburg K et al. (2020) Toxins 12: 221
[3] Buratti et al. (2017) Arch Toxicol 91: 1049
[4] Carmichael WW et al. (1988) Toxicon 26: 971
[5] Chorus I (ed, 2001) Cyanotoxins, Springer Berlin, Heidelberg
[6] Chorus I, Welker M (eds, 2021) Toxic cyanobacteria in water – Second edition, WHO
[7] Dittmann et al. (2013) FEMS Microbiol Rev 37: 23
[8] Dodds WK et al. (1995) J Phycol 31: 2
[9] Francis G (1878) Nature 18:11
[10] Gavelis GS, Gile GH (2018) FEMS Microbiol Lett 365: fny209
[11] Guiry MD, Guiry GM. AlgaeBase, https://www.algaebase.org. accessed 14.07.2025

[12] Huisman J et al. (2018) Nat Rev Microbiol 16: 471
[13] Luckas BLJ et al. (2005) Environ Toxicol 20: 1
[14] Luesch et al (2025) Nat Prod Rep 42: 208
[15] McFadden GI (2014) Cold Spring Harb Perspect Biol 6: a016105
[16] Niedermeyer THJ, Brönstrup M. (2013) Natural product drug discovery from microalgae. In: Posten C, Walter C (eds) Microalgal Biotechnology: Integration and Economy, De Gruyter Brill
[17] Nunnery JK et al. (2010) Curr Op Biotechnol 21: 787
[18] Pearl HW, Huisman J (2009) Environ Microbiol Rep 1: 27
[19] Pearl HW, Paul VJ (2012) Water Res 5: 1349
[20] Pearson L et al. (2010) Mar Drugs 8: 1650
[21] Rinehart et al. (1994) J Appl Phycol 6: 159
[22] Schopf JW, Packer BM (1987) Science 237: 70
[23] Tan LT, Salleh NF (2024) Molecules 29: 5307
[24] Tomitani A, Knoll AH (2006) PNAS 103: 5442
[25] Whitton BA, Potts M (2012) Introduction to the Cyanobacteria. In: Whitton B (eds) Ecology of Cyanobacteria II, Springer, Dordrecht

5.2 Polyketides as Toxins of Cyanobacteria

Among the best studied toxic cyanobacteria are *Moorena producens* and *M. bouillonii*, formerly classified as *Lyngbia majuscula* and *L. bouillonii* (Ph. 5.1 [27, 28]). *Moorena producens*, commonly known as 'mermaid's hair' or 'fireweed', is a filamentous, bloom-forming marine cyanobacterium that occurs in shallow coastal waters in tropical to temperate regions worldwide, e.g., around Hawai'i, Papua New Guinea, Spain, Saudi Arabia, or Australia [1, 2, 7, 8, 17]. According to the 2024 release of the cyanobacterial specialized metabolite database CyanoMetDB [14], approx. 15% of all known cyanobacterial compounds have been isolated from Moorena species, many of which show cytotoxic activity [4, 8]. In addition to polyketides such as aplysiatoxins and alkaloids such as lyngbyatoxins, Moorena species produce a large number of bioactive peptides (see Section 5.5). Most of the Moorena specialized metabolites have only to a very limited extent been studied pharmacologically, and little is known about their toxicity to humans.

The polyether lactones aplysiatoxins (Fig. 5.1) were first isolated from sea hares of the family Aplysiidae, including *Stylocheilus longicauda* [15], which feed on aplysiatoxin-producing cyanobacteria of the genera Moorena or Schizothrix [16, 22]. Structurally related compounds, the oscillatoxins, have been identified in Oscillatoria species [18].

Initially, aplysiatoxins were recognized as causative agents of 'seaweed dermatitis' or 'swimmer's itch' [26]. This irritant dermatological condition, characterized by erythema, blisters, and skin desquamation, is commonly reported among swimmers in the coastal waters of Hawai'i, the Marshall Islands, Okinawa, Florida, and Queensland.

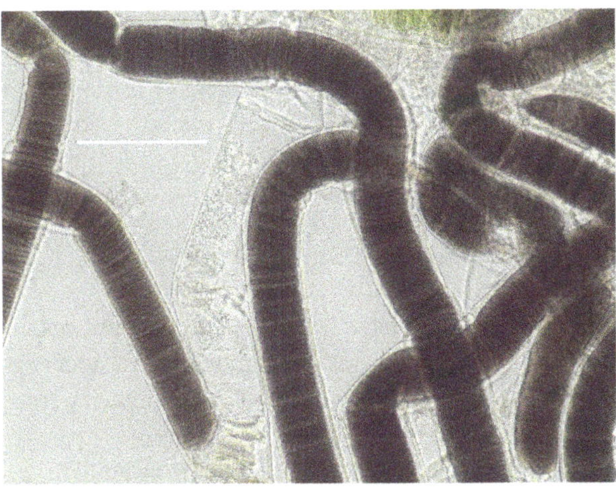

Ph. 5.1: Moorena sp. (source: Evgenia Glukhov).

Aplysiatoxin

Acutiphycin

Scytophycin B

Fig. 5.1: Polyketide toxins of cyanobacteria.

Aplysiatoxins have subsequently been found to be tumor promotors [11], binding to the same cell membrane receptors as the phorbol ester 12-*O*-tetradecanoylphorbol-13-acetate (TPA, see **[159 = Vol. 1]**), leading to activation of protein kinase C and subsequent modulation of cell proliferation and differentiation [10, 19, 24]. Due to this activity, aplysiatoxins are used in experimental models to study carcinogenesis [4]. In mice, aplysiatoxins cause internal hemorrhages after i.p. application of 250 µg/kg BW.

Despite their tumor-promoting activity, analogs of aplysiatoxin have been explored as antineoplastic lead compounds [23, 29].

Human poisoning by aplysiatoxins has also been reported following the ingestion of red algae colonized by epiphytic cyanobacteria producing these toxins [13, 20, 21].

☠ Poisoning by aplysiatoxins occurs after skin contact as 'seaweed dermatitis', characterized by erythema and skin inflammation. Oral ingestion of algae containing aplysiatoxins results in gastrointestinal distress, systemic absorption may lead to neurological symptoms.

Toxin macrolactones are acutiphycin (Fig. 5.1), isolated from *Oscillatoria acutissima* in the coastal waters of Hawai'I [3], scytophycins (Fig. 5.1), obtained from *Scytonema pseudohofmanni* originating from the same coastal waters [12], and tolytoxin from Tolypothrix, which is structurally related to scytophycin B [5]. Acutiphycins, scytophycins, and tolytoxin (LD_{50} scytophycin B 650 µg/kg BW, tolytoxin 1,500 µg/kg BW, i.p., mouse) have a strong cytotoxic and fungicidal effect and show antitumor activity in animal experiments [6, 9, 25].

References

For numbers in bold, see cross chapter literature p. 233.

[1] Albert S et al. (2005) Mar Poll Bull 51: 428
[2] Al-Shehri AM, Mohamed ZA (2007) Ecohydrol Hydrobiol 7: 51
[3] Barchi Jr. JJ et al. (1984) J Am Chem Soc 160: 8193
[4] Burja AM et al. (2001) Tetrahedron 57: 9347
[5] Carmeli S et al. (1990) J Nat Prod 53: 1533
[6] Cavalcante do Amaral S et al. (2023) Mar Drugs 21: 359
[7] Curren E, Leong SCY (2019) Harmful Aglae 86: 10
[8] Curren E et al. (2022) Environ Sci Poll Res 29: 78178
[9] Delawská K et al. (2021) ChemBioChem 23: e202100489
[10] Fujiki H et al. (1984) Biochem Biophys Res Comm 120: 339
[11] Fujiki H, Sugimura T (1987) Adv Cancer Res 49: 223
[12] Ishibashi M et al. (1986) J Org Chem 51: 5300
[13] Iso E, Nagai H (2000) Toxicon 38: 123
[14] Jones MR et al. (2021) Water Res 196: 117017
[15] Kato Y, Scheuer PJ (1974) J Am Chem Soc 96: 2245
[16] Mynderse et al. (1977) Science 196: 538
[17] Martín-García L et al. (2014) Harmful Algae 34: 76
[18] Moore RE et al. (1984) J Org Chem 49: 2484
[19] Moore RE et al. (1986) Carciogenesis 7: 641
[20] Nagai H et al. (1996) Toxicon 34: 753
[21] Nagai H et al. (1997) J Nat Prod 60: 925
[22] Nagai H et al. (2019) Tetrahedron 75: 2486
[23] Nakagawa Y et al. (2009) J Am Chem Soc 131: 7573
[24] Nakamura H et al. (1989) PNAS 86: 9672

[25] Patterson GML, Carmeli S (1992) Arch Microbiol 157: 406
[26] Solomon AE, Stoughton (1978) Arch Dermatol 114: 1333
[27] Tronholm A, Engene N (2019) Notulae algarum 122: 20122
[28] Xu J et al. (2020) Eur J Med Chem 201: 112473
[29] Yanagita RC et al. (2010) Bioorg Med Chem Lett 20: 6064

5.3 Amino Acids as Toxins of Cyanobacteria

β-Methylamino-L-alanine (BMAA; α-amino-β-methylamino-propionic acid) (Fig: 5.2) is a non-proteinogenic amino acid that has been implicated in neurodegenerative diseases (see **[160 = Vol. 2]**). Interest in BMAA arose following observations of an unusually high incidence of a neuropathological disorder among the indigenous Chamorro people of Guam. This disorder, known as amyotrophic lateral sclerosis/Parkinsonism-dementia complex (ALS/PDC), was hypothesized to be linked to traditional dietary practices, specifically, the consumption of flying foxes (fruit bats), which feed on cycad seeds containing cyanobacteria, as well as the direct use of cycad seed flour in cooking [1, 2, 7]. BMAA was subsequently detected in cycad seeds and flying foxes, as well as in post-mortem brain tissue samples of affected patients, suggesting it may play a role in the etiology of the disease [5, 15]. Beyond Guam, BMAA exposure has been proposed as a potential contributor to sporadic ALS cases following consumption of cyanobacteria-contaminated seafood, such as shellfish in France and blue crabs in the United States [11]. Although the exact mode of action of BMAA remains unclear, it is believed to act as an excitotoxin by agonizing glutamatergic receptors, including NMDA, AMPA, and kainate receptors [14, 18]. Neurodegenerative effects of BMAA have been demonstrated in experimental models, including rodents and non-human primates [9, 12, 13, 16].

BMAA has been reported in a wide range of cyanobacterial taxa from marine, freshwater, and terrestrial environments, as well as in cyanobacterial symbionts of lichens and certain plants, potentially providing pathways for human exposure through the food chain. Initially regarded as a widely prevalent cyanotoxin [3, 8], BMAA's significance remains controversial due to inconsistent detection results and methodological challenges [6, 10]. In particular, the reliability of analytical methods used for BMAA quantification, especially in distinguishing it from structurally similar isomers such as DAB (2,4-diaminobutyric acid) and AEG (*N*-(2-aminoethyl)glycine), has been the subject of significant debate [4]. Recent high-specificity analytical studies have failed to confirm BMAA production in many cyanobacterial strains, casting doubt on earlier findings and leaving open questions regarding both its biosynthesis and its definitive role in human neurodegenerative disease [17].

$$H_3C{-}NH{-}CH_2{-}\underset{}{\overset{NH_2}{CH}}{-}COOH$$

β-Methylamino-L-alanine (BMAA)

Fig. 5.2: A toxic amino acid.

References

For numbers in bold, see cross chapter literature p. 233.

[1] Banack AS, Cox PA (2003) 61: 387
[2] Banack AS et al. (2006) J Ethnopharmacol 106: 97
[3] Banack AS et al. (2007) Mar Drugs 5: 180
[4] Bishop SL, Murch SJ (2020) Analyst 145: 13
[5] Bradley WG, Mash DC (2009) Amyotroph Lateral Scler 10: 7
[6] Chernoff N et al. (2017) J Toxicol Environ Health 20: 183
[7] Cox PA, Sacks OW (2002) Neurology 58: 956
[8] Cox PA et al. (2005) PNAS 102: 5074
[9] Cox PA et al. (2016) Proc R Soc B 283: 20152397
[10] Dunlop RA et al. (2921) Neurotox Res 39: 81
[11] Field NC et al. (2013) Toxicon 70: 179
[12] Karlsson O et al. (2009) Toxicol Sci 112: 185
[13] Karlsson O et al. (2015) Arch Toxicol 89: 423
[14 Lobner D et al. (2007) Neurobiol Disease 25: 360
[15] Murch SJ et al. (2004) Acta Neurol Scand 110: 267
[16] Spencer PS et al. (2007) Discovery and Partial Characterization of Primate Motor-System Toxins. In: Bock G, O'Connor M (eds) Ciba Foundation Symposium 126 - Selective Neuronal Death, Novartis Foundation Symposia
[17] Wang ZQ et al. (2023) Water Biol Secur 2: 100208
[18] Weiss JH et al. (1989) Brain Res 497: 64

5.4 Alkaloids as Toxins of Cyanobacteria

5.4.1 Indole Alkaloids

One of the best-studied *Moorena producens* (Ph. 5.1) metabolites is lyngbyatoxin-a (Fig. 5.3), a terpenoid peptide indole alkaloid synthesized via non-ribosomal peptide synthesis [11, 27]. Intriguingly, similar or even identical compounds are also produced by certain actinomycetes, e.g., the teleocidins (by Streptomyces spp. [98]), blastmycetins and olivoretins (both by Streptoverticillium spp. [40, 84]), or pendolmycin (by Nocardiopsis sp. [116]), suggesting plasmid exchange among diverse prokaryotic taxa.

Like with the aplysiatoxins, direct contact with lyngbyatoxin-producing *Moorena producens* can cause 'swimmers itch' [11, 108], and like the aplysiatoxins, lyngbyatoxin-a and its structural analogs act as potent activators of protein kinase C [34, 54], similar to phorbol esters (see [**159** = **Vol. 1**]). They exert strong inflammatory and co-carcinogenic effects in animal experiments, with tumor-promoting activity observed at nanomolar concentrations [35, 44]. Key structural features that are decisive for protein kinase C activation include the hydrophobic substituent at C-7, the lactam ring, and the N-13 methyl group [5, 54]. In vivo, intraperitoneal administration of lyngbyatoxin-a causes injuries in the lung, stomach, and small intestine (lethal dose in mice 250 µg/kg BW [45]).

Lyngbyatoxin-a and other toxins presumably serve as chemical defense against grazers. They are released into the aquatic environment during the autolysis of dying cyanobacterial cells [75]. Bioaccumulation has been observed in fish that feed on *Moorena* spp., with detection of lyngbyatoxin-a, e.g., in the digestive glands, as well as in shellfish [10]. Consequently, these organisms may pose a risk of secondary toxicity to humans.

Skin contact with a blue-green algae blooms, i.e., with mass accumulations of cyanobacteria with a high proportion of *Moorena producens*, can result in intense pruritus, erythema, pustules, conjunctivitis, stomatitis, often accompanied by sore throat, headache, and drowsiness [19, 46]. Wind-driven aerosolization of cyanobacterial toxins may lead to respiratory inflammation [75].

Contact dermatitis caused by bathing in the area of a blue-green algae bloom must be treated symptomatically.

The hapalindoles (Fig. 5.3) from Hapalosiphon spp. and related compounds (ambiguins from Hapalosiphon spp., fischerindoles from *Fischerella ambigua*, welwitindolinones from *Hapalosiphon welwitschii*) are terpenoid indole alkaloids [105]. To date, over 80 members of the hapalindole family have been identified [13]. These compounds, often containing isothiocyanate and isonitrile groups, display a wide range of biological activities. Reported effects include antibacterial, antifungal, antialgal, insecticidal, immunomodulatory, and cytostatic effects [13]. Inhibition of RNA polymerase or sodium channel modulation have been discussed as potential modes of action of these alkaloids [25].

Aetokthonos hydrillicola (Ph. 5.2) has recently become famous as the ‘eagle killer' cyanobacterium. It produces the alkaloid aetokthonotoxin (AETX, Fig. 5.3), a pentabrominated neurotoxin characterized by a rare biindole nitrile substructure. AETX is the causative agent of vacuolar myelinopathy, a wildlife disease that leads to the death of waterfowl like coots, bald eagles, and other birds of prey, amphibians, reptiles, and fish through complex food web interactions [8]. *A. hydrillicola* colonizes the underside of leaves of water thyme, *Hydrilla verticillata*, an invasive aquatic plant widespread in man-made lakes across the southeastern United States [110]. Remarkably, *A. hydrillicola* has been shown to produce two distinct chemically unrelated classes of toxins: aetokthonotoxins, responsible for the neurotoxicity of the cyanobacterium [8], and aetokthonostatins, highly cytotoxic peptides (see Section 5.5 [90]). Given that *H. verticillata* and *A. hydrillicola* are predominantly found in artificial lakes that often serve as drinking water reservoirs, the presence of both toxins may also pose a potential risk to human health. Accordingly, analytical methods for detecting *A. hydrillicola* and aetokthonotoxin have been developed [96, 117].

The tjipanazoles (Fig. 5.3) are halogenated indolo[2,3-a]carbazole alkaloids structurally related to staurosporine, an inhibitor of proteinkinases from Streptomyces species. They are produced by *Tolypothrix tjipanasensis* and *Fischerella ambigua* (recently reclassified as *Symphyonema bifilamentata*) [7,49]. Several tjipanazoles showed antifungal and moderate antibacterial activity, as well as weak cytotoxicity. In addition, some tjipanazoles were found to inhibit ABCG2, an ATP-binding cassette transporter implicated in multidrug resistance [14].

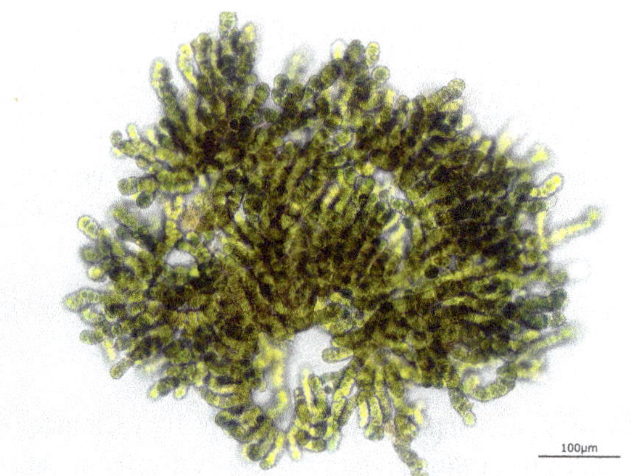

Ph. 5.2: *Aetokthonos hydrillicola* (source: Lenka Štenclová).

Lyngbyatoxin-a

Aetokthonotoxin

Hapalindole A

Tjipanazole D

Fig. 5.3: Indole alkaloids of cyanobacteria.

5.4.2 Homotropane Alkaloids

Anatoxin-a (ATX-a, Fig. 5.4) was first isolated as an acute neurotoxin from the cyanobacterium *Dolichospermum flos-aquae* (formerly *Anabaena flos-aquae*, Ph. 5.3, [21, 24]). Anatoxin-a and its more than 15 derivatives, especially homoanatoxin-a and dihydro- as well as hydroxy-derivatives, were also identified in various other cyanobacteria genera such as Planktothrix [104], Oscillatoria [2], Phormidium [38], or Raphidiopsis [73]. Currently, Microcoleus is considered an increasingly ecotoxicologically relevant genus [50], with extensive studies from Canada [48, 68, 103] and New Zealand [115]. Anatoxin-a producing cyanobacteria have been described from all continents except Antarctica [18, 64].

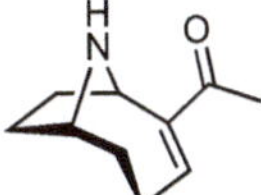

Fig. 5.4: A homotropane alkaloid of cyanobacteria.

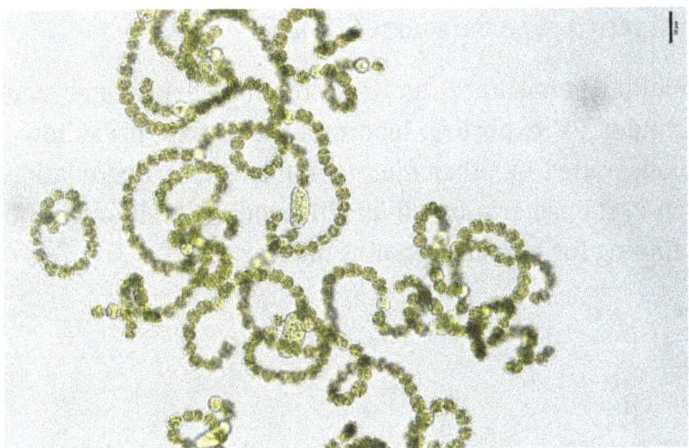

Ph. 5.3: *Dolichospermum flos-aquae* (source: Jan Mareš).

Anatoxin-a is a potent and specific agonist of nicotinic acetylcholine receptors with affinity for both muscular and, to a lesser extent, neuronal subtypes [65, 76, 95, 101]. It is approx. 100 times more potent than nicotine [17, 94, 111] and has 20 times higher receptor affinity than acetylcholine [47]. Only the (+)-enantiomer of anatoxin-a, which is about 150-fold more active than the (–)-enantiomer, occurs in nature [76, 111]. Reported LD_{50} values in mice are approx. 250 µg/kg BW (i.p.), 100 µg/kg BW (i.v.), and 10,000 µg/kg BW (p.o., [17]). Because mice treated i.p. with anatoxin-a were dying within minutes after treatment, the compound was initially dubbed 'very fast death factor' ([VFDF, [24]]). Homoanatoxin-a is similarly potent [113], while the dihydroderivatives exhibit about 10-times reduced toxicity [112].

Blooms of anatoxin-a-producing cyanobacteria, whether pelagic or benthic depending on the strain, are increasing in frequency as a result of rising temperatures, water column stratification, and eutrophication [78]. These blooms pose a growing threat in freshwater systems. Benthic mat samples have reached concentrations as high as 8 mg anatoxin-a per g biomass [64]. Owing to its high water solubility, rapid gastrointestinal absorption, and potent toxicity, its possible occurrence in drinking water poses a serious risk to animal and human health [21]. Water monitoring methods have been established [20, 79] and continue to be improved [117].

🐈 Poisoning by anatoxin-a-producing cyanobacteria species has been documented in livestock, waterfowl, and dogs [21]. Especially in the case of dogs, deaths have often occurred after ingesting detached benthic cyanobacterial mats that had washed ashore [6, 29, 38, 68, 81, 114]. Symptoms of oral intoxication include gastrointestinal distress (nausea, vomiting, abdominal pain), neuromuscular dysfunction (tremors, convulsions, paralysis), and ultimately death due to respiratory paralysis. Notable mass fatalities attributed to anatoxin-a include annual mass deaths of flamingos at Lake Bogoria in Kenya (e.g., 30,000 dead birds in 1999 [55]), devastating the local flamingo population, and over 400 elephant deaths in Botswana, where anatoxin-a-contaminated waterholes are suspected to be the source [63, 106].

Anatoxin-a has low environmental persistence, as it is photolabile in water and undergoes photodegradation under UV exposure, leading to less toxic breakdown products [76]. It can also be biodegraded by other microorganisms [52]. Degradation products have been detected in cyanobacteria-based 'health food' [26], although the toxicological relevance of this finding for humans remains unclear.

5.4.3 Guanidine Alkaloids

Cylindrospermopsin (CYN or CYL, Fig. 5.5) is a polycyclic, zwitterionic, and highly water-soluble alkaloid produced by a range of cyanobacterial genera [37]. It was first isolated from *Raphidiopsis raciborskii*, formerly *Cylindrospermopsis raciborskii* (Ph. 5.4), following an outbreak of a mysterious hepatotoxic illness especially among children on Palm Island, Australia, in 1979. This incident was traced back to a cyanobacterial bloom contaminating the local drinking-water supply [9, 41, 74]. Since then, toxic blooms of cylindrospermopsin-producing cyanobacteria of Cylindrospermopsis and other genera, including Anabaena, Aphanizomenon, Raphidiopsis, and Umezakia spp., have been documented worldwide [16, 37, 80, 97], including in temperate regions.

Cylindrospermopsin's carbon skeleton is mainly biosynthesized via the polyketide pathway, with glycine serving as the starter unit [71]. Structural features essential for its toxicity include the uracil moiety and the hydroxyl group [3].

Cylindrospermopsin is a potent hepatotoxin that irreversibly inhibits protein biosynthesis and glutathione synthesis in hepatocytes, either directly or after metabolic activation by cytochrome P450 enzymes [83, 91, 100]. Although inhibition of protein biosynthesis is a well-documented effect of cylindrospermopsin, it may not be the sole cause of cell death [33]. Covalent modification of DNA and RNA has also been reported [91, 92]. Its precise mode of action, however, remains incompletely understood. Reported LD_{50} values in mice are approx. 5,000 μg/kg BW (p.o.) and 200 μg/kg BW (i.p.) [16, 92].

Ph. 5.4: *Raphidiopsis raciborskii* (source: Jan Mareš).

Cylindrospermopsin is remarkably stable in the absence of sunlight, allowing it to persist for extended periods in turbid and stagnant water bodies [15].

Fatalities after cylindrospermopsin poisoning have been reported in cattle [86]. Cylindrospermopsin has been shown to bioaccumulate in fish during aquaculture, raising concerns for human exposure via consumption of contaminated fish [85]. In Australia, blooms of Cylindrospermopsis have significantly impacted drinking water supplies and caused considerable economic losses [42].

Guanitoxin (GNT, Fig. 5.5), formerly known as anatoxin-a(S), [32] is produced by freshwater cyanobacteria of the genus Dolichospermum (previously classified as *Anabaena flos-aquae*, Ph. 5.3 [12]). Guanitoxin is an unusual cyclic *N*-hydroxyguanidine organophosphate [67]. Like the synthetic organophosphates developed as neurotoxins and used today as insecticides, guanitoxin phosphorylates a serine residue in the active site of acetylcholinesterase, thereby irreversibly inhibiting the enzyme. This inhibition leads to overstimulation of both nicotinic and muscarinic cholinergic receptors [43, 65, 66]. In mice, guanitoxin exposure results in hypersalivation (hence the historical designation '(S)' for 'salivation'), muscle weakness, respiratory distress, and seizures. In rats, chromodacryorrhoea (secretion of bloody tears), muscle weakness, convulsions, and death due to respiratory paralysis occur [30]. The LD_{50} in mice is 20 µg/kg BW, i.p., making guanitoxin approx. 10 times more toxic than anatoxin-a [70].

Animal poisonings by guanitoxin have been reported in various species including cattle, dogs, and birds, especially in the United States and Canada. Monitoring of guanitoxin is complicated due to its chemical instability, which on the other hand reduces its environmental persistence [30].

Saxitoxins are a family of over 50 structurally related tetrahydropurine neurotoxins known for causing paralytic shellfish poisoning (PSP), a severe and potentially life-threatening condition resulting from the consumption of shellfish that have accumulated saxitoxins through filter-feeding on toxic algal or cyanobacterial blooms [22, 59, 102, 63=Vol. 4]. Saxitoxin (STX) is the most prominent representative of this toxin family [87, 88]. Its name derives from the butter clam (*Saxidomus gigantea*) from which the compound was first described. In addition to saxitoxin, the compound family includes neosaxitoxin (NSTX), various gonyautoxins (GTXs), decarbamoylsaxitoxins (dcSTX), and other analogs (Fig. 5.5 [59]). Saxitoxins are not only produced by freshwater cyanobacteria such as Aphanizomenon, Dolichospermum, and Cylindrospermopsis spp. [56], but also by marine dinoflagellates including Alexandrium, Gymnodinium, Pyrodinium spp. [57], which are associated with harmful blooms referred to as 'red tides', [39]). A distinctive feature of saxitoxins is their unusually high nitrogen content, resulting from the incorporation of two arginine molecules during biosynthesis [51].

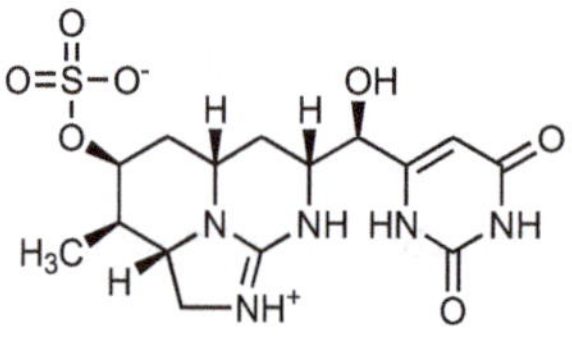

Cylindrospermopsin

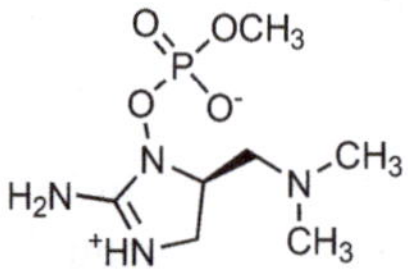

Guanitoxin

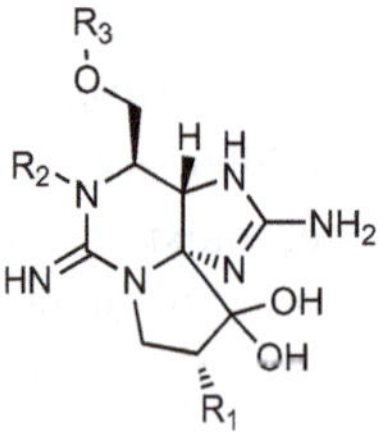

	R_1	R_2	R_3
Saxitoxin	H	H	$CONH_2$
Neosaxitoxin	H	OH	$CONH_2$
Gonyautoxin 1	OSO_3H	OH	$CONH_2$
Descarbamoylsaxitoxin	H	H	H

Fig. 5.5: Guanidine alkaloids of cyanobacteria.

Saxitoxins pose a significant public health and environmental challenge due to their accumulation in both marine and freshwater food webs [82]. Their presence in commercially important shellfish (e.g., mussels, clams, oysters, and scallops) frequently leads to temporary bans on shellfish harvesting, with considerable economic repercussions for coastal communities. Saxitoxins can also accumulate in fish [23, 36, 58], further contributing to food safety concerns. PSP outbreaks have been reported across temperate, subtropical, and tropical coastal regions worldwide [109]. Effective monitoring of saxitoxins in food and environmental samples is essential for public health protection. A range of analytical methods is employed for detection [31]. Regulatory authorities, including the U.S. Food and Drug Administration (FDA) and the European Food Safety Authority (EFSA), have established maximum allowable levels of 80 µg STX equivalents per 100 g of shellfish tissue to prevent human intoxication [1, 107]. Saxitoxins are relatively heat-stable and resistant to cooking and common food processing methods, underscoring the importance of pre-harvest monitoring and toxin surveillance programs.

Saxitoxins potently and reversibly inhibit voltage-gated sodium channels by blocking the channel pore, thereby inhibiting the propagation of action potentials by affected neurons [61, 99]. The blocked neuronal signal transmission results in paralysis of the affected body region. Sublethal exposure can result in nausea, vomiting, diarrhea, abdominal pain, myalgia, and headache [89]. In lethal exposures, affected individuals may remain fully conscious while experiencing progressive paralysis, culminating in respiratory failure. Although no specific antidote exists, supportive care, including artificial respiration, can result in complete recovery. Beyond their toxicological relevance, saxitoxins have found important use as biochemical tools in the study of sodium channels. Although voltage-gated sodium channels were long considered to be selectively targeted, research has shown that other ion channels and enzymes might be involved in saxitoxin bioactivity, as well [60]. In addition, neosaxitoxin has attracted attention as potential therapeutic agent for pain management and long-acting local anesthesia, but no saxitoxin-based drugs have yet received clinical approval [53, 62].

The LD_{50} value of saxitoxin for mice is approx. 3 µg/kg BW (i.v.) and 250 µg/kg BW (p.o., [28]). The oral LD_{50} for humans is about 6 µg/kg BW [77, 109]. Inhalation of aerosolized saxitoxin as well as entry into the body via open wounds can also cause severe intoxications at very low doses [77].

5.4.4 Other Alkaloids

Tubercidin, a 7-desaza purine nucleoside (Fig. 5.6), was isolated from *Tolypothrix byssoidea* and *Scytonema saleyeriense* [4]. It had previously been isolated from Streptomyces spp. in the search for antibiotics. Tubercidin has a stabilizing effect on the microtubules, inhibits DNA and RNA synthesis by disruption of nucleic acid structure following incorporation, and, like some of its synthetic analogues, inhibits adenosine kinase and thus the phosphorylation of ribofuranosylnucleosides in position 5′ [69, 72, 94].

Tubercidin

Fig. 5.6: A 7-desaza purine nucleoside of cyanobacteria.

References

For numbers in bold, see cross-chapter literature p. 233.

[1] Alexander J et al. (2009) EFSA J 1306: 1
[2] Aráoz R et al. (2005) Microbiol 151: 1263
[3] Banker R et al. (2010) J Toxicol Environ Health 62: 281
[4] Barchi JJ Jr. (1983) Phytochem 22: 2851
[5] Basu A et al. (1992) Biochem 31: 3824
[6] Bauer F et al. (2020) Toxins 12: 726
[7] Bonjouklian R et al. (1991) Tetrahedrom 47: 7739
[8] Breinlinger S et al. (2021) Science 371: aax9050
[9] Byth S (1980) Med J Australia 2: 40
[10] Capper A et al. (2005) J Chem Ecol 31: 1595
[11] Cardellina II AJ et al. (1979) Science 204: 193
[12] Carmichael WW, Gorham PR (2017) Proc Int Assoc Theor Appl Limnol 21: 285
[13] Chilczuk T et al. (2020a) Planta Med 86: 96
[14] Chilczuk T et al. (2020b) ChemBioChem 21: 2170
[15] Chiswell RK et al. (1999) Environ Toxicol 14: 155
[16] Chorus I (ed, 2001) Cyanotoxins, Springer Berlin, Heidelberg
[17] Chorus I, Welker M (eds, 2021) Toxic cyanobacteria in water – Second edition, WHO
[18] Christensen VG, Khan E (2020) Sci Total Env 736: 139515
[19] Codd GA, Poon GK (1988) Cyanobacterial Toxins. In: Rogers LJ, Gallon JR (eds) Biochemistry of the Algae and Cyanobacteria, Clarendon Press, Oxford
[20] Codd GA et al. (eds) (1994). Detection Methods for Cyanobacterial Toxins, The Royal Society of Chemistry, Cambridge
[21] Colas S et al. (2021) Environ Res 193: 110590
[22] Cusick KD, Sayler GS (2013) Mar Drugs 11: 991
[23] Deeds JR et al. (2008) Mar Drugs 6: 308
[24] Devlin JP et al. (1977) Can J Chem 55: 1367
[25] Doan NT et al. (2001) FEMS Microbiol Lett 196: 135
[26] Draisci R et al. (2010) Food Add Contam 18: 525
[27] Edwards DJ, Gerwick WH (2004) J Am Chem Soc 126: 11432
[28] EFSA (2009) EFSA J 7: 1019
[29] Fastner J et al. (2018) Toxins 10: 60
[30] Fernandes KF et al. (2024) Chemosphere 352: 141277
[31] Finch SC, Harwood DT (2025) Toxins 17: 105

[32] Fiore MF et al. (2020) Harmful Algae 92: 101737
[33] Froscio SM et al. (2003) Environ Toxicol 18: 243
[34] Fujiki H et al. (1984) Biochem Biophys Res Comm 120: 339
[35] Fujiki H et al. (1988) Cancer Res 48: 4211
[36] Galvão JA et al. (2009) Toxicon 54: 891
[37] Griffiths DJ, Saker ML (2003) Environ Toxicol 18: 78
[38] Gugger M et al. (2005) Toxicon 45: 919
[39] Hackett JD et al. (2013) Mol Biol Evol 30: 70
[40] Hagiwara N et al. (1988) Biosci Biotechnol Biochem 52: 641
[41] Hawkins PR et al. (1985) Appl Environ Microbiol 50: 1292
[42] Herath G (1995) Rev Mark Agric Econ 63: 1
[43] Hyde EG, Carmichael WW (1991) J Biochem Toxicol 6: 195
[44] Irie K et al. (1989) Internat J Cancer 43: 513
[45] Ito E et al. (2002) Toxicon 40: 551
[46] Izumi AK, Moore RE (1987) Clinics Dermatol 5: 1987
[47] James KJ et al. (2007) Anatoxin-a and Analogues: Discovery, Distribution, and Toxicology. In: Botana LM (ed) Phycotoxins: Chemistry and Biochemistry, Blackwell Publishing
[48] Johnston LH et al. (2024) Sci Total Env 917: 170476
[49] Jung P et al. (2021) Microorg 9: 745
[50] Junier P et al. (2024) Water Res X 24: 100252
[51] Kellmann R et al. (2008) Appl Environ Microbiol 74: 4044
[52] Kiviranta J et al. (1991) Arch Hydrobiol 121: 281
[53] Kohane DS et al. (2019) Reg Anesth Pain Med 25: 52
[54] Kozikowski AP et al. (1991) J Med Chem 24: 2420
[55] Krienitz L et al. (2003) FEMS Microbiol Ecol 43: 141
[56] Lagos N et al. (1999) Toxicon 37: 1359
[57] Landsberg JH (2010) Rev Fishery Sci 10: 113
[58] Landsberg JH et al. (2006) Environ Health Perspect 114: 1502
[59] Leal JF, Cristiano MLS (2022) Nat Prod Rep 39: 33
[60] Llewellyn LE (2006) Nat Prod Rep 23: 200
[61] Llewellyn LE (2009) Sodium Channel Inhibiting Marine Toxins. In: Fusetani N, Kem W (eds) Marine Toxins as Research Tools, Springer Berlin/Heidelberg
[62] Lobo K et al. (2015) Anesthesiology 123: 873
[63] Lomeo D et al. (2024) Sci Total Env 957: 177525
[64] Lovin LM, Brooks BW (2019) Mar Freshw Res 71: MF18373
[65] Mahmood NA, Carmichael WW (1986) Toxicon 24: 425
[66] Mahmood NA, Carmichael WW (1986) Toxicon 25: 1221
[67] Matsunaga S et al. (1989) J Am Chem Soc 111: 8021
[68] McCarron P et al. (2023) Toxicon 227: 107086
[69] McGaraughty S et al. (2006) CNS Drug Rev 7: 415
[70] Metcalf JS, Bruno M (2016) Anatoxin-a(S). In: Handbook of Cyanobacterial Monitoring and Cyanotoxin Analysis, Meriluoto J et al. (eds.), John Wiley & Sons
[71] Mihali TK et al. (2008) Appl Environ Microbiol 74: 716
[72] Mooberry SL et al. (1995) Cancer Lett 96: 261
[73] Namikoshi M et al. (2003) Toxicon 42: 533
[74] Ohtani I et al. (1992) J Am Chem Soc 114: 7941
[75] Osborne NJT et al. (2001) Environm Internat 27: 381
[76] Osswald J et al. (2007) Enriron Internat 33: 1070

[77] Patocka J, Stredav L (2002) Brief Review of Natural Nonprotein Neurotoxins. In: Price R (ed) ASA Newsletter, Applied Science and Analysis inc.
[78] Pearl HW, Otten TG (2013) Microb Ecol 65: 995
[79] Plata-Calzado C et al. (2024) Toxins 16: 198
[80] Poniedziałek B et al. (2013) Environ Toicol Pharmacol 34: 651
[81] Puschner B et al. (2008) J Vet Diagn Invest 20: 89
[82] Rossini GP, Hess P (2010) Phycotoxins: chemistry, mechanisms of action and shellfish poisoning. In: Luch A (ed) Molecular, Clinical and Environmental Toxicology, Birkhäuser Verlag, Switzerland
[83] Runnegar MT et al. (1994) Biochem Biophys Res Comm 201: 235
[84] Sakai S et al. (1984) Chem Pharm Bull 32: 354
[85] Saker ML, Eaglesham GK (1999) Toxicon 37: 1065
[86] Saker ML et al. (1999) Environ Toxicol 14: 179
[87] Schantz EJ et al. (1966) Biochem 5: 1191
[88] Schantz EJ et al. (1975) J Am Chem Soc 97: 1238
[89] Schrader A et al. (1992) Tierische Gifte. In: Schrader A (ed) Toxisch bedingte Krankheiten des Nervensystems, Springer, Berlin/New York
[90] Schwark M et al. (2023) PNAS 120: e2219230120
[91] Shaw GR et al. (2000) Ther Drug Monit 22: 89
[92] Shen X et al. (2002) Toxicon 40: 1499
[93] Sirén AL, Feuerstein G (1990) Toxicol Appl Pharmacol 102: 91
[94] Smith CG et al. (1967) Adv Enzyme Regul 5 121
[95] Spivak CE et al. (1980) Mol Pharmacol 18: 384
[96] Štenclová L et al. (2023) Harmful Algae 125: 102425
[97] Stirling DJ, Quilliam MA (2001) Toxicon 39: 1219
[98] Takashima M, Sakai H (1969) Biosci Biotechnol Biochem 24: 647
[99] Tejedor FJ et al. (1988) Biochem 27: 2389
[100] Terano K et al. (1994) Toxicon 32: 833
[101] Thomas P et al. (1993) J Neurochem 60: 2308
[102] Thottumkara AP et al. (2014) Angew Chem Int Ed 53: 5760
[103] Valadez-Cano C et al. (2023) Harmful Algae 124: 102405
[104] Viaggiu E et al. (2004) Environ Tox 19: 191
[105] Walton K, Berry JP (2016) Mar Drugs 14: 73
[106] Wang H et al. (2021) Innovation 11: 100092
[107] Wekell JC et al. (2004) J Shellfish Res 23: 927
[108] Werner KA et al. (2011) Internat J Dermatol 51: 59
[109] WHO (World Health Organization, 2020) WHO/HEP/ECH/WSH/2020.8
[110] Wilde SB et al. (2014) Phytotaxa 181: 243
[111] Wonnacott S, Gallagher T (2006) Mar Drugs 4: 228
[112] Wonnacott S et al. (1991) J Pharmacol Exp Therap 259: 387
[113] Wonnacott S et al. (1992) Biochem Pharmacol 43: 419
[114] Wood SA et al. (2007) Toxicon 50: 292
[115] Wood SA et al. (2018a) Harmful Algae 80: 88
[116] Yamashita T et al. (1988) J Nat Prod 51: 1184
[117] Zamlynny L et al. (2025) Analyt Bioanalyt Chem https://doi.org/10.1007/s00216-025-05829-9

5.5 Peptides as Toxins of Cyanobacteria

5.5.1 General

A large number of cyanobacteria are capable of producing toxic peptides. Toxin producing strains include representatives of the genera Anabaena, Anabaenopsis, Hapalosiphon, Moorena, Microcystis, Nodularia, Nostoc, Oscillatoria, and Planktothrix [68, 97].

These peptide toxins are primarily intracellular and are only released upon cells lysis, for example, in the digestive tract of an animal, or when the cells decompose during or after a cyanobacteria bloom, where they, once released, can contaminate drinking water sources [23, 24].

Many cyanobacterial peptide toxins are cyclic peptides composed of 5–14 amino acid residues, or lipopeptides, which combine peptide and fatty acid moieties. The majority of these compounds are synthesized via non-ribosomal pathways. Their biosynthesis typically involves non-ribosomal peptide synthetases (NRPS), and frequently also polyketide synthases (PKS). The biosynthetic gene clusters (BGCs) responsible for the production of many peptide cyanotoxins have been characterized [35, 36].

Analytical techniques for detecting peptide toxins have advanced significantly in recent years, particularly with the development of novel chromatographic, mass spectrometric, immunologic, and biosensor-based methods [81]. However, current monitoring efforts have disproportionately focused on a small subset of toxins, most notably microcystins, while neglecting many other potentially harmful compounds. Greater analytical attention and regulatory oversight are also needed for food supplements containing cyanobacterial biomass, which may pose a health risk if contaminated [32].

5.5.2 Microcystins

Microcystins are cyclic heptapeptides (Fig. 5.7) containing the unusual amino acid ADDA (3-amino-9-methoxy-10-phenyl-2,3,8-trimethyl-deca-4,6-dienoic acid) or its 9-*O*-acetyl-9-desmethyl derivative. They are the best-studied and most widespread class of peptide toxins produced by cyanobacteria [28, 34, 108, 140, 143]. Microcystins (formerly also referred to as cyanoginosins, cyanoviridins, or 'fast death factor', FDF) were first isolated from *Microcystis aeruginosa* and *M. viridis* (Ph. 5.5). They are widely produced by Microcystis and Planktothrix strains (Ph. 5.6), which are of major toxicological concern in HABs, but production has also been documented in other genera such as Anabaena, Hapalosiphon, Nostoc, and Oscillatoria [23, 24].

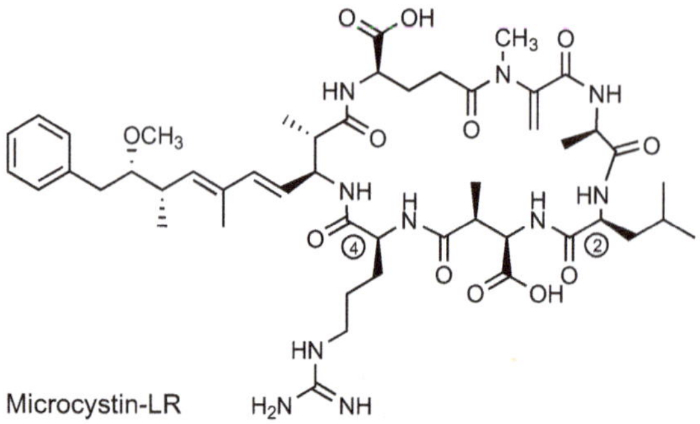

Microcystin-LR

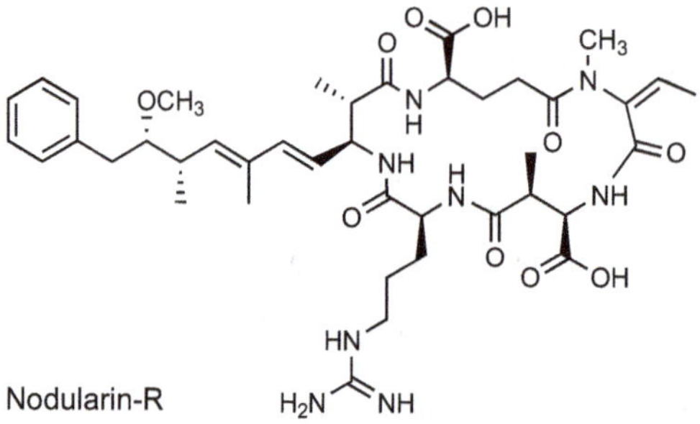

Nodularin-R

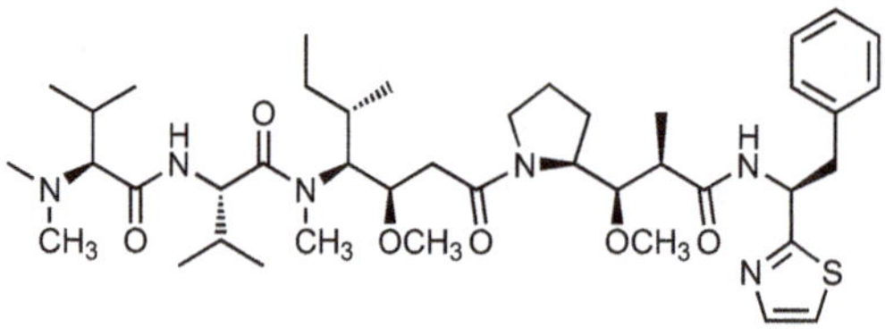

Dolastatin 10

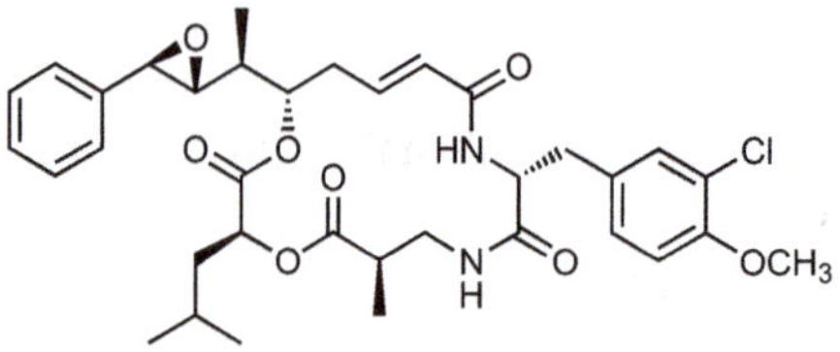

Cryptophycin

Fig. 5.7: Peptide toxins of cyanobacteria, Part 1.

Microcystis aeruginosa is an unicellular organism, 3–9 µm in size, found in planktonic or benthic form in nutrient-rich fresh or brackish water. Colonies are rather irregular and embedded in mucilage. Along with other cyanobacteria, it is often observed in freshwater HABs. *Microcystis viridis* is morphologically similar, but forms more structured colonies with a smooth gelatinous edge. The genus Planktothrix forms unbranched filaments. Toxic species such as *Planktothrix agardhii* and *P. rubescens* have been well studied, as they frequently dominate freshwater and brackish blooms.

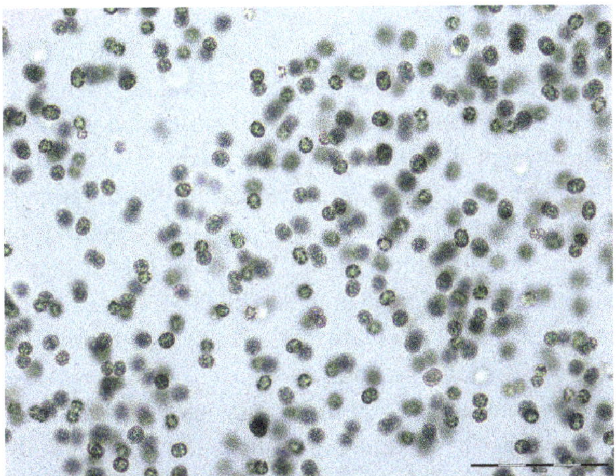

Ph. 5.5: Microcystis sp. CBT 937 (source: Simris Biologics GmbH).

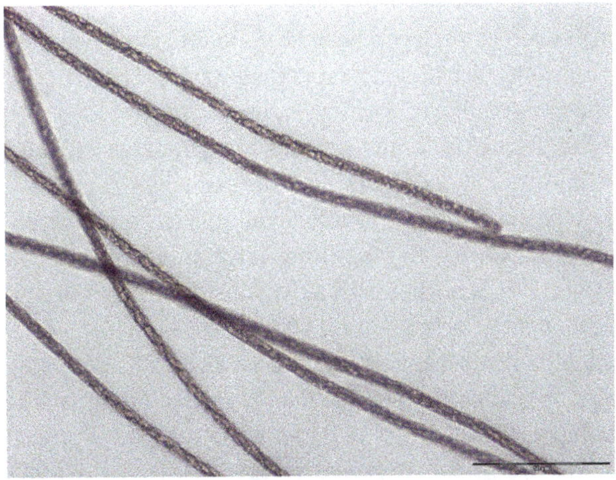

Ph. 5.6: *Planktothrix rubescens* CBT 332 (source: Simris Biologics GmbH).

The widespread distribution of microcystins has been attributed to the ancient evolutionary origin of the microcystin BGC. Over evolutionary time, the capacity for microcystin production has been retained in some lineages and lost in others, suggesting selective pressure for or against toxin synthesis depending on ecological context [113].

Microcystin biosynthesis is non-ribosomal. It is catalyzed by a hybrid non-ribosomal peptide synthetase–polyketide synthase (NRPS–PKS) system encoded by approx. 55 kb of DNA across nine mcy genes. These genes encode peptide synthetases, PKSs (which form the ADDA moiety), hybrid NRPS–PKS enzymes, and tailoring enzymes responsible for structural modifications [25, 58, 136].

To date, over 300 structural variants of microcystins have been identified [7, 69]. These cyclic heptapeptides follow the general structure cyclo(D-alanine^1-X^2-D-erythro-β-methyl-isoaspartic acid^3-Y^4-ADDA^5-D-γ-glutamic acid^6-*N*-methyldehydroalanine^7), where X (position 2) and Y (position 4) are variable L-amino acids. Additional variation arises from modifications to side chains, including demethylations at positions 3 and/or 7, as well as changes to the ADDA moiety or other amino acids [7]. Microcystin nomenclature is standardized [14]. For microcystins with the general structure above, the single-letter code of the amino acids in positions 2 and 4 follow the abbreviation 'MC', e.g., MC-LR for the microcystin containing the L-amino acids leucine (L) and arginine (R). Other modifications are placed in square brackets before the name, e.g., [D-Asp^3]MC-LR indicates the replacement of methyl-isoaspartic acid by isoaspartic acid at position 3.

Microcystin-LR is the most prevalent microcystin in Microcystis strains, but MC-YR and -RR are also well characterized. The dominant variant in a given strain depends on strain type, environmental conditions, and cell age [24, 70, 145, 148]. While most strains produce 3–10 microcystin variants, highly plastic strains may synthesize up to 50 different derivatives [43].

Microcystin-containing HABs are a global problem, and they have been reported from all continents [94]. Very high microcystin levels have been reported in bloom material from Portugal and China, reaching over 7 mg microcystins/g total DW [29, 141, 152, 160]. Eutrophication promotes the proliferation of microcystin-producing taxa [114].

Microcystins potently inhibit serine/threonine protein phosphatases PP1 and PP2A (half-maximal inhibition concentrations (IC_{50}) in the low nanomolar concentration range), disrupting regulation of the cell cycle, cytoskeleton integrity and maintenance, and intracellular signaling [28, 40, 60, 91, 118]. Other targets under investigation include the β-subunit of ATPase [95], mitochondrial aldehyde dehydrogenase [19], calpain, and Ca^{2+}/calmodulin-dependent protein kinase II [33].

The relatively large and amphiphilic microcystins cannot passively diffuse across cell membranes. Instead, cellular uptake occurs via specific organic anion transporting polypeptides (OATP), especially OATP1B1, OATP1B3, and OATP1A2 [54, 74]. Liver cells, where OATP1B1 and OATP1B3 are primarily expressed [54, 72], are particularly vulnerable, resulting in pronounced hepatotoxicity of microcystins [28, 64, 161]. Kidneys, heart, and lungs can be affected by microcystins, as well [11]. Microcystins also have relevant tumor-promoting effects [52, 158].

Interestingly, protein phosphatase inhibition and uptake by OATPs do not correlate [100, 121], resulting in strongly differing in vivo toxicities of individual microcystin variants, e.g., micrrocystin-RR is much less toxic than microcystin-LR [21]. This can complicate risk assessments and water management, as the broadly used immunosorbent assays, in contrast to the more expensive and demanding chromatographic methods, cannot discriminate between microcystin variants. Also, in vivo toxicity of most microcystin variants has not yet been assessed.

In animal models (mice), microcystin exposure results in elevated serum enzymes, prolonged prothrombin time, and intrahepatic hemorrhage due to sinusoidal endothelial damage. Lethality occurs 1–2 h after i.v. or i.p. administration due to hemorrhagic shock. Chronic oral exposure of mice to Microcystis extract in drinking water (750–120,00 µg microcystin/kg/day for 1 year) resulted in liver damage, bronchopneumonia, and reduced lifespan [24].

LD_{50} values of microcystins in mice vary and are, e.g., 25–150 µg/kg BW (i.p.) and 5,000–11,000 µg/kg BW, p.o. [21, 28, 42, 155]. Structural modifications in the ADDA-Glu region, e.g., ADDA isomerization or glutamate acylation, eliminate toxicity, as these two residues are essential for the interaction with the protein phosphatases [2, 50, 53, 91, 134].

Post i.v./i.p. injection of radiolabeled microcystins to mice or rats, about 70% of the toxins localize in the liver. Toxicokinetic data in humans remain limited [1, 32].

Human and animal poisonings by *Microcystis aeruginosa* have been reported globally [29, 143]. Representatives of the other genera mentioned have also been identified as the cause of death in various wild and domestic animals [130]. In humans, exposure can occur via swimming in bloom-affected waters, drinking contaminated water, or consuming fish or shellfish with accumulated microcystins [23]. Uptake by edible plants irrigated with contaminated water is also documented [78]. A particularly notable incident occurred in Brazil in 1996, where water containing 19.5 µg microcystins/L was used for the production of dialysis fluids. It had been obtained from a lake with massive cyanobacteria growth after a dry period. Of the 131 patients who became ill, 100 died of acute liver failure [15, 111]. The detection of microcystins and mcy genes in supposedly health-promoting food supplements based on cyanobacteria is a cause for concern [120]. Approaches to minimize microcystin presence in drinking water are under ongoing development [76, 119].

The WHO has limited free and cell-bound microcystin-LR in drinking water to 1.0 µg/L for long-term, and 12 µg/L for short-term exposure [147].

Symptoms of microcystin poisoning after exposure to recreational waters containing microcystins (2–4 µg/L) or after oral ingestion of the toxins include headaches, myalgia, nausea, diarrhea, and skin irritation [23].

If potentially toxic quantities are ingested, primary removal of toxins must be ensured as quickly as possible. Administration of milk thistle preparations or rifampicin, both blocking microcystin uptake via the OATP, may offer hepatoprotective effects. Experimental therapies include monoclonal antibodies against microcystin LR [23].

Animals mainly fall ill after drinking water contaminated with *Microcystis aeruginosa*. Surprisingly, some species seem to prefer cyanobacteria-laden water over clear water [83]. Documented cases include cattle displaying behavioral disturbances, dehydration, rumenatonia, and mortality after drinking from a stagnant pond in hot, dry weather [47]. Sensitivity of fish is species-specific: While trout and goldfish are insensitive to p.o. intake, microcystins can cause considerable losses in salmon farms ('net pen liver disease' [127]).

Interestingly, microcystins are being explored as anti-cancer agents [100, 101]. Due to their potent PP1/PP2A inhibition, which would be a novel therapeutic mechanism in anti-cancer therapy with unlikely development of resistances, they are attractive as payloads in antibody-drug conjugates. Due to their poor membrane permeability, microcystins are only toxic to cells after transporter-mediated uptake or after uptake by endocytosis when coupled to an antibody. Intriguingly, it might be possible to exploit the poor membrane permeability of microcystins and develop chemically modified derivatives that are not taken up by OAPTs at all, thereby minimizing side effects [121].

5.5.3 Nodularins

Nodularia spumigena is a planktonic, filamentous cyanobacterium that is found in freshwater brackish environments. It was the first cyanobacterium recognized for its toxicity, as observed by Francis in 1878 [45].

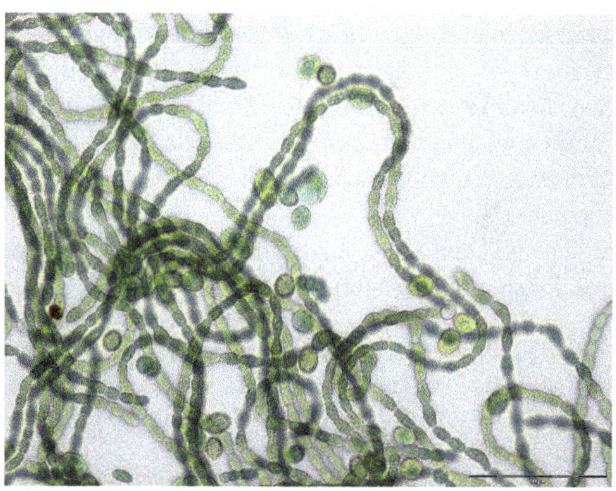

Ph. 5.7: Nodularia sp. CBT 786 (source: Simris Biologics GmbH).

The toxins produced by *N. spumigena* (Ph. 5.7), known as nodularins (NOD, Fig. 5.7), are cyclic pentapeptides structurally related to microcystins. Compared to microcystins, they are a rather small compound family; only 16 variants are known to date [69]. They lack the amino acids typically found at positions 1 and 2 of microcystins, and they usually feature *N*-methyldehydrobutyrine at position 5 [129]. The most commonly detected variant is nodularin-R, which contains arginine. Variants differ in the variable L-amino acid (e.g., motuporin, or [L-Val]-Nodularin, or [L-Har]-Nodularin containing homoarginine), in the ADDA moiety (e.g., desmethylated or the stereoisomeric form 6*Z*-ADDA), or in the degree of methylation on other residues (e.g., desmethyl-Asp -Nodularin). Notably, motuporin has also been detected in the marine sponge *Theonella swinhoei*, likely originating from its cyanobacterial symbionts [144].

In contrast to the widely distributed microcystins, nodularins are most exclusively produced by cyanobacteria of the genus Nodularia [70]. Phylogenetic analyses suggest that the BGC for nodularin evolved from the microcystin BGC [48, 113]. In addition to Nodularia species, some terrestrial Nostoc species (Ph. 5.8) which are cyanobacterial symbionts of plants have also been found to produce nodularins [48].

Blooms of *Nodularia spumigena* are globally observed [20], with extensive occurrences reported, e.g., in Europe in brackish environments in the Baltic Sea. For example, in the hot summer of 2003, plankton samples collected in various Baltic Sea regions contained nodularin concentrations ranging from 150 to 800 µg/L [84].

Toxicologically, nodularins share characteristics with microcystins. They also inhibit the protein phosphatases PP1 and PP2A, and uptake into cells occurs by active transport via OATPs, particularly OATP1B1 [100]. Thus, nodularins are also potent hepatotoxins and tumor-promotors [103, 156].

The LD_{50} of nodularin in mice ranges from 50 to 70 µg/kg BW (i.p.). Histopathological analysis typically reveals extensive hepatic hemorrhaging following exposure to nodularin [13, 39, 117].

Although human intoxications are rare, exposure may occur via ingestion of contaminated drinking water or consumption of fish or shellfish that have bioaccumulated the toxin. There is epidemiological suspicion that chronic exposure to nodularins and microcystins contributes to the elevated liver cancer rates in parts of China, where untreated surface water is commonly used for drinking [157].

Animal poisonings attributed to *Nodularia spumigena* have been reported. Affected animals exhibit dyspnea, opisthotonus, convulsions, pneumonia, and hepatic necrosis upon necropsy. Cases of canine fatalities occur frequently after swimming in lakes with Nodularia blooms, and poisoning of other animals like domestic ducks, sheep, and cattle have also been described [20].

5.5.4 Dolastatins

In 1972, extracts from the sea hare *Dolabella auricularia* were found to exhibit strong cytotoxicity [110]. However, due to the extremely low abundance of the active constituents, it took until 1987 to isolate and determine the structure of the most potent compound, dolastatin 10 (Fig. 5.7), a non-ribosomal linear pentapeptide [71]. Despite the initial assumption that dolastatins were produced by the sea hare itself, subsequent research revealed that these compounds are synthesized by cyanobacteria, e.g., *Symploca hydnoides*, which constitute the sea hare's diet [55, 88]. Chemically related compounds are symplostatin 1 or malevamide D, which are also produced by Symploca strains [56, 61].

At the time of their discovery, dolastatins were the most potent cytotoxins known, with half-maximal effective concentrations (EC_{50}) in the picomolar range. Dolastatin 10 exerts its cytotoxic effects by binding to tubulin near the vinca alkaloid site, thereby disrupting microtubule polymerization and inhibiting mitotic progression [3].

Since their discovery, the dolastatins have been investigated as potential anticancer agents. However, it took nearly four decades before the first synthetic derivative of dolastatin 10, monomethylauristatin E (MMAE), was successfully introduced into clinical use as the cytotoxic payload in the antibody-drug conjugate brentuximab vedotin [99]. Today, six dolastatin-based ADCs have received regulatory approval for human use, with over 30 others in various stages of development [128]. To date, dolastatin 10 remains the only cyanobacterial toxin to have been successfully translated into human medicine.

Dolastatins and related compounds were long considered to be exclusively marine toxins with minimal toxicological relevance to livestock or humans. However, recent findings have identified the freshwater cyanobacterium *Aetokthonos hydrillicola* (Ph. 5.2) as a producer of aetokthonostatins, chemically closely related compounds with comparable toxicity [122]. This cyanobacterium grows epiphytically on the submerged aquatic plant *Hydrilla verticillata*, which is common in the south-eastern United States in artificial lakes frequently used as recreational lakes or drinking water reservoirs. Consequently, the presence of aetokthonostatins, along with a second toxin produced by *A. hydrillicola*, aetokthonotoxin (see Section 5.4), may have implications for both environmental and human health.

5.5.5 Cryptophycins

Cryptophycins are a class of highly cytotoxic cyclic depsipeptides (Fig. 5.7) isolated from Nostoc sp. (Ph. 5.8 [26, 37, 115]). Initially discovered and patented for their antifungal activity, particularly against the human pathogen *Cryptococcus neoformans*, their clinical potential in antifungal therapy was ultimately limited by significant toxicity [59, 123].

Subsequent investigations led to the re-isolation of cryptophycins from a different Nostoc strain, revealing its remarkable cytotoxic potential. Cryptophycins emerged as some of the most powerful known inhibitors of microtubule dynamics, exhibiting cytotoxic effects in the low picomolar concentration range [137]. They exert their action

Ph. 5.8: Nostoc sp. CBT 893 (source: Simris Biologics GmbH).

by binding to a specific site on tubulin and inhibiting microtubule polymerization, effectively arresting the cell cycle at the G2/M transition. Their activity surpasses that of traditional microtubule inhibitors such as paclitaxel (see **[159 = Vol. 1]**) and vinblastine (see **[161 = Vol. 3]**) by factors of 100–1,000 [107].

Extensive structure–activity relationship (SAR) studies have been performed using both semi-synthetic and fully synthetic cryptophycin derivatives [37, 38]. Among these, cryptophycin-52 has attracted interest due to its exceptional potency across a wide range of cancer cell lines, with IC_{50} values in the low picomolar range [142]. Although initial clinical trials of cryptophycin-52 as a standalone chemotherapeutic agent were halted due to dose-limiting toxicities, the compound has since been reconsidered as a cytotoxic payload in antibody-drug conjugates, where targeted delivery may mitigate systemic toxicity [8].

5.5.6 Other Peptide Toxins

The lyngbyastatins (Fig. 5.8) isolated from *Moorena producens* (Ph. 5.1) and other genera are cyclic depsipeptides [31, 57, 149]. They have structural similarity with cyclic dolastatin 15 derivatives from *Symploca hydnoides* (not to be confused with the linear dolastatin 10 analogs discussed above). The lethal dose of lyngbyastatin 2 for mice is 3,000 µg/kg BW [85]. Like the linear dolastatins, they target the cytoskeleton, though not as potently [4]. The cytotoxic and antifungal majusculamide C is a related depsipeptide containing eight amino acids as well as one hydroxy acid residue [16].

Antillatoxin (Fig. 5.8), apratoxin (Fig. 5.8), barbamides (Fig. 5.8), bisebromoamide, curacins (Fig. 5.8), hoiamides, jahanyne, kalkitoxin (Fig. 5.8), kurahyne, lagunamides (Fig. 5.9), lyngbyabellins (Fig. 5.9), malyngamides (Fig. 5.9), and microcolins are exam-

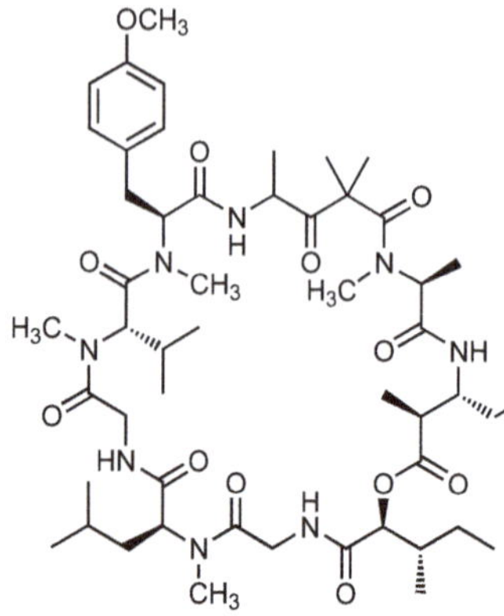

Lyngbyastatin 1

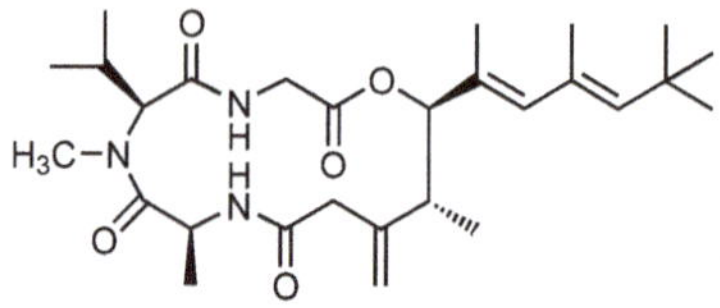

Antillatoxin

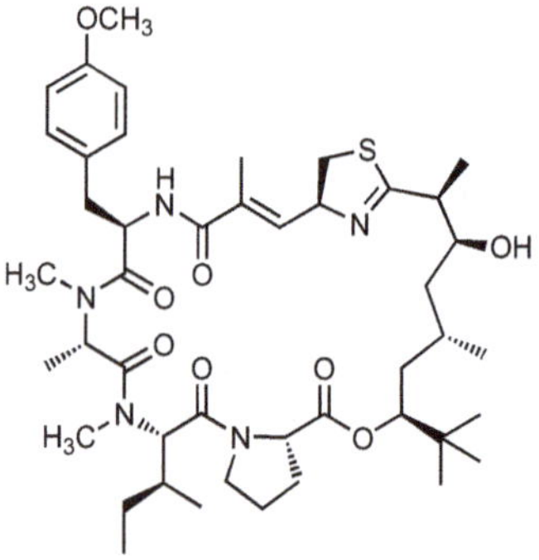

Apratoxin

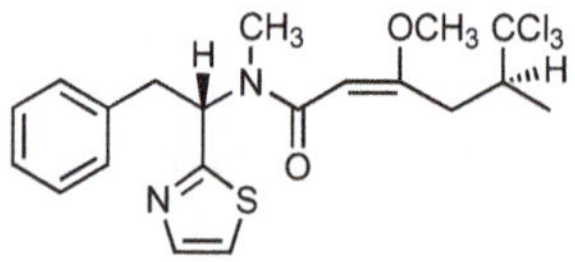

Barbamide

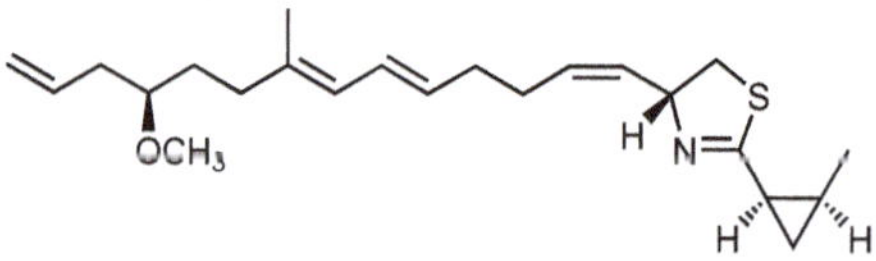

Curacin A

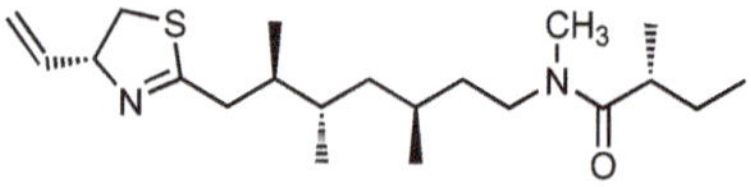

Kalkitoxin

Fig. 5.8: Peptide toxins of cyanobacteria, Part 2.

ples of toxic Moorena sp. metabolites resulting from a mixed PKS/NRPS biosynthetic pathway. The ichthyotoxic antillatoxin A is, similar to the brevetoxins (see **[63 = Vol. 4]**), an activator of voltage-gated sodium channels and has neurotoxic effects [79, 102, 106]. It has been implicated in cases of respiratory irritation, eye inflammation, and contact dermatitis in fishermen [30]. Apratoxin A is a potent cytotoxin with IC_{50} values against human tumor cell lines in the sub-nanomolar concentration range [51, 87, 133]. Its mode of action involves inhibition of the protein translocase Sec61 [62]. Apra-

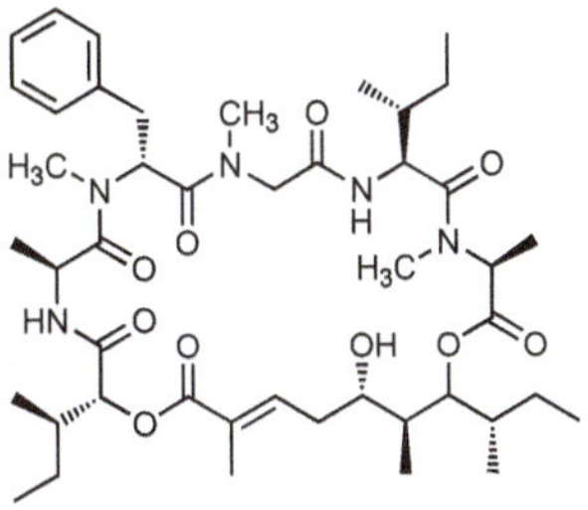

Lagunamide A

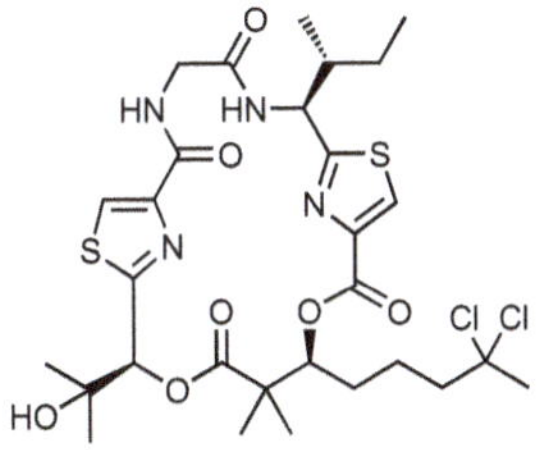

Lyngbyabellin A

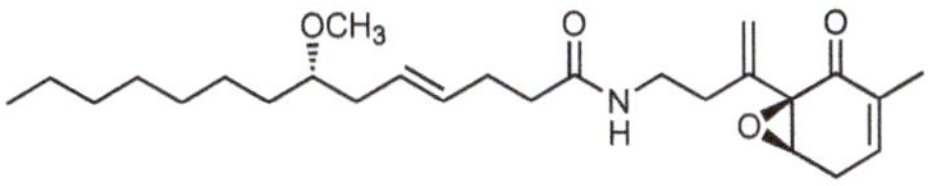

Malyngamide H

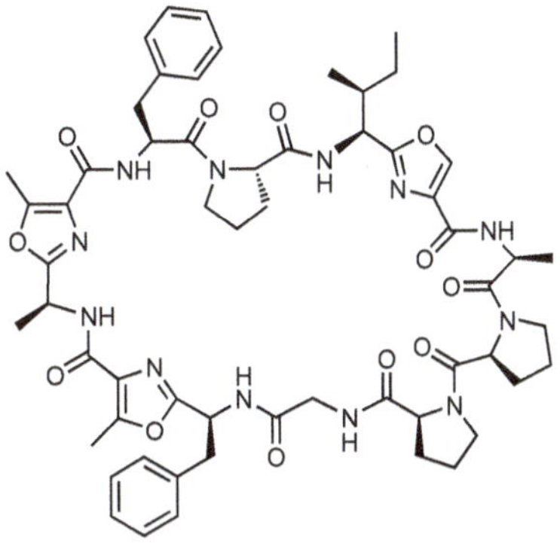

Wewakazole B

Fig. 5.9: Peptide toxins of cyanobacteria, Part 3.

toxin A shows promise as an anticancer drug lead [89]. Barbamide is a chlorinated compound with strong molluscicidal effect [105]. Bisebromoamide is a brominated peptide that is cytotoxic due to protein kinase inhibition and stabilization of actin filaments [131, 135]. Curacin A, a mixed polyketide/non-ribosomal peptide [18], has a strong cytotoxic effect. Like colchicine (see **[161 = Vol. 3]**), it inhibits the formation of microtubules [18, 49] and is discussed for clinical development as anticancer drug [150]. The hoiamides A and B are neurotoxic as partial agonists to voltage-gated sodium channels inducing both apoptosis and necrosis [12, 22, 109]. Jahanyne and the related kurahyne induce apoptosis of human cancer cell lines [66, 67, 104, 153]. Kalkitoxin, an ichthyotoxin (LC_{50} goldfish, *Carassius auratus*, 700 nM), acts by activation of NMDA receptors, inhibition of voltage-gated sodium channels, and inhibition of the electron transport chain complex 1 [5, 77, 98, 139, 151]. Lagunamides are highly cytotoxic due to mitochondrial apoptosis by caspase activation [63, 90, 138]. Lyngbyabellins are cytotoxic, destroying microfilaments [86]. The malyngamides are frequently found lipopeptides with a fatty acid residue of 12 or, more frequently, 14 carbon atoms. Some malyngamides show immunosuppressive, cytotoxic, and antimicrobial activities [27]. The microcolins have antiproliferative and immunosuppressive activities [73, 159].

The ribosomally synthesized and post-translationally modified peptides wewakazoles (Fig. 5.9), produced by *Moorena producens*, were also found to be cytotoxic [82].

In addition to microcystins, Anabaena, Microcystis, Nostoc, and Planktothrix species also often produce peptides with protease-inhibiting activity [68, 99, 112]. Whether they are involved in the toxicity of the respective cyanobacteria or their ecological relevance for the cyanobacteria are unclear, but recent research indicates that these compounds often suppress eukaryotic freshwater organisms like cyanobacteria-feeding Daphnia [124, 125]. Examples for protease-inhibiting cyclic peptides include the chemically related micropeptins and cyanopeptolins (Fig. 5.10) of Microcystis species, the oscillapeptins of Planktothrix species, the nostopeptins of Nostoc species [41, 46, 93, 116, 146], or the anabaenopeptins (Fig. 5.10) from, e.g., Anabaena or Planktothrix sp. [96]. Microviridins are of ribosomal origin and encountered in Microcystis, Planktothrix, and others [17]. Linear protease-inhibiting peptides include the microgi-

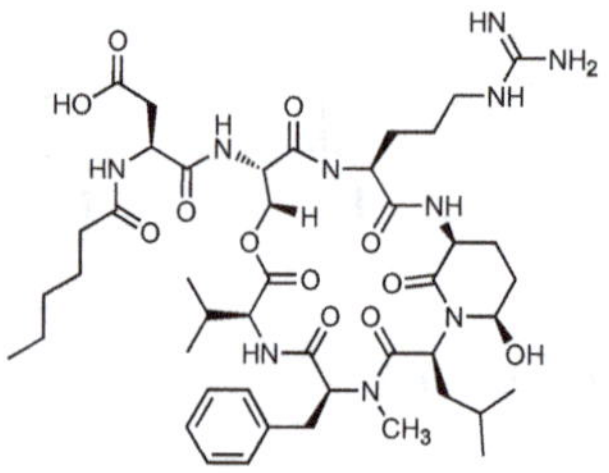

Cyanopeptolin A

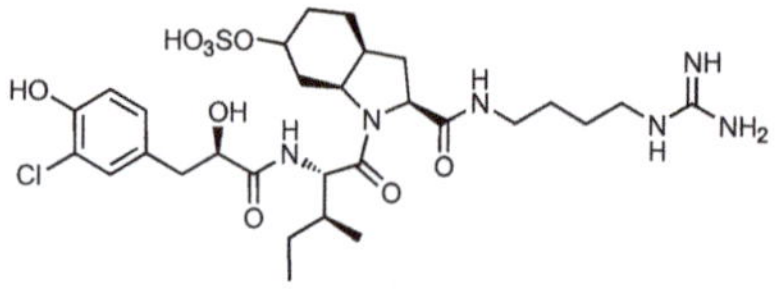

Aeruginosin 98A

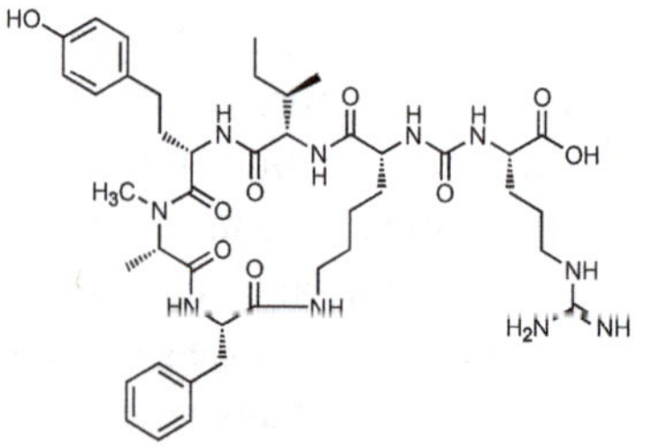

Anabaenopeptin B

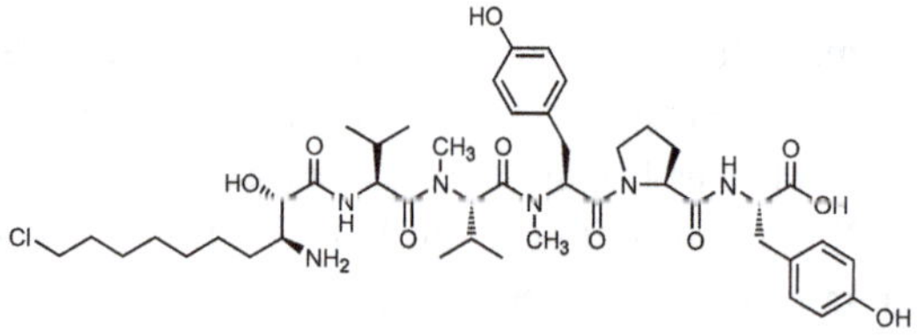

Microginin 299-A

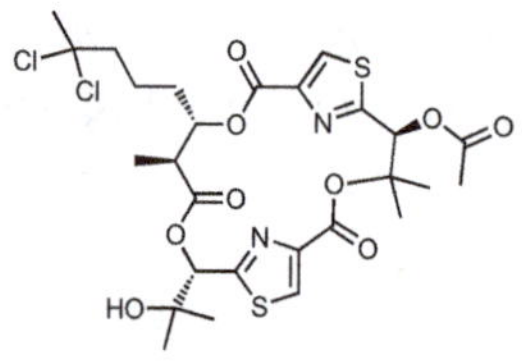

Hectochlorin

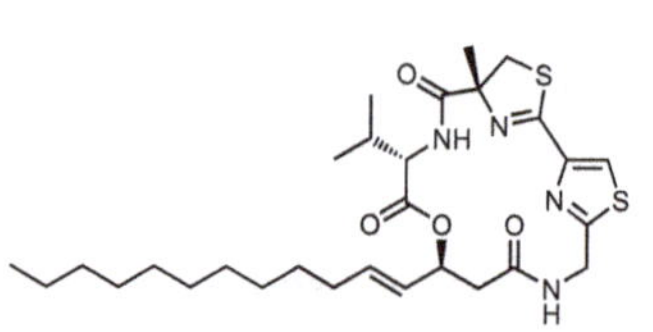

Largazole

Fig. 5.10: Peptides with protease-inhibiting activities and further peptide toxins of cyanobacteria (IV).

nins (Fig. 5.10) and aeruginosins (Fig. 5.10) of Microcystis sp. [65, 75, 80] as well as spumigins of *Nodularia spumigena* [44].

Hectochlorin (Fig. 5.10) and largazole (Fig. 5.10) are cyclic depsipeptides. Hectochlorin, isolated from *Moorena producens*, has been found to be highly cytotoxic due to the induction of hyperpolymerization of eukaryotic actin [92]. Largazole has been isolated from a Symploca sp. It is very toxic to cancer cell lines [132]. It was shown to act as a potent class I, II, and III histone deacetylase (HDAC) inhibitor [9, 154]. Evaluation of structure-activity relationships of synthetic largazole derivatives [6, 126] have led to the synthesis of compounds inhibiting the enzymes in a picomolar concentration range [10].

References

For numbers in bold, see cross chapter literature p. 233.

[1] Arman T, Clarke JD (2021) Toxins 13: 537
[2] Bagu JR et al. (1997) J Biol Chem 272: 5087
[3] Bai R et al. (1990) J Biol Chem 265: 17141
[4] Bai R et al. (1992) Biochem Pharmacol 43: 2637
[5] Berman FW et al. (1999) Toxicon 37: 1645
[6] Bhansali P et al. (2011) J Med Chem 54: 7453
[7] Bouaïcha N et al. (2019) Toxins 11: 714
[8] Bouchard H et al. (2010). Novel Conjugates, Preparations Thereof, and Therapeutic Use Thereof. Patent WO/2011/001052
[9] Bowers A et al. (2008) J Am Chem Soc 130: 11219
[10] Bowers A et al. (2009) Org Lett 11: 1301
[11] Cao L et al. (2019) Toxins 11: 507
[12] Cao T et al. (2015) Mar Drugs 13: 903
[13] Carmichael WW et al. (1988) Appl Environ Microbiol 54: 2257
[14] Carmichael WW et al. (1988) Toxicon 26: 971
[15] Carmichael WW et al. (2001) Environ Health Persp 109: 663
[16] Carter DC et al. (1984) J Org Chem 49: 236
[17] Cavalcante do Amaral S et al. (2021) Mar Drugs 19: 17
[18] Chang Z et al. (2004) J Nat Prod 67: 1356
[19] Chen T et al. (2006) Toxicology 220: 71
[20] Chen Y et al. (2013) ClinChim Acta 425: 18
[21] Chernoff N et al. (2020) Toxins 12: 403
[22] Choi H et al. (2010) J Nat Prod 73: 1411
[23] Chorus I (ed, 2001) Cyanotoxins, Springer Berlin, Heidelberg
[24] Chorus I, Welker M (eds, 2021) Toxic cyanobacteria in water – Second edition, WHO
[25] Christiansen G et al. (2003) J Bacteriol 185: 564
[26] Cragg 2011 et al. (eds. 2011) Anticancer Agents from Natural Products, CRC Press
[27] Curren E et al. (2022) Environ Sci Pollut Res 29: 78178
[28] Dawson RM (1998) Toxicon 36: 953
[29] de Figueirdo DR et al. (2004) Ecotoxicol Environ Safety 59: 151
[30] Dennison WC et al. (1999) Bull Inst Océanogr 632: 501

[31] Dhakal D et al. (2023) J Nat Prod 86: 85
[32] Dietrich D (2005) Toxicol Appl Pharmacol 203: 273
[33] Ding WX, Ong CN (2003) FEMS Microbiol Lett 220: 1
[34] Dittmann E, Wiegand C (2006) Mol Nutr Food Res 50: 7
[35] Dittmann E et al. (2013) FEMS Microbiol Lett 37: 23
[36] Dittmann E et al. (2015) Trends Microbiol 23: 642
[37] Eggen M, Georg GI (2002) Med Res Rev 22: 85
[38] Eißler S et al. (2006) Synthesis 22: 3747
[39] Ericksson JE et al. (1988) Toxin 26: 161
[40] Falconer IR, Yeung DSK (1992) Chem-Biol Interact 81: 181
[41] Faltermann S et al. (2014) Aquat Toxicol 149: 33
[42] Fawell JK et al. (1999) Human Exp Toxicol 18: 162
[43] Fewer DP et al. (2007) BMC Ecol Evol 7: 183
[44] Fewer DP et al. (2009) Mol Microbiol 73: 924
[45] Francis G (1878) Nature 18: 11
[46] Gademann K et al. (2010) J Nat Prod 73: 980
[47] Galey DF et al. (1987) Am J Vet Res 48: 1415
[48] Gehringer MM et al. (2012) ISME J 6: 1834
[49] Gerwick WH et al. (1994) J Org Chem 59: 1243
[50] Goldberg J et al. (1995) Nature 376: 745
[51] Grindberg RV et al. (2011) PLOS One 6: e18565
[52] Gu S et al. (2022) Toxins 14: 715
[53] Gulledge BM et al. (2003) Bioorg Med Chem Lett 13: 2907
[54] Hagenbuch B, Gui C (2007) Xenobiotica 38: 778
[55] Harrigan GG, Goetz G (2002) J Appl Phycol 14: 103
[56] Harrigan GG et al. (1998) J Nat Prod 61 (9): 1075
[57] Harrigan GG et al. (1998) J Nat Prod 61 (10): 1221
[58] Hicks LM et al. (2006) ACS Chem Biol 1: 93
59] Hirsch CF et al. (1990) Antifungal Fermentation Product and Method. US Patent 4,946,835
[60] Honkanen RE et al. (1990) J Biol Chem 265: 19401
[61] Horgen FD et al. (2002) J Nat Prod 65: 487
[62] Huang KC et al. (2016) Mol Cancer Therap 15: 1208
[63] Huang X et al. (2016) Mol Pharmaceutics 13: 3756
[64] Hudnell HK (ed, 2008) Cyanobacterial Harmful Algal Blooms: State of the Science and Research Needs. Springer New York
[65] Ishida K et al. (2000) Tetrahedron 56: 8643
[66] Iwasaki A et al. (2014) RSC Adv 4: 12840
[67] Iwasaki A et al. (2015) Org Lett 17: 652
[68] Janssen EML (2019) Water Res 151: 488
[69] Jones MR et al. (2021) Water Res 196: 117017
[70] Jungblut AD, Neilan BA (2006) Arch Microbiol 185: 107
[71] Kallifidas D et al. (2024) Org Lett 26: 1321
[72] Kalliokoski A, Niemi M (2009) Brit J Pharmacol 158: 693
[73] Koehn FE et al. (1992) J Nat Prod 55: 613
[74] König J et al. (2006) Naunyn Schmiedebergs Arch Pharmacol 372: 432
[75] Kraft M et al. (2006) FEBS Lett 580: 6943
[76] Kulabhusan PK, Campbell K (2024) Sci Total Env 917: 170078
[77] LePage KT et al. (2005) Toxicol Lett 158: 133
[78] Levizou E et al. (2017) Ecotoxicol Environ Safety 143: 193

[79] Li EI et al. (2001) PNAS 98: 7599
[80] Liu J et al. (2023) Mar Drugs 21: 217
[81] Loos R et al. (2017) Cyanotoxins: methods and approaches for their analysis and detection. EU Publications Office
[82] Lopez JAV et al. (2016) J Nat Prod 79: 1213
[83] Lopez Rodas V, Costas E (1999) Res Vet Sci 67: 107
[84] Luckas B et al. (2005) Environ Toxicol 20: 1
[85] Luesch H et al. (1999) J Nat Prod 62: 1702
[86] Luesch H et al. (2000) J Nat Prod 63: 611
[87] Luesch H et al. (2001) J Am Chem Soc 123: 5418
[88] Luesch H et al. (2002) Curr Med Chem 9: 1791
[89] Luesch H et al. (2024) Nat Prod Rep 42: 208
[90] Luo D et al. (2003) Biochem Pharmacol 213: 115608
[91] MacKintosh C et al (1990) FEBS Lett 264: 187
[92] Marquez BLM et al. (2002) J Nat Prod 65: 866
[93] Mazur.Marzec H et al. (2018) Mar Drugs 16: 220
[94] Merder J et al. (2023) Nature Water 1: 844
[95] Mikhailov A et al. (2003) Chem-Biol Interact 142: 223
[96] Monteiro PR et al. (2021) Toxins 13: 522
[97] Moore RE (1996) J Ind Microbiol 16: 134
[98] Morgan JB et al. (2015) Mar Drugs 13: 1552
[99] Niedermeyer THJ, Brönstrup M. (2013) Natural product drug discovery from microalgae. In: Posten C, Walter C (eds) Microalgal Biotechnology: Integration and Economy, De Gruyter Brill, Berlin, Boston
[100] Niedermeyer THJ et al. (2014) PLOS One 9: e91476
[101] Niedermeyer THJ et al. (2018) Modicrocystins and Nodularins. Patent WO/2018/206715
[102] Nogle LM et al. (2001) J Nat Prod 64: 983
[103] Ohta T et al. (1994) Cancer Res 54: 6402
[104] Okamoto S et al. (2015) J Nat Prod 78: 2719
[105] Orjala J, Gerwick WH (2006) J Nat Prod 59: 427
[106] Orjala J et al. (1995) J Am Chem Soc 117: 8281
[107] Panda D et al. (1997) Biochem 36: 12948
[108] Pearson L et al. (2010) Mar Drugs 8: 1650
[109] Pereira A et al. (2009) Chem Biol 16: 893
[110] Pettit GR et al. (1987) J Am Chem Soc 109: 6883
[111] Pouria S et al. (1998) Lancet 352: 21
[112] Radau G (2005) Curr Enzyme Inhib 1: 295
[113] Rantala A et al. (2003) PNAS 101: 568
[114] Rantala A et al. (2006) Appl Environ Microbiol 72: 6101
[115] Rohr J (2006) ACS Chem Biol 1: 747
[116] Rounge TB et al. (2007) Appl Environ Microbiol 73: 7322
[117] Runnegar MT et al. (1988) Toxicon 26: 143
[118] Runnegar MT et al. (1993) Am J Physiol Gastrointest Liver Physiol 265: G224
[119] Sahu N et al. (2025) Sci Total Env 976: 179260
[120] Saker ML et al. (2005) Toxicon 46: 555
[121] Sallandt LL et al. (2025) J Nat Prod 88: 3
[122] Schwark M et al. (2023) PNAS 120: e2219230120
[123] Schwartz RE et al. (1990) J Ind Microbiol Biotechnol 5: 113
[124] Schwarzenberger A (2022) Toxins 14: 770

[125] Schwarzenberger A et al. (2021) Harmful Algae 106: 102062
[126] Seiser T et al. (2008) Angew Chem Int Ed 47: 6483
[127] Shartau RB et al. (2023) Toxins 15: 395
[128] Singh SB (2022) J Nat Prod 85: 666
[129] Sivonen K et al. (1989) Appl Environ Microbiol 55: 1990
[130] Steward I et al. (2008) Cyanobacterial poisoning in livestock, wild mammals and birds – an overview. In: Hudnell HK (ed) Cyanobacterial Harmful Algal Blooms: State of the Science and Research Needs, Springer, New York
[131] Sumiya E et al. (2011) ACS Chem Biol 6: 425
[132] Taori K et al. (2008) J Am Chem Soc 130: 1806
[133] Tarsis EM et al. (2015) Tetrahedron 71: 5029
[134] Taylor C et al. (1996) Bioorg Med Chem Lett 6: 2113
[135] Teruya T et al. (2009) Org Lett 11: 5062
[136] Tillett D et al. (2000) Chem Biol 7: 753
[137] Trimurtulu G et al. (1994) J Am Chem Soc 116: 4729
[138] Tripathi A et al. (2010) J Nat Prod 73: 1810
[139] Umezawa T et al. (2011) J Org Chem 77: 357
[140] van Apeldoorn ME et al. (2007) Mol Nutr Food Res 51: 7
[141] Vasconcelos VM (1994) Arch Hydrobiol 130: 439
[142] Wagner MM et al. (1999) Cancer Cemotherap Pharmacol 43: 115
[143] Watanabe F et al. (eds, 1995) Toxic Microcystis, CRC Press Boca Raton
[144] Wegerski CJ et al. (2007) J Nat Prod 70: 89
[145] Wei N et al. (2024) Water Res 262: 122119
[146] Welker M, von Döhren H (2006) FEMS Microbiol Rev 30: 530
[147] WHO (2017) Guidelines for drinking-water quality, 4th edition
[148] Wilhelm SW et al. (2020) mBio 11: e00529-20
[149] Williams PG et al. (2003) J Nat Prod 66: 1356
[150] Wipf P et al. (2004) Curr Pharm Des 10: 1417
[151] Wu M et al. (2000) J Am Chem Soc 122: 12041
[152] Xu S et al. (2020) Toxins 12: 693
[153] Ye B et al. (2020) Mar Drugs 18: 176 M
[154] Ying Y et al. (2008) J Am Chem Soc 130: 8455
[155] Yoshida T et al. (1998) Natural Toxins 5: 91
[156] Yoshizawa S et al. (1990) J Cancer Red Clin Oncol 116: 609
[157] Yu SZ (1989) In: Tang ZY et al. (eds) Primary Liver Cancer, Springer, Berlin
[158] Zeguro B (2016) Mini-Rev Med Chem 16: 1042
[159] Zhang LH et al. (1997) Life Sci 60: 751
[160] Zhang QX et al. (1991) Environ Toxicol Chem 10: 313
[161] Zurawell RW et al. (2011) J Toxicol Enriron Health 8: 1

Cross-Chapter Literature and Further Reading

(Numbers inside the chapters in bold)

[1] Aktories K et al. (eds.) (2022) Allgemeine und Spezielle Pharmakologie und Toxikologie, 13th ed, Urban & Fischer, München, Jena and Elsevier, Amsterdam. Formerly: Forth W et al. (2001) Allgemeine und spezielle Pharmakologie und Toxikologie. Urban & Fischer, München

[2] Alouf JE, Freer JH (eds.) (1999) The Comprehensive Sourcebook of Bacterial Protein Toxins. Academic Press, London, Boston, New York, Sydney, Tokyo, Toronto

[3] Amberger-Lahrmann M, Schmähl D (1988) Gifte-Geschichte der Toxikologie. Springer, Berlin

[4] Ammon HPT (ed.) (2001) Arzneimittelneben- und Wechselwirkungen. 4th ed., WVG, Stuttgart

[5] Armstrong J, Pascu O (2015) Toxicology Handbook. 3rd ed., Elsevier, Amsterdam

[6] Arora D (1986) Mushrooms Demystified. Ten Speed Press, Emeryville

[7] Barceloux DG (2008) Medical Toxicology of Natural Substances, Foods, Fungi, Medicinal Herbs, Plants, and Venomous Animals. John Wiley & Sons Inc., Hoboken, New Jersey

[8] Barile FA (2010) Clinical Toxicology. 2nd ed., CRC Press, Boca Raton

[9] Barton S, Nakanishi K (eds.) (1999) Comprehensive Natural Products Chemistry. Vol. 1–9, Elsevier, Amsterdam

[10] Beasley V (2004) Veterinary Toxicology. Int Vet Information Service. https://libguides.libraries.wsu.edu/veterinarymedicine/IVIS

[11] Belitz HD, Grosch W (1992) Lehrbuch der Lebensmittelchemie. 4th ed., Springer, Berlin

[12] Bentz H (1969) Nutztiervergiftung. Erkennung und Verhütung. Fischer-Verlag, Jena

[13] Berndt S (2016) Wenig bekannte Pilzvergiftungen. Biol Unserer Zeit 46(3): 170

[14] Betina V (1989) Mycotoxins, Chemical, Biological and Environmental Aspects. Elsevier, Amsterdam

[15] BfR (2017) Risiko Vergiftungsunfälle bei Kindern. www.bfr.bund.de/cm/350/risiko-vergiftungsunfaelle-bei-kindern.pdf

[16] Blaschek W et al. (eds.) (CD ROM since 2015) Hager ROM/Hagers Enzyklopädie der Arzneistoffe und Drogen. Springer, Berlin, Heidelberg, New York, 2019, formerly Hagers Handbuch der Pharmazeutischen Praxis. Last printed edition Hänsel R et al. (eds.) (1992-1998), Drogen und Folgebände Drogen, 5th ed.

[17] Boit HG (1961) Ergebnisse der Alkaloidchemie bis 1960. Akademie-Verlag, Berlin

[18] Botana LM (ed.) (2024) Environmental Toxicology. Non-bacterial Toxins. De Gruyter, Berlin, Boston

[19] Bresinsky A, Besl H (1985) Giftpilze. Ein Handbuch für Apotheker, Ärzte und Biologen. WVG, Stuttgart

[20] Brugsch H, Klimmer OR (1966) Vergiftungen im Kindesalter. Ferdinand Enke, Stuttgart

[21] Buckingham J (ed.) (2020) Dictionary of Natural Products. Vol 1–6, Suppl 1, Taylor and Francis Ltd., London

[22] Burns DL et al. (eds.) (2003) Bacterial Protein Toxins. ASM Press, Herndon

[23] Canadian Biodiversity Information Facility (2015) Poisonous Plants Information System. http://www.cbif.gc.ca/eng/species-bank/canadian-poisonous-plants-information-system/poisonous-plants-sites/?id=1370403265037

[24] Centers for Disease Control and Prevention. Poisonous plants. http://www.cdc.gov/niosh/topics/plants/

[25] CliniTox. www.clinitox.ch. www.giftpflanzen.ch

[26] Cole RJ, Cox RH (1981) Handbook of Toxic Fungal Metabolites. Academic Press, New York

[27] Dabrowski WM, Sikorski ZE (eds.) (2004) Toxins in Food. CRC Press, Boca Raton

[28] Daunderer M (1990) Drogenhandbuch für Klinik und Praxis. Diagnose, Therapie, Nachweis, Prophylaxe, Recht, Drogenprofile. Vol. III, Pharmakologie, ecomed, Landsberg/Lech

https://doi.org/10.1515/9783110728576-006

[29] Daunderer M (1995) Lexikon der Pflanzen- und Tiergifte, Diagnostik und Therapie. Nikol Verlagsges, Hamburg
[30] De´Mello JFP (ed.) (1997) Handbook of Plant and Fungal Toxicants. CRC Press, Boca Raton
[31] DGfM (Deutsche Gesellschaft für Mykologie). www.dgfm-ev.de
[32] Dingermann T et al. (2004) Arzneidrogen, 5th ed., Spektrum Akad Verlag, München
[33] Döpke W (1976) Ergebnisse der Alkaloidchemie, Vol I, Part 1 and 2: 1960–1968. Akademie-Verlag, Berlin
[34] Döpke W (1978) Ergebnisse der Alkaloidchemie, Vol II, Part 1 and 2: 1969–1970. Akademie-Verlag, Berlin
[35] Dunkelberg H et al. (2007) Handbuch der Lebensmitteltoxikologie. Vol. 1–5, Wiley, Chichester
[36] Durrant M (2013) Handbook of Clinical Toxicology. Hayle Medical, New York
[37] Efferth T (2006) Molekulare Pharmakologie und Toxikologie. Springer, Berlin
[38] Eisenbrandt G et al. (2005) Toxikologie für Naturwissenschaftler und Mediziner. 3rd ed., Wiley-VCH, Weinheim
[39] European Union (EU, 2023) Commission regulation (EU) 2023/915 of 25 April 2023 on maximum levels for certain contaminants in food and repealing Regulation (EC) No 1881. https://eur-lex.europa.eu/legal-content/EN/TXT/?uri=CELEX%3A32023R0915
[40] FDA. www.fda.gov/food/natural-toxins-food/mycotoxins
[41] Flament E et al. (2020) Human poisoning from poisonous higher fungi: focus on analytical toxicology and case reports in forensic toxicology. Pharmaceuticals 13: 454
[42] Flammer R (2014) Giftpilze, Pilzvergiftungen. Ein Nachschlagewerk für Ärzte, Apotheker, Biologen, Mykologen, Pilzexperten und Pilzsammler, AT Verlag, Aarau
[43] Frohne D, Pfänder HJ (2005) Poisonous Plants. 2nd ed., Manson Publishing LTD
[44] Geisslinger G et al. (2020) Mutschler Arzneimittelwirkungen. Pharmakologie – Klinische Pharmakologie–Toxikologie. 11th ed., WVG, Stuttgart. Formerly: Mutschler E et al.: Arzneimittelwirkungen. Lehrbuch der Pharmakologie und Toxikologie. WVG, Stuttgart
[45] Gessner O (1931: 1st ed., 1953: 2nd ed., 1974: 3rd ed.) Published and reedited by Orzechowski G: Gift- und Arzneipflanzen von Mitteleuropa. Carl-Winter-Universitäts-Verlag, Heidelberg
[46] GGIZ. www.GGIZ-Erfurt.de (Gemeinsames Giftinformationszentrum der Länder Mecklenburg-Vorpommern, Sachsen, Sachsen-Anhalt und Thüringen, Germany)
[47] Gonçalves J et al. (2021) Psychoactive substances of natural origin: Toxicological aspects, therapeutic properties and analysis in biological samples. Molecules 26: 1397
[48] Gossel TA, Bricker JD (2019) Principles of Clinical Toxicology. 3rd ed., CRC Press, Boca Raton
[49] Govorushenko S et al. (2019) Poisoning associated with the use of mushrooms: A review of the global pattern and main characteristics. Food Chem Toxicol 128: 267
[50] Graeme KA (2014) Mycetism. A review of the recent literature. J Med Toxicol 10: 173
[51] Grandjean P (2016) Paracelsus revisited: the dose concept in a complex world, Basic Clin Pharmacol Toxicol 119 (2): 126
[52] Gupta RC (ed.) (2012) Veterinary Toxicology: Basic and Clinical Principles. 2nd ed., Elsevier INC, San Diego
[53] Guthmann J (2024) Heilende Pilze. 3rd ed., Quelle & Meyer, Wiebelsheim
[54] Hahn A et al. (eds.) (2016) Ernährung. 3rd ed., WVG, Stuttgart
[55] Handbuch der Experimentellen Pharmakologie. Vol 1 (1935) – Vol 65 (1983), Springer, Berlin
[56] Hapke HJ (1988) Toxikologie für Veterinärmedizin. 2nd ed., Ferdinand Enke, Stuttgart
[57] Harborne JB et al. (1977) Dictionary of Plant Toxins. Wiley, Chichester
[58] Hartwig A et al. (2020) Mode of action-based risk assessment of genotoxic carcinogens. Arch Toxicol 94: 1787
[59] Hausen BM, Vieluf IK (1997) Allergiepflanzen, Handbuch und Atlas. 2nd ed., Nikol, Hamburg

[60] He MQ et al. (2022) Potential benefits and harms: a review of poisonous mushrooms in the world. Fungal Biol Rev 42: 56
[61] Helfrich W, Winter CK (eds.) (2000) Food Toxicology. Taylor and Francis Ltd., London
[62] Herrmanns-Clausen M et al. (2019) Akzidentelle Vergiftungen mit Gartenpflanzen und Pflanzen in der freien Natur. Bundesgesundheitsbl 62: 73. https://doi.org/10.1007/s00103-018-2853-5
[63] Hildebrandt JP et al. (2023) Natural Poisons and Venoms (Vol. 4): Animal Toxins. De Gruyter, Berlin, Boston
[64] Hodgson E, Roe M (2014) Dictionary of Toxicology, 3rd ed., Elsevier, Amsterdam
[65] IARC Monographs on the Identification of Carcinogenic Hazards to Humans: https://monographs.iarc.who.int/list-of-classifications
[66] ImazekiR, Hongo T (1998) Coloured illustrations of Fungi of Japan. Hoikusha, Osaka
[67] Index Fungorum. https://indexfungorum.org
[68] Japanese Mushroom Picture Book. Visual Dictionary Fungi of Japan, Yamato Gorgesha
[69] Jo WS et al. (2014) Toxicological profiles of poisonous, edible, and medicinal mushrooms. Mycobiology 42(3): 215
[70] Karow T, Lang-Roth R (2022) Allgemeine und spezielle Pharmakologie und Toxikologie 2023/2024. 31st ed., Karow Verlag, Berlin
[71] Karrer W et al. (1958) Konstitution und Vorkommen der organischen Pflanzenstoffe (exclusive Alkaloide). Suppl Vol 1 (1977), Suppl Vol 2, part 1 (1981), part 2 (1985), Birkhäuser, Basel
[72] Keeler RF, Tu AT (eds.) (1983 and 1991) Handbook of Natural Toxins. Vol I: Toxicology of Plant and Fungal Compounds. Marcel Dekker, New York, and CRC Press, Boca Raton
[73] Kell V (1991) Giftpilze und Pilzgifte. Ziemsen Verlag, Wittenberg Lutherstadt
[74] Khan IA, Abourashed EA (2010) Leung´s Encyclopedia of Common Natural Ingredients. 3rd ed, Wiley, Hoboken, New Jersey
[75] Kibby G (2023) Mushrooms and Toadstools of Britain and Europe. Vol 1–4. Privately published by G Kibby
[76] Kirk PM, Cannon PF, Minter DW, Stalpers JA (eds.) (2008) Dictionary of the Fungi. 10th ed., CABI Europe, UK
[77] Klaassen CD (ed.) (2019) Casarett and Doull´s Toxicology–The Basic Science of Poisons. 9th ed., McGraw-Hill, New York
[78] Knudsen H, Vesterholt J (eds.) (2012) Funga Nordica. Nordsvamp, København
[79] Kort RP, Teng SC (1996) Fungi of China. Mycotaxon, Oxford, UK
[80] Kreisel H (2014) Ethnomykologie. Weissdorn-Verlag, Jena
[81] Krienke EG et al. (1986) Vergiftungen im Kindesalter. Ferdinand Enke, Stuttgart
[82] Laessoe T, Petersen JH (2019) Fungi of Temperate Europe. Vol. 1+2. Princeton University Press, Princeton, New Jersey
[83] Lander DG (2017) Veterinary Toxicology. Agri Bio Vet Press, Delhi
[84] Laux HE (2001) Der große Kosmos-Pilzführer. Alle Speisepilze mit ihren giftigen Doppelgängern. Franckh-Kosmos, Stuttgart
[85] Lendac Data Systems Ltdex System. www.pharmaceuticalonline.com/doc/toxicology-database-0001
[86] Lewin L (1992) Gifte und Vergiftungen–Lehrbuch der Toxikologie. 6th ed., Karl F Haug Verlag, Heidelberg, reprint of the original edition from 1929
[87] Lewin L (2007) Die Gifte in der Weltgeschichte. updated and expanded new edition of Sorgegfrey C, Carl Ueberreuter, Wien
[88] Liebenow H et al. (2005) Risiko Pilze – Einschätzung und Hinweise, BfR Pressestelle. Berlin
[89] Lindner E (1990) Toxikologie der Nahrungsmittel. Thieme, Stuttgart, New York
[90] List PH, Hörhammer L (1967–1980) Hagers Handbuch der Pharmazeutischen Praxis. 4th ed., Vol 1–8, Springer, Berlin

[91] LiverTox: Clinical and Research Information on Drug-Induced Liver Injury [Internet]. Bethesda (MD): National Institute of Diabetes and Digestive and Kidney Diseases (2012) https://www.ncbi.nlm.nih.gov/books/

[92] Luch A (ed.), Molecular, Clinical and Environmental Toxicology. Vol 1: Molecular Toxicology (2009), Vol 2: Clinical Toxicology (2010), Vol. 3: Environmental Toxicology (2012). Birkhäuser, Basel

[93] Luckner M (1984) Secondary Metabolism in Microorganisms, Plants and Animals. Fischer-Verlag, Jena

[94] Ludewig R, Regenthal R (2015) Akute Vergiftungen und Arzneimittelüberdosierungen. 11th ed., WVG, Stuttgart

[95] Lüllmann H et al. (2016) Pharmakologie und Toxikologie. 18th ed., Georg Thieme, Stuttgart

[96] Mabberley DJ (2017) Mabberley´s Plant-Book, a Portable Dictionary of Plants, their Classification and Uses. 4th ed., Cambridge University Press, Cambridge

[97] Machholz R, Lewerenz HJ (eds.) (1989) Lebensmitteltoxikologie. Akademie Verlag, Springer, Berlin

[98] Madaus G (1979) Lehrbuch der biologischen Arzneimittel. Olms-Verlag, Hildesheim

[99] Manske RHF, Holmers HL (1950, Founder of the book series; continued 1979 by Manske RHF and Rodrigo RGA, 1983 by Brossi A and 1992 by Cordell GA) The Alkaloids-Chemistry and Physiology. Vol 1 (1950) – Vol 63 (2006). Academic Press, New York

[100] Marquardt H et al. (2019) Toxikologie, 4th ed., WVG, Stuttgart

[101] Martinez D, Lohs KH (1986) Giftmagie und Realität, Nutzen und Verderben. Nikol Verlagsges, Hamburg

[102] Maurer H, Brandt SD (2018) New Psychoactive Substances. Handbook Exp Pharmacol: 252

[103] May TW, Wood AE (1957) Fungi of Australia. Vol 2A. Catalogue and Bibliography of Australian Macrofungi 1. Basidiomycota CSIRO Publishing, Clayton South, VIC

[104] May TW et al. (2003) Fungi of Australia. Vol 2B. Catalogue and Bibliography of Australian Macrofungi 2. Basidiomycota p. p. & Myxomycota. CSIRO Publishing, Clayton South, VIC

[105] McKenzie R (2012) Australia´s Poisonous Plants, Fungi and Cyanobacteria: A Guide to Species of Medical and Veterinary Importance. CSIRO Publishing, Clayton South, VIC

[106] Melzig FM, Hiller K (2023) Lexikon der Arzneipflanzen und Drogen. 3rd ed., Springer Spektrum, Berlin

[107] Michael E, Hennig B, Kreisel H (1981–1988) Handbuch für Pilzfreunde. Vol 1–6, Fischer, Jena

[108] Moeschlin S (1986) Klinik und Therapie der Vergiftungen. Thieme, Stuttgart

[109] Morton JF (1977) Poisonous and Injourios Higher Plants and Fungi. In: Forensic Medicine III. Tedeschi CG (ed.), WB Saunders Company, Philadelphia, PA, p. 1456

[110] Moser M (1940–1995) Kleine Kryptogamenflora, Vol 2, Part A (1963), B/2 (1983) Gustav Fischer Jena, Stuttgart

[111] Moshobane MC et al. (2020) Plants and mushrooms associated with animal poisoning incidents in South Africa. Vet Rec Open 7: e000402

[112] Mothes K, Schütte HR (1969) Biosynthese der Alkaloide. Deutscher Verlag der Wissenschaften, Berlin

[113] Mothes K et al. (1985) Biochemistry of Alkaloids. Deutscher Verlag der Wissenschaften, Berlin

[114] MSD Manual Veterinary Manual. www.msdvetmanual.com/toxicology/mycotoxicoses

[115] Mtewa AG et al. (eds.) (2021) Poisonous Plants and Phytochemicals in Drug Development. Wiley & Sons, Inc. Hoboken, New Jersey

[116] Mücke W, Lemmen C (2004) Schimmelpilze – Vorkommen, Gesundheitsgefahren, Schutzmaßnahmen, ecomed, Landsberg/Lech

[117] Müller U (1989) Grundlagen der Lebensmittelmikrobiologie. Fachverlag Leipzig

[118] Mycobank.www.mirri.org/upcp_product/mycobank/

[119] National Poison Data System. www.aapcc.org/national-poison-data-system (NPDS)

[120] Nelson LS, Balick MJ (2020) Handbook of Poisonous and Injurious Plants. Springer, New York

[121] Neuwinger HD (2000) African Traditional Medicine. A Dictionary of Plant Use and Application. Medpharm Scientific Publ., Stuttgart
[122] Nuhn P (2006) Naturstoffchemie: mikrobielle, pflanzliche und tierische Naturstoffe. Hirzel, Stuttgart
[123] Oberdisse E, Hackenthal E (eds.) (2002) Pharmakologie und Toxikologie. 3rd ed., Springer, Berlin
[124] Petejova N et al. (2019) Acute toxic kidney injury. Renal Failure 41(1): 576
[125] Peterson ME et al. (2006) Small Animal Toxicology. W.B. Saunders, Philadelphia
[126] Pharmacopoea Europaea (Ph. Eur., European Pharmacopeia, 2023), 11th ed., European Pharmacopeia Commission (EPC), Strasbourg
[127] Phillips R (2010) Mushrooms and Other Fungi of North America. Firefly Books LTD
[128] Pilzliste (2020). Pilzliste_1_Auflage_10_2020.xlsx. www.bvl.bund.de.
[129] Plumlee KH (2004) Clinical Veterinary Toxicology. Mosby, St. Louis
[130] Pilze der Schweiz. Vol 1–6. Mykologia, Luzern
[131] POISINDEX®System, IBM Micromedex, access via WVG, Stuttgart
[132] Proft Th (ed.) (2005) Microbial Toxins. Molecular and Cellular Biology. Horizon Bioscience, Wymondham, Norfolk
[133] Quattrocchi U (2012) CRC World Dictionary of Medicinal and Poisonous Plants. Routledge, London
[134] Queensland Health, Poison Information Centre (2015) Plants and Mushrooms (Fungi) Poisonous to People in Queensland. https://www.childrens.health.qld.gov.au/chq/our-services/queensland-poisons-information-centre/plants-mushrooms/
[135] Rätsch Ch (1998) Enzyklopädie der psychoaktiven Pflanzen – Botanik, Ethnopharmakologie und Anwendungen. AT-Verlag, Aarau, and WVG, Stuttgart
[136] Rätsch Ch, Müller-Ebeling C (2006) Lexikon der Liebesmittel. AT Verlag Aarau 2003, and WVG, Stuttgart
[137] Rechcigl JR (2020) CRC Handbook of Foodborne Diseases of Biological Origin, CRC Press, Boca Raton
[138] Reiss J (1981) Mykotoxine in Lebensmitteln. Gustav Fischer, Stuttgart
[139] Reiss J (1998) Schimmelpilze, Lebensweise, Nutzen, Schaden, Bekämpfung. Springer, Berlin
[140] Riet-Correa F et al. (2011) Poisoning by Plants, Mycotoxins and related Toxins. CABI, Wallingford
[141] Robinson L (2018) A Practical Guide to Toxicology and Human Health Risk Assessment. Wiley, Hoboken, New Jersey
[142] Roth L, Frank H, Kormann K. (1990) Giftpilze, Pilzgifte. ecomed, Landsberg/Lech
[143] Rumack BH, Salzman MD (1978) Mushroom Poisoning, Diagnosis and Treatment. CRC Press, Boca Raton
[144] Schäfer C, Marschall-Kunz B (2014) Gifte und Vergiftungen in Haushalt, Garten, Freizeit. WVG, Stuttgart
[145] Schmidbauer W, vom Scheidt J (2004) Handbuch der Rauschdrogen. 11th ed., Nymphenburger in der FA Herbig Verlagsbuchhandlung, München
[146] Schmidt RJ, Botanical Dermatology Database. http://www.botanical-dermatology-database.info
[147] Schultes RE, Hofmann A (1981) Pflanzen der Götter–Die magischen Kräfte der Rausch- und Giftgewächse. Hallwag Verlag, Bern
[148] Seeger R, Neumann HG (1990) Giftlexikon. Ein Handbuch für Ärzte, Apotheker und Naturwissenschaftler. last actualization 2015 by Schrenk D et al., WVG, Stuttgart
[149] Singh BR, Tu AT (eds.) (2011) Natural Toxins 2: Structures, Mechanism of Action and Detection. Springer, Berlin
[150] Soyka M (2010) Drogennotfälle: Diagnostik, Klinisches Erscheinungsbild, Therapie. Schattauer, Stuttgart
[151] Spoerke DG, Rumack BH (1994) Handbook of Mushroom Poisonig & Treatment. CRC Press, Boca Raton
[152] Sticher O et al. (eds.) (2015) Hänsel/Sticher Pharmakognosie–Phytopharmazie. 10th ed., WVG, Stuttgart

[153] Stickel F, Shouval D (2015) Hepatotoxicity of herbal and dietary supplements: an update. Arch Toxicol 89(6): 851

[154] Tang W, Eisenbrand G (2011) Chinese Drugs of Plant Origin, Chemistry, Pharmacology, and Use in Traditional and Modern Medicine. Springer, Berlin

[155] Tedeschi CG et al. (1977) Forensic Medicine – a Study in Trauma and Environmental Hazards. III. Environmental Hazards. WB Saunders Company, Philadelphia

[156] Teuscher E (2006) Medicinal Spices, A Handbook of Culinary Herbs, Spices, Spice Mixtures and their Essential Oils. Medpharm Scientific Publishers, Stuttgart

[157] Teuscher E (ed.) (2018) Gewürze und Küchenkräuter. WVG, Stuttgart

[158] Teuscher E, Lindequist U (2010) Biogene Gifte, Biologie–Chemie–Pharmakologie–Toxikologie. 3rd ed., WVG, Stuttgart

[159] Teuscher E, Lindequist U (2023) Natural Poisons and Venoms (Vol. 1). Plant Toxins: Terpenes and Steroids. De Gruyter, Berlin, Boston

[160] Teuscher E, Lindequist U (2024) Natural Poisons and Venoms (Vol. 2). Plant Toxins: Polyketides, Phenylpropanoids and Further Compounds. De Gruyter, Berlin, Boston

[161] Teuscher E, Lindequist U (2025) Natural Poisons and Venoms (Vol. 3). Plant Toxins: Alkaloids and Lectins. De Gruyter, Berlin

[162] Teuscher E et al. (2020) Biogene Arzneimittel. 8th ed., WVG, Stuttgart

[163] The Merck Veterinary Manual (2008) http://www.merckvetmanual.com/mvm/index.jsp?cfile=htm/bc/210800.htm&word=prussic%2cacid

[164] Tox Info Suisse Zürich (2019) Giftige Garten- und Wildpflanzen. https://toxinfo.ch/customer/files/28/Giftige-Garten-und-Wildpflanzen-fuer-Tox-Info-Suisse.pdf

[165] TOXLINE (National Library of Medicine bibliographic database for toxicology). www.nlm.nih.gov/databases/download/toxlinesubset.html

[166] Tu A (ed.) (2019) Handbook of Natural Toxins. Vol. 7: Food Poisoning. Routledge, New York

[167] USDA Agricultural Research Service Poisonous Plant Res. www.ars.usda.gov

[168] Uter W (2020) Contact allergy – emerging allergens and public health impact. Int J Environ Res Public Health 17: 2404

[169] Van Wyk BE (2005) Food Plants of the World. WVG, Stuttgart

[170] Van Wyk BE et al. (2002) Poisonous Plants of South Africa. Briza Publ, Pretoria

[171] Vergiftungs-Informations-Zentrale Freiburg (2019) Liste ausgewählter Giftpflanzen. https://www.uniklinik-freiburg.de/giftberatung/liste-ausgewaehlter-giftpflanzen.html

[172] Von Mühlendahl KE et al. (2003) Vergiftungen im Kindesalter. 4th ed., Thieme, Stuttgart

[173] Wagstaff DJ (2008) International Poisonous Plant Checklist. CRC Press, Boca Raton

[174] Watt JM, Breyer-Brandwijk MG (1982) The Medicinal and Poisonous Plants of Southern and Eastern Africa. E und S Livingstone LTD, Edinburgh

[175] Weidenbörner M (2001) Encyclopedia of Food Mycotoxins. Springer, Berlin

[176] Wendt S et al. (2022) Poisoning by plants. Dtsch Ärztebl Int 119: 317

[177] Wennig R et al. (2020) Mushroom poisoning. Dtsch Ärztebl Int 2020; 117: 701 and additions: Berndt S (2021) Dtsch Ärztebl Int 2021: 118, Haufs MG (2021) Dtsch Ärztebl Int 2021: 118

[178] White J et al. (2019) Mushroom poisoning: A proposed new clinical classification. Toxicon 157: 53

[179] Wexler P (2014) Encyclopedia of Toxicology. 3rd ed., Elsevier, Amsterdam

[180] Wirth W, Gloxhuber C (1994) Toxikologie. 5th ed., Thieme-Verlag, Stuttgart, New York

[181] www.toxinfo.org

[182] Yin X et al. (2019) Mushroom toxins: chemistry and toxicology. J Agric Food Chem 67 (18): 5053

[183] Young T, Young AM (2005) A Field Guide to the Fungi of Australia. New South Publishing. Dictionary Wallingford

Poison Information Centers

(Selection of PI numbers below, without guarantee of correctness)

Dial the emergency number of the country you are in (EN) (for numbers, see https://www.taschenhirn.de/allgemeinbildung-reisen/internationale-notrufnummern) **or from abroad** (**PI**, see international dialing code in the internet file above) and you will be connected with the relevant poison information center in this country.

Worldwide, PI: (+45) 82 12 12 12
All European countries. EN: 112
Austria, PI: +44(1) 406 43 43
Australia, PI: 131 126 or EN: 000
Belgium, PI: +32(70) 245 245
Brazil, EN: 192
Bulgary, PI: +359(2) 515 32
Denmark, PI: +45 (35) 316-060
Egypt, EN: 123
England, PI: +44 (171) 635 91 91, 999 works in England and in all former British colonies
Finland, PI: +358(9) 472 977
France, PI: +33 (3) 883 737 37
Germany, EN: 112, this number also works in India, Great Britain, and in all European countries
Greece, PI: +30 (1) 799 37 77
Hungaria, PI: +36 (1) 215 215
India, PI: 1800 345 033
Indonesia, EN: 112
Italy, PI: +39 (6) 490 663
Norway, PI: +47(22) 591 300
People's Republic of China, EN: 120, works also in Japan
Poland, PI: +48 (42) 657 99 00
Russian Federation, PI: +7 (95) 928 16 47
South Africa, EN: 10 177
Spain, PI: +34 (91) 562 84 69
Sweden, PI: +46 (8) 736 03 84
Switzerland, PI: +41(1) 251 51 51
The Netherlands, PI: +31 (39) 274 88 88
Turkey, PI: +90(312)433 70 01
USA, PI: 1-800-222-1222 or 11911 or EN: 911, works also in Philippines

Other numbers use EN, see above.

https://doi.org/10.1515/9783110728576-007

Index

The page numbers with the photos of the corresponding fungi or bacteria and the structural formulas of the corresponding compounds are printed in bold.

https://doi.org/10.1515/9783110728576-008

Index Volumes 1–4

Red=Volume 1, Green=Volume 2, Blue=Volume 3, Orange=Volume 4

Please note that due to technical limitations, the different colours used in the index cannot be displayed in the EPUB version of the book. For accurate colour representation, kindly refer to the print or eBook (PDF) version.

https://doi.org/10.1515/9783110728576-009

www.ingramcontent.com/pod-product-compliance
Lightning Source LLC
Chambersburg PA
CBHW080914220326
41598CB00034B/5573

* 9 7 8 3 1 1 0 7 2 8 5 6 9 *